REVIEWS in MINERALOGY

(Formerly: "Short Course Notes")

Volume 1

SULFIDE MINERALOGY

PAUL H. RIBBE, Editor

Authors: B. J. WUENSCH
Department of Metallurgy & Materials Science
Massachusetts Institute of Technology
Cambridge, Massachusetts

C. T. PREWITT
Department of Earth & Space Sciences
State University of New York
Stony Brook, New York

V. RAJAMANI AND S. D. SCOTT
Department of Geology
University of Toronto
Toronto, Ontario, Canada

J. R. CRAIG
Department of Geological Sciences
Virginia Polytechnic Institute & State University
Blacksburg, Virginia

P. B. BARTON
United States Geological Survey
Reston, Virginia

Series Editor:

P. H. RIBBE
Department of Geological Sciences
Virginia Polytechnic Institute & State University
Blacksburg, Virginia

MINERALOGICAL SOCIETY OF AMERICA

PRINTED BY

BookCrafters, Inc.
Chelsea, Michigan 48118

REVIEWS in MINERALOGY

(Formerly: SHORT COURSE NOTES)

ISSN   0275-0279

VOLUME 1:   SULFIDE MINERALOGY

ISBN 0-939950-01-4

Additional copies of this volume as well as those
listed below may be obtained at moderate cost from

Mineralogical Society of America
2000 Florida Avenue, NW
Washington, D.C.   20009

Vol.		No.of pages
1	SULFIDE MINERALOGY. P.H. Ribbe, Editor (1974)	284
2	FELDSPAR MINERALOGY. P.H. Ribbe, Editor (1975; revised 1982)	∿350
3	OXIDE MINERALS. Douglas Rumble III, Editor (1976)	502
4	MINERALOGY and GEOLOGY of NATURAL ZEOLITES. F.A. Mumpton, Editor (1977)	232
5	ORTHOSILICATES. P.H. Ribbe, Editor (1980; revised 1982)	∿410
6	MARINE MINERALS. R.G. Burns, Editor (1979)	380
7	PYROXENES. C.T. Prewitt, Editor (1980)	525
8	KINETICS of GEOCHEMICAL PROCESSES. A.C. Lasaga and R.J. Kirkpatrick, Editors (1981)	391
9A	AMPHIBOLES and Other Hydrous Pyriboles - Mineralogy. D.R. Veblen, Editor (1981)	372
9B	AMPHIBOLES: Petrology and Experimental Phase Relations. D.R. Veblen and P.H. Ribbe, Editors (1982)	∿375

TABLE OF CONTENTS

ERRATA AND ADDENDA FOR THE SECOND PRINTING . . . . (v)
    (continued at the back of the book on pages R-33ff.)

FOREWORD . . . . (vi)

DISCLAIMER . . . . (vi)

Chapter 1. DETERMINATION, RELATIONSHIPS, AND CLASSIFICATION OF SULFIDE MINERAL STRUCTURES

(B. J. Wuensch)

Introduction. . . . . W- 1

Crystal Structure Determination 2

Problems in Sulfide Structure Determination 5

    Suitable specimens   5
    Chemical and structural complexity   6
    X-ray absorption   7

Packing Considerations. 8

Polymorphism and Polytypism 13

Derivative Structures 16

"Composite" Structures. 17

Classification Schemes. 18

Chapter 2. SULFIDE CRYSTAL CHEMISTRY

(B. J. Wuensch)

Introduction. W-21

Elemental Sulfur. 21

Ionic Sulfides. 23

Disulfides and Derivatives. 23

Monosulfides and Derivatives. 25

The Nickel Sulfides 31

Silver and Copper Sulfides. 32

Rocksalt Derivatives. 34

Group V Metal Sulfides and Related Sulfosalts 34

    The Group V sulfides   35
    Stibnite derivatives   36
    The Bi sulfosalts   38

Lead-Antimony Sulfosalts. 39

Lead-Arsenic Sulfosalts 41

Additional Sulfosalt Structures 43

Chapter 3. ELECTRON INTERACTIONS AND CHEMICAL BONDING IN SULFIDES

(C. T. Prewitt and V. Rajamani)

Introduction. . . . . PR- 1

Electronic Structures of the Elements 2

    Wave equations   2
    Quantum numbers and the hydrogen atom   3
    The periodic system   5
    Filling of the *d* orbitals of the transition metals   5

Group Theory. 6

    Basic principles   6
    Classes of symmetry operations   9
    Representations of groups   10
    Orbitals and energy levels   13

Chemical Bonding in Sulfides. 14

    Crystal field theory   14
    Molecular orbital theory   18
    Band theory   22

Some Aspects of Metal-Metal Bonding in Fe, Ni, and Ni Sulfides. . . . 23

Physical Properties of Transition-Metal Sulfides and Band Theory. . . 27

    Reflectivity   31

Interatomic Distances and Covalency in Sulfides 32

Application of Bonding Theories to Specific Sulfides. 34

    Pyrite   35
    Thiospinels   36
    Pentlandite   38
    Conclusions   41

Chapter 4. EXPERIMENTAL METHODS IN SULFIDE SYNTHESIS

(S.D. Scott)

Introduction . . . . S- 1

Some Principles of Experimentation 1

    Appearance-of-phase method   2
    Equilibrium   3
    Applications to nature   4

Methods. 5

    Evacuated silica tube   5
    Differential thermal analysis   9
    Salt flux   10
    Hydrothermal recrystallization   11

Analysis of Run Products 17

    Microscopy   17
    X-ray diffractometry   18
    Microprobe analysis   23

Presentation of Experimental Results 24

Sulfur Activity: Its Measurement and Control. 26

    S_{L-V} buffer   29
    Dew point   30
    Gas mixing   30
    Electrum tarnish   34
    Pyrrhotite indicator   34
    Electrochemical cells   35

Growth of Large Single Crystals. 36

    Growth from melts   36
    Sublimation   37
    Chemical vapor transport   37
    Growth in halide fluxes   37
    Growth in gels   37
    Hydrothermal growth   38

Chapter 5. SULFIDE PHASE EQUILIBRIA

(Craig and Scott)

Introduction CS-1

Survey of Data Sources on Sulfide Phase Equilibria *(Craig)* 3

Major Sources of Thermochemical Data on Sulfides 20

The Fe-S System *(Scott)*. 21

    Phases   24

        Troilite 24, Mackinawite 25, "Hexagonal" pyrrhotite 26, Monoclinic pyrrhotite 28, "Anomalous" pyrrhotite 29, γ iron sulfide 29, Smythite 29, Greigite 30, Pyrite 30

    Pyrite-Pyrrhotite Solvus   31

    Thermochemistry   32

    Pyrite + Hexagonal Pyrrhotite Solvus   33

        Pyrrhotite field 34, Pyrite field 36, Free energy and heat of formation of pyrite 36

    Effect of Pressure on Phase Relations   36

        FeS activity in pyrrhotite 36, Pyrite + pyrrhotite solvus 37

The Fe-Zn-S System *(Scott)* 41

    FeS Activity and Sphalerite Composition   43
    Sphalerite + Troilite + Iron   44
    Sphalerite + Pyrrhotite   47
    Sphalerite + Pyrrhotite + Pyrite   47
    Iron-rich patches   48
    Effect of Pressure on the Sphalerite + Pyrite + Pyrrhotite
      Solvus: The Sphalerite Geobarometer   51
    Phase Relations Below 300°C   54
    Partial Molar Volume of FeS in Sphalerite (V_{FeS}^{-sp}) 56

The Cu-S System *(Craig)*. 58

    Chalcocite 60, Djurleite 61, Anilite 61, Covellite and "Blue-Remaining" Covellite 61, CuS_2 62

    Thermochemistry   62

The Cu-Fe-S System *(Craig)* 64

    Digenite 70, Bornite 71, X-, Anomalous, or Sulfur-rich Bornite 71, Cubanite 71, Idaite 72, Fukuchilite 73, Chalcopyrite 73, Talnakhite, Intermediate Phase I, Intermediate Phase II 73, Mooihoekite and Intermediate Phase A 73, Haycockite 74, $Cu_{0.12}Fe_{0.94}S_{1.00}$ Phase 74

    Thermochemistry   74

The Ni-S System *(Craig)*. 77

    Heazlewoodite 79, Godlevskite 79, Millerite 79, Polydymite 80, Vaesite 80

    Thermochemistry   80

The Fe-Ni-S System *(Craig)* 82

    Pentlandite 87, Violarite 88, Bravoite 89, $(Fe,Ni)_{1-x}S$ Monosulfide Solid Solution (mss) 89

    Thermochemistry   89

Sulfosalts *(Craig)*. CS-91

    Thermochemistry   93

Stoichiometry of Sulfides *(Scott)* 99

    Silver Sulfide   102
    Molybdenite   102
    Mercury Sulfide   103
    Zinc Sulfide   104

      Nomenclature 104, Nonstoichiometry of zinc sulfide 105,
      Sphalerite-wurtzite inversion 106, Wurtzite polytypes 109.

Chapter 6. SULFIDE PETROLOGY

(P. B. Barton)

(Paper reprinted from *Mineralogical Society of America Special
Paper 3*, 187-198 (1970))

Introduction. . . . . B- 1

Complexity of the Problem of Interpreting Mineral Associations.   1

Thermodynamic Approach to Phase Diagrams.   4

General Factors Influencing the Compositions of Minerals.   7

The Sulfidation State of Natural Environments   9

Goals of Current Research 11

References. 11

REFERENCES (Unified bibliography except for Chapter 6). . . . . R- 1

ACKNOWLEDGMENTS R-32

SUPPLEMENTAL REFERENCES (added for the second printing) R-33

ERRATA AND ADDENDA (continued from page (o)). R-34

SUPPLEMENT TO *Survey of Data Sources on Sulfide Phase Equilibria*
    by J. R. Craig R-36

ERRATA and ADDENDA

compiled for the second printing of *Sulfide Mineralogy*

    This list is continued on the last pages of this book, and it includes, in
particular, an up-dating of the *Survey of Data Sources on Sulfide Phase Equilibria*
by J. R. Craig which begins in Chapter 5, page CS-3.

Page No.

W-27   Table W-1. Sphalerite supercell for haycockite should be "2x2x6."

W-31   Line 8. For pentlandite add reference: Hall and Stewart (1973a).

W-40   Table W-2. Zinckenite formula should be $Pb_6S_{14}S_{27}$.

PR-3   5 lines from bottom. Should read: "by s, p, d, f instead of 0,1,2,3..."

PR-10 C_4 basis vector designation should be $\begin{bmatrix} -y \\ x \\ z \end{bmatrix}$

PR-13   Line 5. $\bar{1}$ should be "T."

PR-13   Lower matrix. $(2x^2-x^2-y^2, x^2-y^2)$ should be "$(2z^2-x^2-y^2, x^2-y^2)$."

PR-35   11 lines from bottom. Should read "Therefore, iron in pyrite is *in a
        low spin state and pyrite is* a diamagnetic semiconductor."

PR-35   3 lines from bottom. Repulaion should read "repulsion."

S-4   Line 18. (1973) should be "(1963)."

S-18   Line 1. Ferromagnetic should be "ferrimagnetic."

S-18   Line 10. 0.05% should be "0.5%."

S-23   7 lines from bottom. Re Desborough *et al.* (1971) references, see also
        critique by Reed (1972) *Am. Mineral.* 57, 1551.

S-24   Line 1. Emmission should be "emission."

S-27   Line 4. Should read $-\dfrac{d \log k}{d(1/T)}$.

S-27   Line 10. Formula should read

$$\frac{d \log a_{S_2}}{d(1/T)} = \frac{\Delta H^\circ}{2.303R} + \frac{2d \log a_{MS_2}}{d(1/T)} - \frac{2d \log a_{MS}}{d(1/T)}$$

S-28   6 lines from bottom. Table S-2 should be "Table S-3."

S-32   Line 10. Fig. S-18 should be "Fig. S-19."
CS-4   As-S. Realgar stability is 265°C, high form of AsS is 307°C.
             Orpiment stability is 170°C, high form of As_2S_3 is 315°C.
        Cr-S. Cr_7S_8 should be "Cr_7S_8."

CS-5   Hg-S. Cinnabar and metacinnabar formulae are both $Hg_{1+x}S$.

CS-6   Mn-S. Haverite should be "hauerite."

CS-6   Os-S. Reference should be Ying-chen Jen & Yu-jen Teng (1973).

CS-8   Ag-Cu-S. Mckinstiyite should be "mckinstryite."

             Stromeyerite formula should be "$Ag_{1+x}Cu_{1-x}S$."

CS-9   As-Pb-S. Add "see also page W-42."

CS-11,12 Cu-Fe-S. Idaite formula should be "Cu_3FeS_4."

             See Table CS-6 should read "see Table CS-7."

             Add references of MacLean *et al.* (1972) and Cabri *et al.* (1973).

CS-13   Fe-Ni-S. Add reference of Harris and Nickel (1972).

CS-14   Pb-Sb-S. Add "see also page W-40."

CS-15   Pd-Pt-S. Reference should be as for Os-S on page CS-6.

CS-15   Add Ag-As-Pb-S. "Marrite = $PbAgAsS_3$, Wuensch (1967)."

CS-15   Add Ag-Bi-Cu-S. Chen and Chang (1974), *Can. Mineral.* 12, 404-410.

CS-15   Ag-Fe-Ni-S. Hall and Steward should be "Hall and Stewart." Add
             Mandziuk and Scott (1975) *Geol. Soc. Am. Abstr. Progr.* 7,
             1187.

FOREWORD

Short courses of mineralogical interest were begun in 1965 in conjunction with the annual meetings of the Geological Society of America. Sponsored by the American Geological Institute Committee on Education and directed by J.V. Smith of the University of Chicago, short courses of feldspars (1965), pyroxenes and amphiboles (1966), sheet silicates (1967), and resonance spectroscopy (1968) were presented by mineralogists with expertise in these subject areas. With each succeeding year the course notes became more comprehensive and formalized, and AGI published these at low cost for distribution in the geological community. Unfortunately, AGI has been financially unable to continue this service.

In 1973 President J.V. Smith surveyed members of the Mineralogical Society of America concerning the desireability of renewing this sort of effort, and in response the M.S.A. Council appointed a committee to initiate a series of short courses under their sponsorship. The mineralogy of sulfides was selected as the topic for the first of these courses, and the primary lecturers, J.R. Craig, C.T. Prewitt, S.D. Scott, and B.J. Wuensch, gathered in Blacksburg, Virginia in May 1974 to organize their presentation. The assistance of V. Rajamani as both co-author and lecturer and P.B. Barton as special lecturer was enlisted, and the work of writing and editing this volume began. The Short Course on Sulfide Mineralogy was given November 15-17, 1974, preceding the annual meetings of the affiliated societies of the Geological Society of America, at the Sheraton Four Ambassadors Hotel in Miami, with eighty persons in attendance.

The editor and committee chairman is particularly grateful to Professors J.V. Smith and S.W. Bailey for their strong support of this effort. As Treasurer of M.S.A., Dr. Philip M. Bethke volunteered much helpful financial advice, and Mrs. Mary Holliman, Managing Editor of *The American Mineralogist*, gave considerable time in technical editing. Cheryl Crum was responsible for most of the typing, assisted by Ramonda Haycocks, Lyn Groover, and Margie Strickler. John B. Higgins proof-read the final copy, and S.A. Kissin kindly provided data from his unpublished dissertation.

Paul H. Ribbe
Blacksburg, Virginia
October 18, 1974

FOREWORD to the FOURTH PRINTING

This book was first published under the title *SHORT COURSE NOTES, VOLUME 1: Sulfide Mineralogy*. Fifteen hundred copies were printed in 1974, 1500 in 1976, and 1500 in 1979. This, the fourth printing (2000 copies), was completed in August 1982 under the new serial title, *REVIEWS in MINERALOGY*, which was adopted in 1980 by the Mineralogical Society of America. The 1976 lists of errata, addenda, supplemental references, and a supplement to the table beginning on page CS-3 are included on page v and pages R-33ff. at the back of this volume.

DISCLAIMER

This book was prepared by the authors for the use of participants in the Sulfide Mineralogy Short Course and for others who desire an introduction to the mineralogy of sulfides. No claim is made that they are complete; they have not been openly reviewed.

Ch. I

DETERMINATION, RELATIONSHIPS, AND CLASSIFICATION

OF SULFIDE MINERAL STRUCTURES

Bernardt J. Wuensch

INTRODUCTION

Sulfur occurs in nature as a major constituent of three mineral groups: the sulfides, the sulfosalts and the sulfates. Many other classes of sulfur compounds exist, of course, and a few are known as mineral species. Julienite, $NaCo[NCS]_4 \cdot 8H_2O$, (Preisinger, 1952) is a thiocyanate; kermesite, Sb_2S_2O (Kupcik, 1967) is an oxysulfide. Voltzite was long described as Zn_5S_4O, but has been shown (Frondel, 1967) to be a mixture of wurtzite and a zinc-bearing organometallic compound. Such minerals are of interest because of the rarity of these compounds in nature. Sulfates are the only salts important in mineralogy because $SO_4{}^{2-}$ represents the final product of oxidation.

The present series of lectures is concerned with sulfides, minerals in which one or more metals are combined with sulfur[*]. Minerals in which more than three metals have a distinct structural role (e.g., hatchite $PbTlAgAs_2S_5$) are rare. The sulfosalts are a sub-group of this family of minerals in which one of the metals is a Group V element such as As, Sb or Bi. Relatively few metals are involved as the second metallic component in sulfosalts - usually Pb, Cu, or Ag, and less frequently Zn, Hg, or a transition metal. Separation of the sulfosalts from sulfides is based upon early chemical concepts and is somewhat artificial. Many sulfosalts are closely related to sulfides, and are accordingly included in the present discussion. Any modern distinction is probably best based upon crystalchemical characteristics. The Group V metal in those structures commonly regarded as sulfosalts has either three S neighbors which, together with the metal, form a trigonal pyramid or, alternatively, a square [1+2+2] pyramidal arrangement of neighbors. In contrast, the coordination of sulfur about the Group V metal in a few minerals is tetrahedral (e.g., enargite, Cu_3AsS_4) or involves metal-metal bonds -- e.g., parkerite, $Ni_3Bi_2S_2$ (Fleet, 1973) or lautite, CuAsS (Kulpe, 1961; Marumo, and Nowacki, 1964). The latter minerals are best included among the sulfides if a distinction is to be made.

Sulfur has the s^2p^4 electron configuration common to all Group VI elements. Its electronegativity of 2.5 is moderate in comparison to the value of 3.5 assigned to oxygen. Still heavier elements in Group VI become progressively more metallic as their atomic number increases. In sulfur and subsequent Group VI elements d orbitals are available for hybridization.

------

[*]Dana's system also includes, under "sulfides", the selenides and tellurides, and minerals in which a Group V element is the electronegative element.

The crystal chemistry of the sulfides thus differs markedly from that of oxygen and is more complex. Several types of bonds might be formed to complete the s^2p^4 subshell: Two electrons might be added to form S^{2-}; one electron might be added and a single electron-pair bond formed (e.g., SH^-); two electron-pair bonds might also be formed or, alternatively, a larger number of bonds involving a variety of possible hybrid orbitals. Close metal-metal separations may give sulfides appreciable metallic character. All of these modes of bond formation are known for sulfur (see Ch. 3). Its coordination in structures is thus extremely variable. Regular and symmetric tetrahedral or octahedral coordination occurs about sulfur in many simple sulfides. In more complex sulfides and sulfosalts, however, extremely distorted coordination polyhedra are encountered. Interatomic distances often range almost continuously over an interval of 1Å or more. The designation of nearest- and second-nearest neighbors and even the specification of a coordination polyhedron thus becomes somewhat arbitrary.

Our discussion of the structural aspects of sulfide mineralogy is divided into two chapters. In this chapter we will review some of the basic concepts of atomic packing and relationships between structures, and we will outline present-day procedures in crystal structure determination. The latter discussion is not intended to be comprehensive. Instead it is presented as background with which the experimental difficulties associated with sulfide structure determination may be contrasted. Chapter 2 is a discussion of sulfide crystal chemistry. The organization of the presentation of this material presupposes some basis for the structural classification of sulfides. The concluding section of this chapter will accordingly discuss the various schemes which have been proposed.

CRYSTAL STRUCTURE DETERMINATION

The following data specify the arrangement of atoms in a crystalline material: lattice constants (the dimensions and interaxial angles of the unit cell); the crystallographic space group (which embraces information about the space lattice type, crystal class, and crystal system); the coordinates (referred to unit cell edges as a basis) of each symmetry-independent atom contained within the unit cell; the occupancy of each site in cases where a position is only partially occupied or statistically occupied by two or more species in disorder; and parameters which describe the thermal vibration of each atom about its equilibrium position. The description of thermal motion sometimes involves an assumption of isotropy -- that is, the root-mean-square displacement is assumed to be independent of the direction of displacement. A more general representation assumes an ellipsoidal variation. In this case a symmetric second-order tensor is required to describe thermal motion. When the atom occupies a position of symmetry, the shape and orientation of the thermal ellipsoid must conform to the point symmetry of the site. This condition requires equalities among certain tensor elements or may require some elements to be identically zero. Along with parameters which depend on the perfection and microstructure of a specimen ("primary" and "secondary" extinction), the above data define completely the diffraction effects which will be produced by a crystal of a given mineral. Correspondence between observed and computed diffraction effects provides the usual basis for establishing the veracity of a structural model.

The position of the diffraction maxima produced by a crystal depend only upon the size and shape of the unit cell. The space group is established (though usually with an ambiguity which may be resolved by recourse to crystal morphology, physical properties, or the statistical distribution of diffracted intensities) from the symmetry of the diffraction effect and systematic absences among the set of possible intensities.

The relative intensities of the diffraction maxima depend on the types and positions of the atoms contained within the unit cell. Measured values for the diffracted intensities thus constitute the raw data from which a crystal structure is determined. Before analysis, however, the intensities must be corrected for physical and geometric factors which depend on the type of diffraction experiment and the shape of the specimen used. The former include an effect due to X-ray polarization, which depends upon the value of the diffraction angle, and the Lorentz factor, essentially a measure of the relative time available for diffraction as the crystal is rotated through the diffracting position. The factor dependent upon specimen shape is due to absorption of both the incident and diffracted beam by the specimen. After making the physical and geometric corrections, the square root of each corrected intensity is proportional to the "structure factor", F, a sort of normalized amplitude for each diffracted beam which depends only upon the diffraction angle and the type and arrangement of atoms in the unit cell.

The process of structure determination from these data may be compared, both literally and figuratively, to the process of formation of a magnified image of a periodic object with a lens. In the optical process, rays scattered from the object combine to form a diffraction pattern in the focal plane of the lens but then proceed on without interruption to combine to form an image. The X-ray diffraction pattern produced by a crystal is strictly analagous to the optical diffraction pattern. Specifically, it is the *Fourier transform* of the scattering density of the object. In the X-ray experiment, however, the process of "image formation" is interrupted at the stage of the diffraction pattern. The intensities of the diffraction pattern cannot, either experimentally or analytically, be routinely allowed to act as a subsidiary source for image formation because the maxima in the pattern do not have the same relative phase. Only the *amplitudes* of the diffraction maxima are recorded in an experiment. The data on relative phases is lost. This is the infamous "phase problem" of crystal structure determination.

Two basic approaches have evolved for solution of the phase problem. A first concerns itself with the information which is contained in the magnitude of the structure factor alone. A Fourier synthesis using each of the structure factors

(*with* their appropriate phase) as coefficients would be equivalent to image formation from the diffraction pattern. Summation of the series provides the number of
electrons per unit volume at all points within the unit cell. Summation of an
analogous series which uses instead the squares of the *magnitudes* of the structure factors as coefficients may be shown to provide a function which displays
maxima at positions which correspond to the vectors between all atoms in the
structure translated to a common origin. The result of this summation is known
as the *Patterson function*. A crystal which contains N atoms per cell will contain N^2 maxima in its Patterson map, of which N represent the interaction of an
atom with itself and thus occur at the origin. Systematic procedures have been
developed which permit the unraveling of the Patterson function to obtain the
crystal structure. The difficulty of this procedure increases rapidly with N.

A second body of procedures is based on the fact that the phases of the
structure factors are not independent of their magnitudes. Certain inequalities
and probability relations may be derived. These permit assignment of phases
through a "boot strap" procedure. Eventually, probable phases may be assigned
to a sufficient number of the larger structure factors that a Fourier series may
be employed to reveal the atomic positions.

The final stage in the determination of a crystal structure consists of
improvement of the trial parameters which have been obtained through either
Patterson or direct-phasing procedures. The refinement was done in early days
through successive Fourier syntheses. With the advent of large high-speed com-
puters in the past decade, refinement is more customarily performed through
least-squares methods. The measured structure factors are used as observations,
and the trial parameters (e.g., atomic positions, thermal motion parameters, scale
factor, and extinction parameters) are adjusted to give a least-squares fit between
observed and calculated structure factors. A moderately complicated sulfide
structure may involve two or three thousand measured structure factors and of
the order of 200 adjustable parameters.

The degree of agreement between the observed and calculated structure factors
provides the measure of the correctness and precision of a crystal structure
determination. A convenient and commonly-used measure of the overall agreement
is the "residual", "reliability index" or "R-factor": $\Sigma\left||F_{obs}|-|F_{cal}|\right|/\Sigma F_{obs}$.
Rearrangement of this definition, $R = \Sigma F_{obs}(|\Delta F|/F_{obs})/\Sigma F_{obs}$, shows R to be a
weighted average of the fractional difference between F_{obs} and F_{cal}, where the
magnitude of F is used as the weight.

A typical value of R in a well-refined structure is today of the order of 2-
7%. This provides atomic coordinates with standard deviations in the range
0.001 to 0.0001 (in terms of fractions of a cell edge). The precision, however,
depends on the scattering power of the atom in question as well as the level of
refinement. The precision of the atomic positions, in turn, translates to a
precision in interatomic distances which is typically ±0.003 to ±0.007Å.

The above discussion is only a qualitative outline of the process of crystal structure determination. The interested reader will find several books devoted to its practical aspects (e.g., Buerger, 1960; Stout and Jensen, 1968).

PROBLEMS IN SULFIDE STRUCTURE DETERMINATION

The procedures for structure determination which were outlined in the preceding section obviously apply to analysis of any type of crystalline material, and are not dependent upon whether the specimen is organic or inorganic, biological or mineralogical in origin. Structural study of sulfides, however, often involves experimental difficulties which make these materials much less amenable to study than other inorganic compounds. The acquisition of detailed structural information has therefore lagged in comparison to study of, for example, the silicates. The gaps in our understanding of sulfide crystal chemistry stem in part from a lack of data, as well as the complexity of these minerals.

Suitable Specimens

A relatively perfect single-crystal specimen of a mineral must be isolated before a detailed structural analysis may procede. The lack of a suitable specimen often impedes a study which, in other respects, would be relatively straight forward. Unlike many other minerals, sulfides are often massive or poorly crystallized. Crystals of some species are almost invariably bent or possess low-angle grain boundaries. Certain sulfides and sulfosalts have pronounced acicular habit, occurring only as bundles of fibers which have their axes of elongation in common.

As will be seen later, many simple sulfides have structures which are based upon close-packed arrays of sulfur atoms. Small displacements of the metal atom positions often cause the true symmetry of the structure to be lower than that of the ideal close-packed array. These structures are pseudosymmetric and thus are commonly and sometimes invariably twinned. Monoclinic pyrrhotite, Fe_7S_8, for example, is markedly pseudo-hexagonal. To the writer's knowledge, only a single untwinned fragment has ever been discovered. The problem of twinning is compounded if, in addition to being pseudosymmetric, the mineral displays a rapid phase transformation to a more symmetric form at a temperature which is lower than that of deposition. For example, chalcocite (Cu_2S) transforms to a hexagonal form at $\sim 105°C$. The low temperature form, usually twinned, was thought to be orthorhombic and pseudo-hexagonal. Low-chalcocite is now known to be monoclinic (Evans, 1971). Twinning need not prevent a crystal structure determination. The X-ray intensities which are measured may be proportioned to the separate members of a twin provided that the members are of unequal volume, or if not all reflections superpose. Such analysis requires the collection of much redundant data and is tedious to apply. Thus this procedure has been attempted in only a small number of problems.

Yet another problem encountered in isolating a suitable specimen may be caused by exsolution following deposition of a mineral. The presence of the second phase may be difficult to detect if the intergrowth is coherent. More insidious types of intergrowths are known in sulfides. Many of the Pb-As sulfosalts, for example, have structures based upon layers of Pb in 9-fold coordination alternating with a Pb-As layer of distorted rocksalt-like structure whose thickness and composition differs slightly among the different minerals of this group. "Single crystals" have been observed (Marumo and Nowacki, 1967b) in which the rocksalt layer may fluctuate between compositions corresponding to more than one unique mineral. Similarly a large number of minerals with slightly different (Pb+Cu):Bi ratios appear to exist as superstructures intermediate to aikinite, $PbCuBiS_3$, and bis-muthinite, Bi_2S_3. Coherent intergrowths of several of these phases have been observed (Welin, 1966).

Chemical and Structural Complexity

Solid solution is present in most of the minerals found in nature. In primarily ionic materials the extent to which solid solution occurs is deter-mined by the radius of the ions involved and is subject to the restriction that (+) and (-) ionic charges balance. This often makes it possible to deduce an ideal composition for an ionic material even though extensive replacement has occurred.

Sulfides and sulfosalts have predominantly covalent or metallic character. The relative number of atoms in a formula unit need not be simple ratios, and furthermore, departures from stoichiometry may occur (see Ch. 5). Many examples might be given. Tetrahedrite, one of the most common sulfosalts, had frequently been described as Cu_3SbS_3 or Cu_3SbS_{3+x}. The mineral was shown through structure determination (Pauling and Neumann, 1934; Wuensch, 1964) to have the rather unlikely composition $Cu_{12}Sb_4S_{13}$. Even this composition is approximate! Recent studies of phase equilibria in the system (Skinner et al, 1972; Tatsuke and Morimoto, 1973) have shown a wide solid solution range $Cu_{12+x}Sb_{4+y}S_{13}$ with $0.11 < x < 1.77$ and $0.03 < y < 0.30$. The stability field does not include either Cu_3SbS_3 or $Cu_{12}Sb_4S_{13}$.

Small amounts of impurity are often problematical in sulfides. It is often not clear whether they represent substitution, whether they play a structural role, or whether they are necessary to stabilize the structure. For example, in hutchinsonite, $(Tl,Pb)_2As_5S_9$, 0.05-0.09 Ag or Cu is normally present. This has been shown not to be an essential component of the phase (Takéuchi et al, 1965). In contrast, similar amounts of the same element have been shown to be necessary to the formation of other minerals. Hall (1967) demonstrated that polybasite, $Ag_{16}Sb_2S_{11}$, is stable $only$ if an amount of Cu greater than 3.1 wt. %, but less than 7.6 wt. % has substituted for Ag.

Even in the absence of solid solution and nonstoichiometry, analytic tech-niques such as wet-chemical or electron microprobe analysis may not be sufficiently

accurate to distinguish between alternative formulae for a sulfide. The
uncertainty of the weight fraction for sulfur in the presence of other much
heavier elements may span several atoms in a corresponding formula unit. Careful
determination of the composition of nuffieldite through a number of analytic
techniques provided excellent fit to $Pb_{10}Cu_4Bi_{10}S_{27}$. Neither the number of metal
atoms nor the number of sulfur atoms, however, were permitted by the space group
of the mineral. A structure determination (Kohatsu and Wuensch, 1973) suggests
$Pb_2Cu(Pb,Bi)Bi_2S_7$ as the appropriate compotision.

A structure determination is frequently the means by which questions of composition are resolved. This information is often obtained with difficulty. The
chemical complexity of a sulfide is often paralleled by structural complexity.
As noted above, the difficulty of a structure determination increases rapidly as
the number of atoms in the unit cell increases. Large unit cells seem the rule
rather than the exception, particularly for sulfosalts. As will be seen later,
six of the 17 known Pb-Sb sulfosalts have one lattice constant in excess of 40Å!
Some of the long-period structures are superstructures, a term which will be
defined below. Analysis of such structures is especially difficult, in part
because information about structural detail is contained in the very weak X-ray
intensities which are known with poor precision, and in part because large
numbers of alternative structural models fit the data equally well.

X-ray Absorption

Present-day counter diffractometer techniques provide means for measuring
diffracted X-ray intensities to an accuracy of a few percent. Before structural analysis with these data may procede, however, it is necessary to reduce
the intensities to a set of structure factors by correction, with comparable
accuracy, for absorption of radiation by the specimen. Many sulfides and sulfosalts commonly involve elements of high atomic number (Pb or Bi, for example)
which have high mass absorption coefficients. Linear absorption coefficients
of 1000 to 1500 cm^{-1} are common for Pb-Bi sulfosalts in CuKα radiation. This
means that as little as 3×10^{-7} of the intensity of an incident X-ray beam may
be transmitted through a specimen only 0.1 mm thick.

The high absorption of X-radiation by sulfides which contain heavy metal
atoms remains a serious obstacle in precise determination of their crystal structures. The magnitude of the diffracted intensities decreases exponentially with
increasing crystal diameter. Small specimens must therefore be used to keep
absorption within manageable bounds. On the other hand, the intensities are proportional to sample volume. The intensity of a reflection thus decreases as the
cube of specimen size in an equi-dimensional sample. Further, as specimen size
decreases, small-scale features of the specimen shape, which would be negligible
in a less-absorbing specimen, become increasingly difficult to measure, and the
precision of the absorption correction decreases correspondingly. A compromise
in specimen size must be sought but, in any case, the magnitudes of the

diffracted intensities are small and are subject to increased standard errors as a result of poorer counting statistics.

A decade ago, sulfide crystal structure refinements were reported with disagreement indices which were often as high as 25 to 40%. Improved corrections for absorption have increased precision, but the level to which refinement procedes in present-day studies is often in the range of 10% to 15%, a factor of 2 to 3 higher than that which is almost routinely available for organic compounds and less-absorbing minerals.

Considerable effort and computation expense are involved in making precise absorption corrections and in performing least-squares refinement of a structure. One may justifiably ask whether such efforts are necessary. Very often the objective of a structure refinement is the study of subtle effects such as site occupancies, thermal motion, or the details of bonding as reflected in minor variations in interatomic distances. Data of the highest precision are required if the results of such studies.are to be meaningful. But, if the objective of a study is the determination of an unknown structure toward the end of establishing relations between minerals in a given family, does refinement to a level of 5% relative to a refinement of 15%, say, provide any advantages other than reduced standard deviations in atomic positions and bond lengths? Sometimes not. In many cases, however, data of the highest quality are required if the identification of metals and interpretation of the linkage of coordination polyhedra is to be made. The scattering of X-rays in heavy-metal sulfides is completely dominated by the metals. In a Pb-Bi sulfide, for example, the metals atoms account for as much as 80% of the scattering matter in the unit cell. Determination of sulfur atom locations in such structures is almost comparable to the problem of locating hydrogen atoms in an organic structure. Accurate data are required if the sulfur atoms are to be located and their positions refined. Accurate positions, in turn, are necessary if the bonding scheme or type of metal is to be ascertained in certain sulfides. Several of the species simultaneously present in sulfides have virtually identical scattering power. Examples are Pb and Bi ($Z = 82$ and 83, respectively) Ag and Sb (47 and 51), or Fe, Cu, and As (26, 29 and 33). Such atoms must be distinguished through subtle differences in coordination geometry and bond distances. The literature contains several examples of instances in which recent refinement of an early structure determination has provided a rather different crystal-chemical interpretation of a structure or assignment of chemical species, even though the general nature of the atomic arrangement proved to be correct: galenobismutite, $PbBi_2S_4$ (Iitaka and Nowacki, 1962); lorandite, $TlAsS_2$ (Zemann and Zemann, 1959; Knowles, 1965; Fleet, 1973) and gratonite, $Pb_9As_4S_{15}$ (Rösch, 1963; Ribar and Nowacki, 1969).

PACKING CONSIDERATIONS

The cohesive forces which maintain atoms in the solid state arise because the energy of an atom or ion pair at infinite separation decreases and passes through a minimum as the atoms approach one another (see p. PR-19 ff.). The depth of the potential at equilibrium separation represents the binding energy of the pair. If a third neighbor X is brought into the proximity of an A atom in an A-X pair, a comparable amount of energy is released, although it may not be as large as that of the initial A-X bond if a repulsive X-X interaction exists. If the bonding between the atoms or ions is non-directional, one can obtain the lowest energy in an array of atoms if the maximum possible number of bonds is formed about each atom.[*] The atomic configuration of minimum energy is thus the array with maximum packing density.

The strategy in obtaining maximum packing density of an array of atoms of different size is not unlike that in organizing a suitcase with maximum efficiency. One carefully arranges the larger items; then the smaller things -- socks, toothbrush and the like -- are put in the remaining holes.

The maximum packing density for a collection of identical spheres in two dimensions is the close-packed layer (Fig. W-1). The unit net is hexagonal with a periodicity a equal to twice the radius of a sphere. A subsequent equivalent layer will be in closest approach to the initial layer if the spheres occ γ one of two sets of "hollows" among the spheres of the initial layer. These sites are mutually exclusive in that they are too close to be simultaneously occupied. If the origin of the unit net is taken at the position of a sphere in the original layer, and this location is denoted by A, the coordinates of the two possible locations for the next layer are B, 2/3, 1/3, and C, 1/3, 2/3. Assume that location B is selected for the second layer. Figure W-1 shows that the hollows for the location of a third layer are (relative to the original net) A, 0, 0, or C, 1/3, 2/3. That is, the third layer may be placed directly over the first, or in the position corresponding to that which was not used for the second. The stacking sequence of successive layers in a close-packed array of spheres may thus be specified by a sequence of symbols A, B, or C. The only restriction, if the array is to be close-packed, is that a given symbol cannot directly follow itself in the sequence.

Two special stacking sequences lead to simple, highly symmetric arrays. It may be noted that the three alternative positions for spheres in successive layers all fall above the long diagonal of the hexagonal net -- that is, above

------

[*] Sulfides have appreciable covalent character, and the bonds about a given atom will be decidely directional. The development presented here thus cannot strictly apply. As will be seen, however, the interstices among close-packed arrays involve neighbors at locations consistent with the orientations of important hybrid orbitals — tetrahedral coordination for sp^3 hybridization. Close-packed arrays are thus encountered in covalent structures even though the bonding is directional.

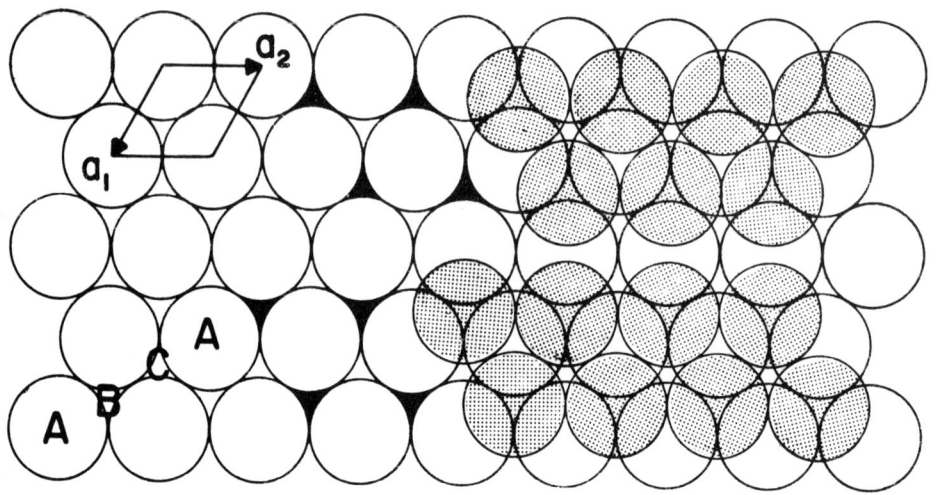

Fig. W-1. A close-packed layer of identical spheres. Alternative positions
for a subsequent equivalent layer are indicated.

[$\bar{1}$ 1]. If the offset of neighboring layers is always in the same sense, either
+1/3 [$\bar{1}$ 1] or −1/3 [$\bar{1}$ 1], a sequence which may be denoted ABCABC... results.
This array has cubic symmetry, a face-centered cubic lattice, and is known as
the *cubic close-packed* configuration (ccp). If the spheres are in contact, the
lattice constant of the array is determined by the fact that a face diagonal is
equal to four times the radius of the spheres. Therefore $a = 2\sqrt{2}r$.

If the direction of offset of successive layers alternates between +1/3 [$\bar{1}$ 1]
and −1/3 [$\bar{1}$ 1], a sequence of layers results which may be denoted ABABAB... (or
by any alternating pair of symbols). This array, which has a primitive hexagonal
lattice, and space group $P6_3/mmc$, is known as the *hexagonal close-packed* (hcp)
array. The lattice constants are $a = 2r$, and $c = \sqrt{32/3}r^{*}$.

Two types of interstices exist among the spheres in the ccp and hcp arrays.
The coordination about one type of site is tetrahedral, that about the second
is octahedral. The location of the tetrahedral and octahedral sites within a
unit cell of the ccp array is depicted in Fig. W-2. The coordinates of these
positions are:

— — — — — — —

*This result is easily derived by reference to the ccp array. The former is a
three-layer structure, so that c_{hcp} equals 2/3 the body diagonal of the ccp cell.
Therefore $c_{hcp} = 2/3a_{ccp} \sqrt{3} = 2/3 (2\sqrt{2})r\sqrt{3}$. $a_{hcp} = 2r$, from which $c_{hcp}/a_{hcp} = \sqrt{8/3}$.

<u>Cubic close-packed array</u>   tetrahedral interstices: octahedral interstices:

1/4	1/4	1/4	1/4	1/4	3/4	1/2	0	0
1/4	3/4	1/4	1/4	3/4	3/4	0	1/2'	0
3/4	1/4	1/4	3/4	1/4	3/4	0	0	0
3/4	3/4	1/4	3/4	3/4	3/4	1/2	1/2	1/2

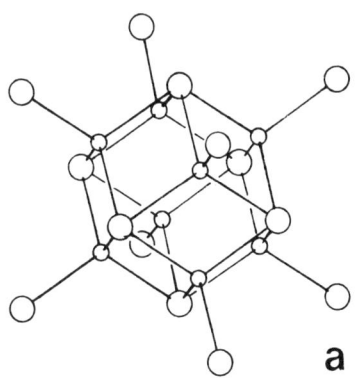 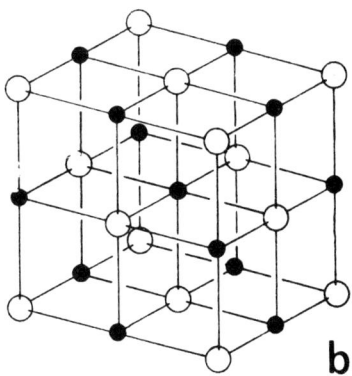

Fig. W-2.   The location of (a) tetrahedrally and (b) octahedrally coordinated
            interstices within a unit cell of a cubic close-packed array of spheres.

These positions correspond to equipoints 8 c $\bar{4}$ 3m and 4 b m3m of space group
$Fm3m$, respectively. From the point symmetry of these sites it follows that both
the coordination tetrahedron and octahedron are *regular* if the array of close
packed spheres is truly face-centered cubic.

    Figure W-3 illustrates the positions of the tetrahedral and octahedral inter-
stices within a unit cell of an hcp array of spheres. The coordinates of these
positions are (if the interior close-packed sphere is placed at 1/3 2/3 1/2):

<u>Hexagonal close-packed array</u>   tetrahedral interstices: octahedral interstices:

0	0	3/8	2/3	1/3	1/4
0	0	5/8	2/3	1/3	3/4
1/3	2/3	1/8			
1/3	2/3	7/8			

These positions correspond to equipoints 4 f 3m and 2 a $\bar{3}m$ in space group $P6_3/mmc$.
Neither tetrahedral or octahedral sites are required by symmetry to be regular.

    The tetrahedral **and** octahedral sites in a ccp array are linked by sharing of
edges. In the hcp array the tetrahedral sites are linked by alternate sharing of
vertices and faces along c, and by sharing of edges normal to c. The octahedral
sites in the hcp array are linked by sharing of faces along c and sharing of edges
normal to c. A ccp arrangement of anions is therefore more favorable in ionic
structures. The ccp array has four spheres, four octahedral sites, and eight

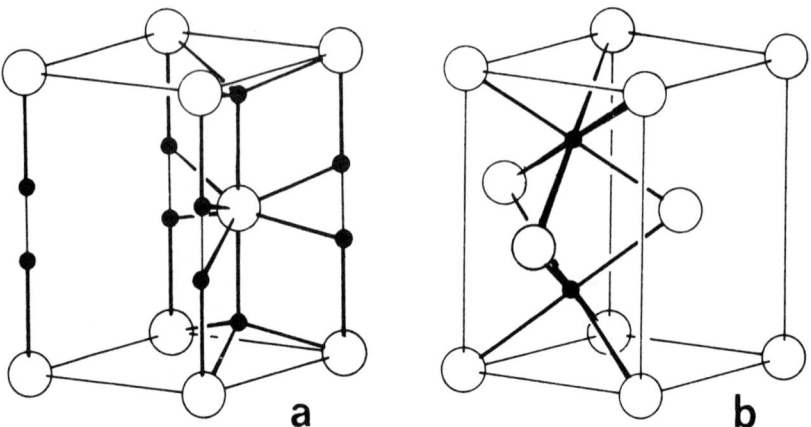

Fig. W-3. The location of (a) tetrahedrally and (b) octahedrally-coordinated
interstices within a unit cell of a hexagonal close-packed array of
spheres.

tetrahedral sites per unit cell. The hcp array has two spheres, two octahedral
sites, and four tetrahedral sites per cell. The number of octahedral sites is therefore equal to the number of spheres in the close-packed array, while the number of
tetrahedral interstices is equal to twice the number of spheres. This relation is
independent of the details of the stacking sequence and may be demonstrated for the
general case.

Figure W-4 depicts two adjacent close-packed layers of spheres. About each
sphere in the lower layer there exist three tetrahedral interstices, plus a

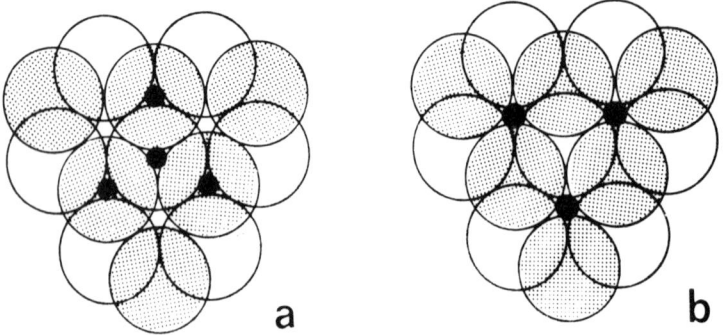

Fig. W-4. The position of (a) tetrahedral sites and (b) octahedral sites between
adjacent layers in an arbitrary close-packed sequence of spheres.

fourth directly above. Each interstice is coordinated by four spheres, so that 1/4 of each interstice "belongs to" each of the spheres. The number of tetrahedral interstices per sphere is thus $4(1/4) = 1$. A similar geometry pertains to the close packed layer directly below the first layer regardless of the stacking sequence. Thus there are always two tetrahedral sites per close-packed sphere. Figure W-4b shows that in the first layer there exist three octahedral sites adjacent to each sphere, and one-sixth of each octahedral site "belongs to" that sphere. A similar situation exists below the first layer, independent of the stacking sequence. The total number of octahedral sites per sphere is thus $2 \times 3 \times 1/6 = 1$.

Quite generally, then, if the sulfur atoms in a sulfide $A_m B_n S_p$ are arranged in a close packed array, regardless of the stacking sequence a fraction $m/2p$ of the available tetrahedral interstices are occupied if the A atoms are tetrahedrally coordinated, and a fraction n/p of the available octahedral interstices are occupied when the B atoms are octahedrally coordinated. The notion of type and occupied fraction of available interstices plays a prominent role in most of the schemes which have been proposed for the classification of sulfide structures.

The sulfur atom array in a few high-temperature sulfide phases is body-centered cubic. This configuration is relatively rare, and the geometry of the interstices will consequently not be examined in detail. Both octahedral and tetrahedral interstices may again be located, but the coordination about neither site is regular. The octahedral sites are located at the mid-points of cell faces and at the mid-points of cell edges. The six neighboring spheres are located in mutually orthogonal directions, but they are not equidistant (four are at $a\sqrt{2}/2$, and two at $a/2$). Tetrahedral interstices are located on cell faces at positions such as 0, 1/4, 1/2. All four neighbors are equidistant from this site (at $a\sqrt{5}/4$) but the interatomic angles are not those of a regular tetrahedron.

POLYMORPHISM AND POLYTYPISM

The alternative stacking sequences discussed in the preceding section would represent different crystallographic forms of a compound containing but one chemical specie. This represents an example of *polymorphism* -- literally "many forms" -- the phenomenon of a given chemical compound existing with two or more distinct crystal structures. Sulfides have provided many of the early and well-known examples of polymorphism. Examples to be discussed below are the wurtzite and sphalerite forms of ZnS and the pyrite and marcasite forms of FeS_2 (*cf*. Ch.5). *Allotropy* is a term having similar meaning except that its use is confined to the description of polymorphism in elements.

Polymorphs have different crystal structures; the differences may be in nearest-neighbor configurations (the graphite and diamond forms of carbon, for example), but often the nearest-neighbor coordination remains the same and only second- and higher-order neighbors are differently arranged. Wurtzite and sphalerite and marcasite and pyrite are polymorphs in which the later situation exists. The difference in energy between alternative configurations may therefore be quite small. Nevertheless, since the atomic arrangements are distinct, the energies cannot be precisely the same, however close, and it follows that at a given temperature and pressure only one polymorph is the stable structure. Remaining polymorphs, if they exist, must be metastable. The existence of a metastable polymorph does not imply a small difference in energy between two structures, but rather that the phase transformation is hindered by kinetics. The slogan "Diamonds are forever" has its basis in kinetics and is thermodynamically false.

The rigorous specification of polymorphism in some systems is difficult and subject to subtle questions. A phase may be stabilized by small concentrations of impurities. In other cases, as will be discussed in Chapter 5 on sulfide phase equilibria, the stoichiometry of one "polymorph" may be slightly different. Such phases are not strictly polymorphous since they do not have precisely the same chemical composition. In this discussion of crystal chemistry the concept of polymorphism will be somewhat loosely used to describe a relation between structures for which no gross chemical differences exist. There is clearly a grey area in which use of the term may be disputed.

The permitted stacking arrangements for layers of close-packed spheres represent a special type of polymorphism. A two-dimensional structural unit is common to all the configurations, and they differ only in stacking sequence, that is, in their arrangement in a third direction. This special type of polymorphism is termed *polytypism*. The phenomenon is possible in layer structures in which two or more alternate positions exist for the placement of subsequent layers. A total of three positions are involved for close-packed structures. Four positions occur, however, in silicides and the class of intermetallic compounds known as Laves phases.

The literature on polytypism is complicated by the fact that several different schemes for denoting a polytype have come into use. For close-packed structures the sequence of letters *ABC*... etc. serves not only to identify the polytype but also to completely specify its crystal structure. An alternative description, involving triangles and deltas to indicate the direction of the displacement of subsequent layers has come into use, primarily in literature concerned with stacking faults. The information conveyed is the same. The triangle-delta notation has a slight advantage in that its symbolism is independent of the choice of origin of the cell. That is, the same sequence of symbols results for both close-packed sequences *ABABAB*... and *BCBCBC*..., which are obviously equivalent. Both conventions have liabilities. The detailed crystal structure of a polytype must be determined before a sequence of symbols may be assigned. Secondly, the sequences of symbols may become extraordinarily unwieldy. This may be appreciated by realizing that there exists a polytype of SiC which has a lattice constant of approximately 1500Å; 594 layers must be placed before the stacking sequence repeats!

Ramsdell (1947) proposed a compact notation for designation of close-packed polytypes which avoids these problems. The notation consists of two parts: A number specifying how many layers are contained within the periodicity (that is, the c axis dimension divided by the thickness of one layer) is given first. This is followed by a letter giving the lattice type (H = hexagonal, R = rhombohedral, C = cubic). The ccp sequence is 3C and the hcp arrangement 2H. The 1500Å SiC structure referred to above is 549R. While the crystal structure is *not* specified by this notation, the symbol may readily be established from the information available in a diffraction pattern. One might ask, however, whether the notation is unique. Might several polytypes exist with the same number of layers and lattice type? This problem does indeed occur. Two polytypes each are known for SiC 10H, 36H, and 51R. The distinction between each is made by the addition of arbitrary subscripts a and b to the Ramsdell symbol.

Two additional closely-related symbolisms have come into common use for denoting the structures of close-packed polytypes. A system introduced by Jagodzinski (1954a) assigns either the letter h or c to a layer depending upon whether the layers adjacent on either side are displaced in the opposite directions along [$\bar{1}10$] (h for hcp-like) or displaced in the same direction (c for ccp-like). Thus the A layer in the sequence *BCACB* would be given the symbol h, while the A layer in the sequence *BCABC* would be given the sumbol c. The advantage of this notation is that, again, it is independent of the choice of origin of the cell. Further, an economy in symbols may result (by a factor of up to 1/3). To illustrate, the stacking sequence in the close-packed structure of Sm is

> *ABABCBCAC*   *ABAB* -----   (9R)
> hhchhchh   c --------   hhc

It might appear that details of the structure are lost through this condensation, but Jagodzinski (1954b) has given rules for resurrecting this information.

(a) If the number of h symbols is odd, the structure is hexagonal (space group $P6_3/mmc$) and contains within c a number of layers equal to twice the total number of symbols.

(b) If the number of h symbols is even, assign ±1 to each symbol, changing sign after each h. (For example, for Sm, hhc, one could write either +1-1+1 or -1+1-1).

(b-1) If the sum of these integers is not a multiple of 3 the structure is trigonal, space group $R3m$ with three times as many layers within c as symbols (e.g., Sm).

(b-2) If the sum of these integers is equal to a multiple of 3, then the structure is trigonal, space group $P3m$, with the same number of layers as symbols.

A somewhat more compact notation, closely related to that of Jagodzinski, has been proposed by Zhdanow (1945). The stacking sequence is represented by a series of integers each of which represents the number of symbols following each h in the Jagodzinski notation. Thus Sm, hhc, becomes (12); SiC 21R, hcchccc becomes (34). It should be noted that the Zhdanow notation specified the *total* number of layers within c -- for example SiC 10H$_b$, hcccc is (55) and not (5), since, using Jagodzinski's rules, the number of layers in the cell is twice the number of symbols.

As with the notion of polymorphism, there exist fine points in connection with a precise definition of polytypism. We have seen, for example, that the configuration about the tetrahedral sites among a ccp array is constrained by symmetry to be regular ($\bar{4}3m$) while the symmetry about tetrahedral sites in an hcp array is lower ($3m$). Should structures formed by placing atoms in the tetra-hedral sites in the two arrays be considered polytypes if minor distortions cause the structure of the tetrahedral layer to be slightly different in the two cases? Again, we will adopt a rather loose definition and consider the struc-tures to be polytypes if there is close structural similarity in the two-dimensional unit.

There are many interesting questions associated with polytypes. How can a layer sense its correct position when the translation-equivalent layer may be as far away as 1500Å? Many complex stacking sequences are believed to be repli-cated by growth on screw dislocations with large Burgers vectors, but this inter-pretation is not completely successful. Additional discussion and a general survey of polytypism may be found in a book by Verma and Krishna (1966) speci-fically devoted to the topic.

DERIVATIVE STRUCTURES

Quite frequently a complicated crystal structure may bear a close relation-ship to a simpler atomic array. A relation is often intuitively recognized when examining models of crystal structures and verbalized in statements such as "Sphalerite resembles diamond, except that half of the black spheres are replaced by red spheres" or that an overall arrangement of spheres is essentially close-packed, but distorted. Buerger (1947) termed these more complex arrays *derivative structures*. In derivative structures the symmetry of a simpler parent array is degraded through perturbations which may sometimes suppress certain symmetry operations in the basic structure. The symmetry of the derivative structure may thus be a subgroup of that of the parent configuration. If one of the operations which is suppressed is translation, the derivative structure has a unit cell with a larger volume than that of the basic structure. The lattice constants of the derivative structure may then be expressed as some vector sum of the translations in the parent structure. This special class of derivative is termed a *superstructure*. The relation is manifested in diffraction patterns by the presence of a set of intense reflections (*substructure reflections*, which closely resemble the diffrac-tion effects produced by the parent structure) plus a collection of weak reflec-tions which indicate the presence of a larger cell and which contain information on the nature of the perturbation which produced the increased periodicity.

There are four basic mechanisms through which derivative structures may be formed:

(a) substitution of one or more types of atom for another

(b) ordered omission of atoms

(c) addition of atoms to a site which is unoccupied in the parent structure (termed a "stuffed" derivative by Buerger)

(d) distortion of an array.

Two, three, or sometimes four mechanisms may be operative in a given derivative.

Sulfide and sulfosalt structures provide a rich collection of examples of derivative structures. Among the structures to be discussed in the following chapter, the following examples may be selected to illustrate each mechanism.

Substitution:

$ZnZnS_2$ (sphalerite) $FeSS$ (marcasite)

$CuFeS_2$ (chalcopyrite) $FeAsS$ (arsenopyrite)

Omission and distortion:

Fe_8S_8 (troilite)

Fe_7S_8 (pyrrhotite)

Addition ("stuffing") and substitution:

$\square BiBiS_3$ (bismuthinite)

$CuPbBiS_3$ (aikinite)

Distortion:

MS (NiAs-type, hexagonal)

CrS (monoclinic)

Substitution, omission, addition and distortion:

$Zn_{12}Zn_4S_{16}\square$ (sphalerite)

$Cu_{12}Sb_4S_{12}S$ (tetrahedrite)

"COMPOSITE" STRUCTURES

Another type of relationship occurs in sulfide and especially sulfosalt structures which is less frequent in the crystal chemistry of other minerals. Very complex structures are not infrequently found to contain blocks or domains of simpler structure. These are, in turn, joined in an irregular fashion. The coordination of atons is quite regular within such regions (the rocksalt structure type is a common example), but highly irregular arrangements occur about the atoms at the boundaries of the domains. Such structures are not derivative structures, and no terminology has been proposed to describe them. In the literature of other minerals, one sometimes encounters expressions such as "twinning of unit cells", but these are not satisfactory. In the present discussion such arrays will be denoted as *composite structures*.

Many examples of composite structures exist among the sulfosalts which contain Pb. The plagionite group $Pb_{3+2n}Sb_8S_{15+2n}$ (with n = 0 to 3) consists of a family of homologous monoclinic structures in which two lattice constants remain invariant while the third increases with n. The structures consist of slabs of rocksalt structure alternately parallel to (112) and ($1\bar{1}2$) as one procedes along the variable lattice constant (Cho and Wuensch, 1970, 1974; Kohatsu and Wuensch, 1974). Successive members of the series differ only in the width of the rocksalt slab. Other Pb sulfosalts such as cosalite, $Pb_2Bi_2S_5$ (Weitz and Hellner, 1960), ramdohrite (Kawada and Hellner, 1971), lillianite, Pb_3BiS_6 (Takagi and Takeuchi, 1972) and several synthetic Pb-Bi sulfides (Otto and Strunz, 1968) have large regions of rocksalt structure. In the latter structures these domains are "polysynthetically twinned" on (311).

Other more subtle composites exist in more complex structures. Kobellite is a disordered Bi-Sb lead sulfosalt of composition $Pb_{6-x}(Bi,Sb)_{7+x}(Fe,Cu)S_{17.5}$ with 0<x<1, which displays two large lattice constants a = 22.5, b = 34.0Å. The structure (Miehe, 1971) is a composite of two regions, one of which resembles the structure of cosalite, $Pb_2Bi_2S_5$, and the other jamesonite, $FePb_4Sb_6S_{14}$ (Niizeki and Buerger, 1957). The two domains partially overlap. Each domain is not only extraordinarily similar to the parent structure, but the disordered Bi and Sb are partitioned according to the nature of the domain: The sites in the cosalite region contain primarily Bi, those in the jamesonite region primarily Sb, while the Group V metal sites common to both the cosalite and jamesonite regions contain comparable fractions of both Sb and Bi.

CLASSIFICATION SCHEMES

It is convenient to have a basis upon which to classify the many minerals found in nature. A scheme used to establish a set of cubby holes may be quite arbitrary. One could use chemical composition as a basis, crystallographic symmetry, or even some physical characteristic such as color. Ideally, however, a classification scheme should possess some predictive power. (Color provides a satisfactory basis for classifying paint, but has obvious inadequacies for other materials). The mode of linkage of SiO_4 tetrahedra and the reflection of this linkage in chemical composition provides a powerful basis for classifying silicates, and the resulting organization is one of the great successes of crystal chemistry. A comparable and equally satisfactory scheme for sulfides has proven more elusive.

Early reviews of the crystal chemistry of sulfides and sulfosalts (e.g., Gruner, 1929), when the structures of only a few simple minerals were known, recognized that many of these phases were based upon close-packed arrangements of sulfur atoms and that the distinction between sulfides and sulfosalts was somewhat artificial. Important differences from oxides and silicates were also noted, however, such as the presence of S_2 "radicals" in minerals such as pyrite.

The availability of additional crystallographic data (although not structures) prompted Berry (1943)[*] to propose a classification based on cell dimensions. Relationships were apparent, but the arrangement was not completely satisfactory, and many apparent similarities have not been supported by subsequent structure determinations. The early 1950's saw a spurt of activity in the determination of sulfide and sulfosalt structures. Results for about 18 minerals were reported although, as noted by Berry (1965), fewer than half of these studies provided refinement to a residual under 20%. Nevertheless the results gave support to the observation that even the more complicated sulfides and sulfosalts had structures that, at least in part, consisted of close-packed arrays of sulfur.

Hellner (1958) provided a review of known sulfide structures and classification on the basis of a "formula factor", $f_1 = f_1^O + f_1^T$, where f_1^O is the ratio of the number of octahedrally coordinated atoms to the number of S atoms and f_1^T is a similar ratio for tetrahedrally coordinated atoms. As shown in preceding sections, this "factor" provides a measure of the fraction of available sites occupied; $f_1^O < 1$ or $f_1^T < 2$ implies either that less than the maximum number of T or o sites is filled, or that certain close-packed S atom locations are "split" (e.g., at the boundary between domains in a composite structure). Structures were divided into five groups, each of which were then subdivided according to the nature of the S atom packing (ccp, hcp, or other cp sequence) and further subdivided according to the magnitude of the "formula factor". The major groups were I - metals in tetrahedral sites, II - metals in both tetrahedral sites and octahedral sites, III - metals in octahedral sites, IV - metals in tetrahedral and/or octahedral sites with only a portion of the sulfur atoms in close-packing, and V - structures which did not fit the classification. The disadvantage of the classification is that an atom in a nominally octahedral site may, in fact, have only three or four nearest neighbors. Details of the structure and some correlative features are therefore lost. Further, a significant fraction of the structures known even then fell into categories IV and V.

Most attempts at classification after 1960 have been concerned primarily with the sulfosalts since these minerals have been the most problematic. In addition, a large number of precise studies have been successfully conducted in recent years. Nowacki (1969) presented an extensive compilation of the crystal data, bond distances and angles, and literature for sulfosalt structure determinations which provided a complete summary of studies undertaken up to that date. A sulfosalt is regarded as having general structural formula $[B_m S_n | S_p | A(1)_q | A(2)_r]$. The B atoms represent the Group V metals, and the $B_m S_n$ unit was taken to represent the structural network built of BS_3 pyramids or BS_4 tetrahedra. The p remaining S atoms are considered "extra" atoms which are not a part of the grouping

- - - - - - -

[*]Berry, L. G. (1943) Studies of mineral sulpho-salts, VII. A systematic arrangement on the basis of cell dimensions. *Univ. Toronto Studies, Geol. Ser.* 48, 9-41.

of Group V polyhedra. The A(1) atoms are metals of low (2 to 4) coordination number, while A(2) represents metals atoms with coordination numbers in the range 6 to 9. Classification is based upon a factor, ϕ, defined as the ratio of the number of S atoms to the total number of Group V metal atoms. Each range of ϕ is then subdivided according to linkage of Group V metal pyramids or tetrahedra (e.g., isolated units, ribbons, rings, etc.). The classification system thus focuses attention on the Group V metal coordination polyhedra and their linkage, an interpretation which ascribes to these polyhedra a role similar to that played by SiO_4 tetrahedra in determining the crystal chemistry of silicates. The system possibly should be extended and refined through distinction between the trigonal pyramid formed with As and the [1+2+2] square pyramid formed about Bi and, in some cases, Sb.

Concurrent with Nowacki's classification, Takéuchi and Sadanaga (1969) proposed a scheme which also focussed upon the Group V metal polyhedra and their linkage. An important distinction was made between the geometry of bridging and non-bridging B-S-B bonds. The structural feature dominant in controlling sulfosalt structures was considered to be the relative sizes of the BS_3 groups and the polyhedra formed about the other atoms in the structure. In view of the irregular nature of coordination polyhedra in sulfides, the assignment of radii is difficult. Takéuchi and Sadanaga therefore take the principal quantum number of the valence electron shell of these metals to be a measure of "size". When more than one metal atom is present, an average is taken over atoms in a formula unit. A second quantity (the ratio of the sum of principal quantum numbers of the metal atoms to a corresponding sum for Group V metals in the composition) was taken as a measure of the relative volumes occupied by the two species. These two quantities were used to define a two-dimensional field. When the compositions of sulfosalts were used to place them in this diagram it was found that four major areas occurred: I - sulfosalts containing Cu as a major component, II - sulfosalts containing Ag, III - the Pb sulfosalts and other sulfosalts known to be closely related to the rocksalt structure, IV - the majority of Pb sulfosalts and those having relatively high Group V metal content. Area IV could be further subdivided on the basis of the type of Group V metal, providing subfields containing related structures with similar Pb coordination groups. The classification of Takéuchi and Sadanaga is not only based on geometric criteria which control structure, but would also appear to have some predictive power.

A very recent summary and classification of simple sulfide structures has been prepared by Zoltai (1974). The classification is again based upon packing considerations, but the concept of packing has been generalized to include incomplete and corrugated layers and the larger variety of interstices associated with such generalization. A notation is developed to permit effective specification of these variations in structures. The result is a classification which now successfully includes many structures which could not be accomodated in a system based upon ideal close-packing. Copies of the compilation are available at cost from the author.

Ch. 2

SULFIDE CRYSTAL CHEMISTRY

Bernhardt J. Wuensch

INTRODUCTION

This section will present a broad survey of the crystal chemistry of sulfides and sulfosalts. The space available is obviously not sufficient for either an exhaustive survey or even a complete description of the varied and complex structures to which reference will be made. Although this review must be incomplete, it will be addressed toward three main objectives. An attempt will be made to describe structures and cite literature for the majority of minerals discussed in subsequent chapters concerned with bonding and phase equilibria. We will also attempt to present examples of the frequent close relationships between very complex structures and those which are the simple and fundamental structure types. This thread of unity between the simple and complex is one of the most interesting and important themes of sulfide crystal chemistry. Finally, it is hoped that this chapter will provide at least a partial guide to the literature on sulfide crystallography.

No uniform system of classification will be used to organize the discussion. At some points it is convenient to separate structures according to bond type or on the basis of chemical composition. At other junctures it is valuable to use derivative or composite structure relationships to bring together on structural grounds a group of minerals of diverse chemistry. At yet other points it is of interest to separately discuss a specialized subfamily of minerals which illustrate in striking fashion a particular relation between composition and structure. The organization follows the general outlines of a presentation given elsewhere (Wuensch, 1972) and, to some extent, the present discussion is an expansion and update of that review.

ELEMENTAL SULFUR

The ability of sulfur to form covalent bonds is reflected in the ease with which repeated S-S bonds may form to create chains or rings. One example is provided by the polysulfide ions, S_n^{2-}. Ions are known in which n ranges from 2 up to at least 6. The thionate ions $S_nO_6^{2-}$ have analogous structure but instead terminate with a pyramid of oxygen atoms. Further examples are provided by the large number of complex structures assumed by elemental sulfur.

The stable form of sulfur at normal temperature and pressure is orthorhombic. The structure (Fig. W-5a) contains a puckered, crown-like S_8 molecule (Abrahams, 1955, 1961). The molecules are efficiently packed, and the intermolecular sulfur-sulfur contacts range from 3.31 to 3.83Å. Above 95.4°C sulfur assumes a monoclinic structure in which similar S_8 molecules are packed in an arrangement which has orientational disorder (Sands, 1965). The monoclinic gamma form of sulfur (Watanabe, 1974) also contains this molecule. The average S-S bond distances in the S_8 molecules are 2.048, 2.06 and 2.044Å for α-(orthorhombic), β-, and γ-S_8, respectively. These values are all somewhat less than the 2.08Å separation proposed for a single bond by Pauling. It has been suggested that this reflects some double-bond character due to the involvement of d orbitals.

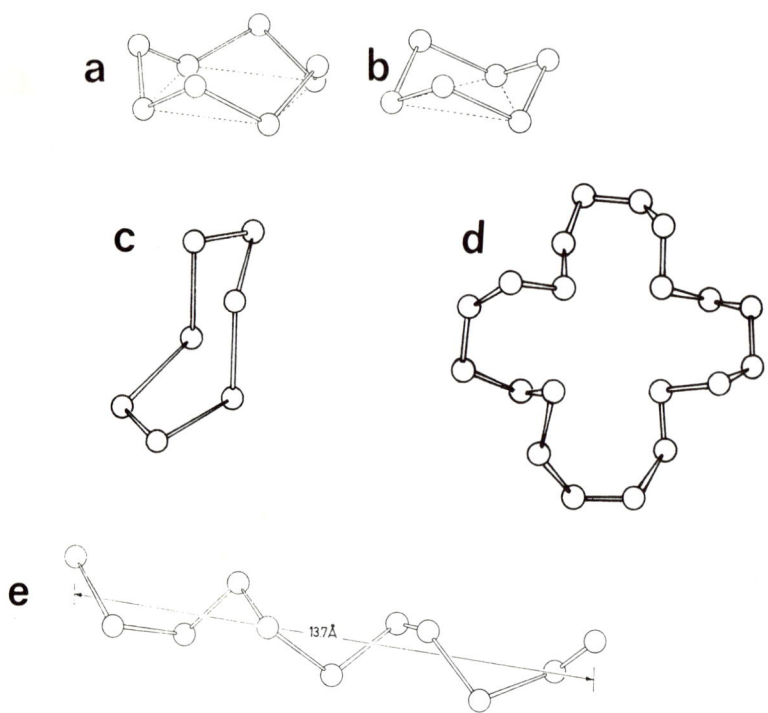

Fig. W-5. Atomic arrangements in polymorphs of elemental sulfur (a) S_8 ring found in α, β, and γ sulfur (b) S_6 ring in rhombohedral sulfur (c) S_7 (d) S_{20} (3) helical chain found in fibrous forms of sulfur.

A rhombohedral form of sulfur may be crystallized from solution. This modification has been found (Donohue et al, 1961) to contain an S_6 molecule (Fig. W-5b). These molecules are arranged normal to c in a cubic close-packed sequence with intermolecular contacts (3.20 to 3.75Å) which are comparable to those in orthorhombic sulfur. More complex molecules have recently been prepared. A form has been synthesized which contains an S_7 molecule with chair configuration (Kawada and Hellner, 1970 -- Fig. W-5c), while a large clover-leaf-like molecule (Fig. W-5d) has been found in S_{20} (Debaerdemarker et al, 1973). Two different configurations are found for molecules S_{18} (Debaerdemaker and Kutoglu, 1973). A purple paramagnetic form of sulfur may be condensed from a vapor. It is believed to contain S_2 pairs (Rice and Sparrow, 1953). A green form is thought to consist of S_8 chains (Rice and Ditter, 1953).

Three forms of sulfur have been synthesized under the influence of pressure (Geller, 1966). These structures are closely related and appear to consist of helical chains containing ten atoms and three turns within a characteristic 13.7Å period (Fig. W-5e). These structures bear close resemblance to the fibrous sulfur which may be prepared by stretching quenched amorphous sulfur (Tuinstra, 1966; Geller and Lind, 1969; Donohue et al, 1969).

The ability of sulfur to form S-S bonds has no counterpart in oxide or silicate mineralogy. The occurrence of S_2 pairs has also been found in a few sulfides and sulfosalts. Among these are the pyrite and marcasite forms of FeS_2 and covellite (CuS) which are discussed in detail below. Patronite $V(S_2)_2$ (Allmann et al, 1964; Kutoglu and Allmann, 1972); livingstonite, $HgSb_4S_8$ (Niizeki and Buerger, 1957a) and $Cu_4Bi_4S_9$ (Ozawa and Takéuchi, 1972) also contain such groups.

IONIC SULFIDES

Sulfur forms purely ionic bonds only with very electropositive ions of low valence. These compounds assume simple structures characteristic of ionic materials. The alkaline-earth sulfides assume the rocksalt structure type in which all of the octahedral interstices are occupied among a face-centered cubic array of anions (Fig. W-2b). The alkali sulfides assume the antifluorite structure type (Fig. W-2a). The cations are arranged in face-centered cubic packing; sulfur ions occupy all available tetrahedral interstices.

DISULFIDES AND DERIVATIVES

Disulfides assume structures which do not have analogues among the more ionic oxides. Compounds such as TiS_2, ZrS_2, and SnS_2 assume the hexagonal CdI_2 structure type depicted in Fig. W-6a. The metal atoms are octahedrally coordinated by S. These octahedra share edges to form a continuous layer. Such layers are stacked directly above one another in a simple hexagonal sequence. The structure may be viewed as a hcp array of sulfur atoms, with half of the octahedral interstices -- those on alternate layers -- vacant. Tungstenite, WS_2, and molybdenite, MoS_2 (Dickinson and Pauling, 1923), assume a somewhat similar but distinct layer structure in which the 6-fold coordination about the metal atoms is prismatic rather than octahedral. A rhombohedral three-layer polytype is known for molybdenite (Takéuchi and Nowacki, 1964). The latter structure is also assumed by NbS_2.

Two additional disulfide structure types are important in sulfide mineralogy. Pyrite, FeS_2 (Bragg, 1914; Elliott, 1960) has the cubic structure shown in Fig. W-7. Iron atoms are in a face-centered cubic array. A covalently-bonded S_2 pair (S-S = 2.14Å) occupies positions corresponding to that of the anion in the rocksalt structure type. The cell contains four such pairs, each oriented along one of the four distinct body-diagonals of the cell. Pyrite may thus be regarded as a derivative of the rocksalt structure type. The Fe atoms accordingly form an octahedron about the S_2 pair. Each sulfur atom is coordinated by three Fe and one S atom in a distorted tetrahedral configuration; each Fe is octahedrally coordinated by S.

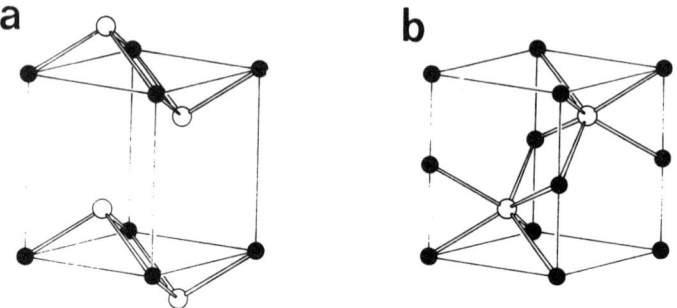

Fig. W-6.   (a) The CdI_2-type layer structure.   (b) The NiAs structure type. (Solid circles represent metals atoms, the open circles sulfur.)

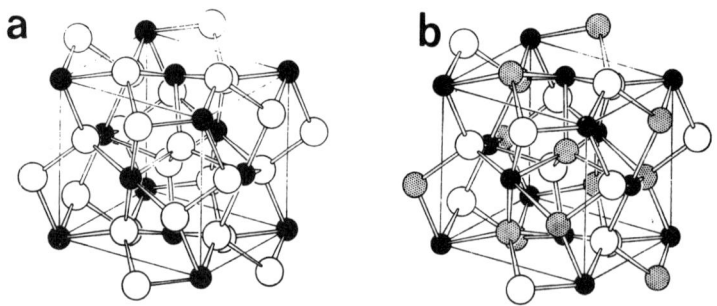

Fig. W-7.   (a) The structure of pyrite, FeS_2 (solid circles represent Fe, open circles S atoms).   (b) The structure of ullmannite, NiSbS, a derivative of the pyrite structure (solid circles represent Ni, shaded circles Sb, open circles S atoms). After Wuensch (1972).

Members of the cobaltite group (cobaltite, CoAsS; gersdorffite, NiAsS; and ullmannite, NiSbS) have structures closely related to that of pyrite. The Group V metal replaces one of the sulfur atoms in the S_2 pair. There was much disagreement in the early literature on whether the substitution occurs in an ordered fashion or in a random manner which would cause the structures to be statistically isostructural with pyrite. It is now clear that three distinct forms exist for at least some members of this group. All three minerals exist in disordered forms (Giese and Kerr, 1965; Bayliss, 1968) isostructural with pyrite. The structures of ordered isometric forms have been determined for ullmannite, NiSbS (Takéuchi, 1957) and gersdorffite (Bayliss and Stephenson, 1967). These structures are sub-stitution derivatives of pyrite; the structure of ullmannite is depicted in Fig. W-7b. A third form exists, however, in which atomic displacements cause a dimen-sionally isometric structure to have orthorhombic symmetry (Giese and Kerr, 1965; Bayliss and Stephenson, 1968).

Marcasite is an orthorhombic polymorph of FeS_2 (Buerger, 1931; Brostigen et al, 1973) whose structure is depicted in Fig. W-8a. The nearest-neighbor con-figuration is the same as that in pyrite (S-S = 2.21Å). Arsenopyrite (Buerger, 1936) is a derivative obtained from the marcasite structure type in a fashion analogous to the manner in which ordered members of the cobaltite group are derived from pyrite. Gudmundite, FeSbS has essentially the same structure (Buerger, 1939). The crystal chemistry of arsenopyrite, however, is subtle. The As:S ratio may vary from 1.22 to 0.82. The structure is triclinic (Morimoto and Clark, 1961) but the As-rich phases approach monoclinic symmetry and a cell which is dimensionally orthorhombic.

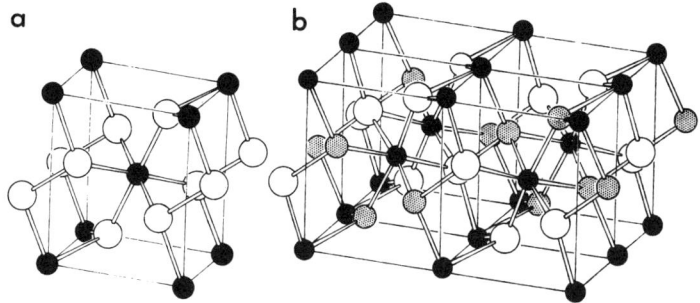

Fig. W-8. (a) The crystal structure of marcasite FeS_2 (solid circles represent Fe, open circles S atoms) (b) The structure of arsenopyrite, FeAsS (shaded circles represent As). After Wuensch (1972).

MONOSULFIDES AND DERIVATIVES

The monosulfides, as a group, behave in two dissimilar fashions. Many assume a common structure type which is characteristic of sulfides. Alternatively, a few assume unexpectedly complex and unique structures which are shared by no other mineral. Covellite, CuS, is one of the minerals whose simple composition gives no indication of an unexpectedly complex structure, Fig. W-9. The structure (Oftedal, 1932; Berry, 1957) contains one type of copper in tetrahedral coordination. The tetrahedra share corners to form a continuous layer. Two such layers share apices; a second type of copper is located in trigonal interstices among these apices to build a planar CuS layer. The resulting structure consists of a sheet

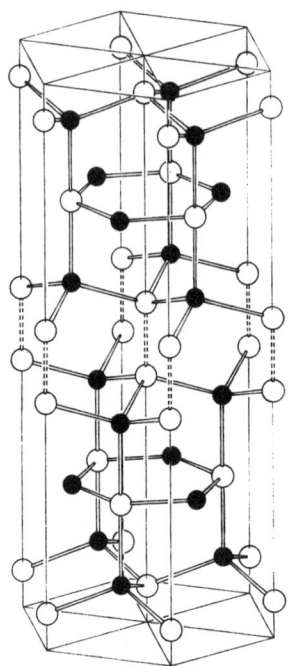

Fig. W-9. The structure of covellite, CuS (solid circles represent Cu, open circles S. The dotted lines indicate covalent S-S bonds.) After Wuensch (1972).

of CuS_3 triangles, sharing corners, sandwiched between a double layer of CuS_4 tetrahedra. Sulfur-sulfur bonds link the layers.

Cinnabar, HgS (Aurivillus, 1950; Auvray and Genet, 1973) has a hexagonal structure which is derived from the rocksalt structure by distortion. Six S neighbors coordinate Hg, but they are grouped in pairs at distances of 2.36, 3.10, and 3.30Å, reflecting the tendency of Hg to form sp bonds. If only the short 2.36Å bonds are considered, the structure consists of helical Hg-S chains with an Hg-S-Hg bond angle of 105.2° which is comparable to that observed in elemental sulfur. Cooperite, PtS (Bannister and Hey, 1932) has a unique structure which arises from the ability of Pt to form square planar dsp^2 bonds; the sulfur atom has tetrahedral coordination. The structure may be visualized in terms of the sphalerite configuration (shown below in Fig. W-10).

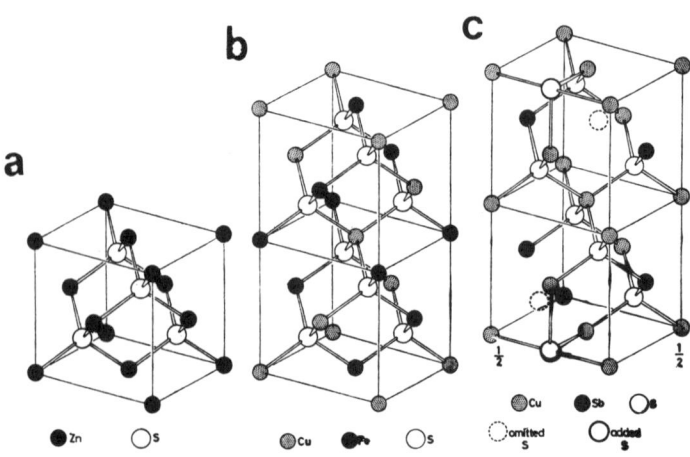

Fig. W-10. The sphalerite structure and some derivatives (a) sphalerite, Zns (b) chalcopyrite, $CuFeS_2$ (c) tetrahedrite, $Cu_{12}Sb_4S_{13}$. After Wuensch (1972).

Imagine S in a fcc array and Pt occupying one-half of the tetrahedral interstices. If the layer of S atoms which occupy the centers of the sides of the cell is replaced by a face-centered arrangement equivalent to the base of the cell, the tetragonal structure of cooperite results, Fig. W-11. It is of interest to note that the S-Pt-S bond angle is 97.5°, intermediate between the value of 90° for a regular square coordination for Pt, and the angle of 109.5° required for a tetrahedral structure.

Tetrahedral Structures and Derivatives

The Group IIB monosulfides crystallize with either the sphalerite structure type (Fig. W-10a) -- Be, CdS, metacinnabarite HgS, MnS, ZnS -- or with the hexagonal wurtzite structure type (Fig. W-12a) -- greenockite CdS, MnS, ZnS. Both are close-packed structures with metal atoms occupying one-half of the available tetrahedral interstices. The two structures may be regarded as polytypes which differ in the stacking of a layer of tetrahedra sharing vertices - a ccp sequence in the case of sphalerite, an hcp sequence in the case of wurtzite. Many additional polytypes of ZnS are known.

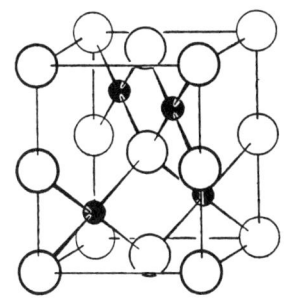

Fig. W-11. The structure of PtS (solid circles represent Pt)

A large number of derivatives of the sphalerite structure type are known. Chalcopyrite, $CuFeS_2$ (Pauling and Brockway, 1932; Hall and Stewart, 1973) is a tetragonal superstructure based upon sphalerite (Fig. W-11b) with an ordered arrangement of Cu and Fe. Stannite, Cu_2FeSnS_4 (Brockway, 1934) has a supercell of similar geometry in which layers of ordered Fe and Sn alternate with layers of Cu. A number of new minerals closely related to chalcopyrite have been very recently described. The compositions and lattice geometries are summarized in Table W-1.

Table W-1. Stuffed derivatives related to chalcopyrite

Mineral	Space Group	Lattice constants (Å)	Sphalerite supercell	Reference
Mooihoekite, $Cu_9Fe_9S_{16}$	$P\bar{4}2m$	a = 10.585 c = 5.383	2x2x1	Hall and Rowland (1973)
Talnakhite, $Cu_9Fe_8S_{16}$	$I\bar{4}3m$	a = 10.593	2x2x2	Hall and Gabe (1972)
Haycockite, $Cu_4Fe_5S_8$	$P222$	a = 10.705 b = 10.734 c = 31.63	2x2x3	Rowland and Hall (1973)

All of these minerals are superstructures derived from sphalerite by the addition of interstitial metal atoms and possible Cu-Fe ordering.

The sulfosalts provide more complex examples of derivatives of the sphalerite structure type. Lautite (Kulpe, 1961; Marumo and Nowacki, 1964) is an ortho-rhombic derivative with $[a,b,c]$ = [3/2, 3/2, 0 / -1/2, 1/2, 0 / 001] $[a_1,a_2,a_3]$ of sphalerite. The coordination of all atoms is tetrahedral, but Cu has 3 S and 1 As as neighbors, As has 1 Cu, 2As + 1S, while the sulfur atoms are coordinated by 3 Cu and 1 As. The As-As separation is the same as found in elemental arsenic. Lautite should be grouped with the sulfides in view of the As-As bonds which are present. Luzonite, Cu_3AsS_4, is a tetragonal derivative of sphalerite with $[a_1,a_2,c]$ = [100/010/002] $[a_1,a_2,a_3]$ (Marumo and Nowacki, 1967). Layers of tetra-hedrally coordinated Cu alternate with ordered layers of Cu and As. Enargite (discussed below) is a polymorph of luzonite derived, instead, from wurtzite.

Lautite, luzonite and enargite (along with members of the cobaltite and arsenopyrite groups) are atypical in that the Group V metal has tetrahedral coordination. (In *only* luzonite and enargite, however, is the group V metal tetrahedrally coordinated only by S.) In most sulfosalts the Group V metal forms either three nearly orthogonal bonds or, alternatively, a [1+2+2] square pyramidal coordination with, perhaps, sp^3d^2 hybridization. Many sulfosalts are therefore derived from the sphalerite structure by omission of the fourth S atom which would have been bonded to the Group V metal. An example is provided by the rhombohedral mineral nowackiite, $Cu_6Zn_3As_4S_{12}$ (Marumo, 1967).

Tennantite, $Cu_{12}As_4S_{13}$ (Pauling and Neuman, 1934; Wuensch *et al*, 1966) and tetrahedrite, $Cu_{12}Sb_4S_{13}$ (Wuensch, 1964) are more complex derivatives which are superstructures with double the cell edge of sphalerite (Fig. W-11c). One-fourth of the metal atom sites are replaced by a Group V metal which forms only three bonds with S. One-fourth of the S atoms in each cell are therefore omitted. This leaves one-half of the Cu atoms with but two sulfur atoms as neighbors and two additional S are therefore added to an interstitial site in the sphalerite array to complete a triangular coordination for these atoms. (The presence of an interstitial *anion* is unique among close-packed structures.) The resulting array may be expressed with the structural formula $Cu_6^{[4]}Cu_6^{[3]}Sb_4^{[3]}[S_{12}^{[3Cu+Sb]}S^{[6Cu]}]$.

Tetrahedrite minerals are capable of incorporating large amounts of other elements in solid solution. Schwatzite, a variety containing Hg, has been shown by Kalbskopf (1971) to preferentially accommodate the mercury atoms in the tetra-hedrally-coordinated Cu site. Freibergite, an Ag-rich specie, incorporates the impurity in the 3-coordinated Cu site (Kalbskopf, 1972), and there is evidence that the interstitial sulfur site is not fully occupied. Much remains to be learned about minerals of the tetrahedrite group. These minerals display a range of stoichiometry but,, as has been previously mentioned, the field of stabi-lity does *not* include either $Cu_{12}Sb_4S_{13}$ or Cu_3SbS_3 (Skinner *et al*, 1972; Tatsuka and Morimoto, 1973). *Pure* tetrahedrites decompose upon cooling to two distinct tetrahedrites with slightly different lattice constants. (The solid solution in naturally-occurring minerals prevents this decomposition.) A new phase, called *pseudotetrahedrite* by Tatsuka and Morimoto (1973), has been observed which has twice the lattice constant of tetrahedrite, i.e., it is a 4x4x4 superstructure based upon sphalerite.

Derivatives of wurtzite are relatively rare, but enargite (Fig. W-12b) provides an example of an orthorhombic superstructure (Pauling and Weinbaum, 1934; Adiwidjaja and Löhn, 1970). Structures in which all atoms have tetrahedral coordination may occur whenever the number of bonding electrons available in the structure is equal to four times the number of atoms. Many synthetic sulfides are among the compounds which meet this requirement. Such compositions have been systematized in detail by Parthe (1964).

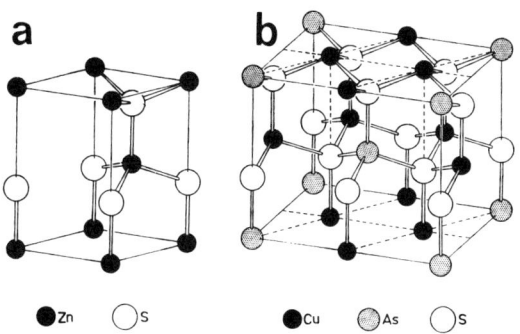

Fig. W-12. (a) The wurtzite structure type. (b) Enargite, Cu_3AsS_4, an orthorhombic derivative of wurtzite. After Wuensch (1972).

The Nickel-Arsenide Structure Type and Derivatives

The NiAs structure type (Fig. W-6b) is a hexagonal structure in which the more electro-negative atoms are arranged in hcp. The remaining atoms occupy all available octahedral interstices. The coordination of the electronegative atom is thus trigonal-prismatic. The octahedral sites are linked along c by the sharing of faces. This configuration would not be assumed by an ionic material. Metal-metal bonding along c undoubtedly contributes to the stability of the structure, and the ratio c/a is almost always smaller than the value of $\sqrt{8/3}$ = 1.633 which obtains for an ideal close-packed configuration. (Values of c/a as low as 1.2 have been noted, but ratios in excess of the ideal value are occasionally found.)

The NiAs structure type is assumed by most of the transition metal sulfides. Non-stoichiometry (usually metal atom vacancies) and derivative structure are very common. CrS, for example, is a distorted monoclinic derivative (Jellinek, 1957). Alternatively, metal atoms may be preferentially removed from certain layers of octahedral sites. Fig. W-6 shows that the CdI_2 structure type results if every-other layer of octahedral sites is ommitted. Solid solution may thus occur between these two structure types, and this is the case in melonite,

NiTe–NiTe$_2$. If the concentration of vacancies is large, the structure of lowest energy will contain an ordered array of vacancies. Jellinek (1957) has shown that what had been believed to be a broad, homogeneous field of non–stoichiometric composition Cr$_{1-x}$S, actually consisted of a closely-spaced series of discreet phases whose compositions (Cr$_7$S$_8$, Cr$_5$S$_6$, Cr$_3$S$_4$, and Cr$_2$S$_3$) reflect different numbers and ordering schemes of vacancies.

Pyrrhotite, Fe$_{1-x}$S, is an extremely complex derivative of the NiAs structure type which is important in mineralogy. Troilite, stoichiometric FeS, has a structure derived from NiAs through small displacements of Fe normal to c and small displacements of S parallel to c (Evans, 1970). A bewildering variety of ordered structures occur among the non–stoichiometric iron sulfides. The equilibrium is discussed in detail in Ch. 5, and is complicated by changes in magnetic structure as well as in the atomic configuration. The crystal structures of two forms of pyrrhotite are known in detail.

One variety has a and b equal to twice the dimensions of the ortho-hexagonal double-cell of NiAs, while the c axis is four times larger. According to current nomenclature this form is loosely referred to as "4c" pyrrhotite. The vacant sites (Bertaut, 1952; Tokonami et al, 1972) are confined to alternate layers of octahedral sites in the NiAs configuration. The arrangement of vacancies with these layers is indicated schematically in Fig. W-13, where the larger open circles indicate the vacant sites. Sulfur atoms as well as the intervening layers of completely filled octahedral sites are not indicated. It may be noted that the arrangement of vacancies is the same in each level. The detailed structure of a second form of Fe$_7$S$_8$ is also available (Fleet, 1971). This variety of pyrrhotite ("3c") is hexagonal with a and c equal to twice and three times the lattice constant a and c of the NiAs substructure. The structure is somewhat analogous to that of the 4c structure in that the array is slightly distorted and the vacant sites occur again on alternate layers of sites.

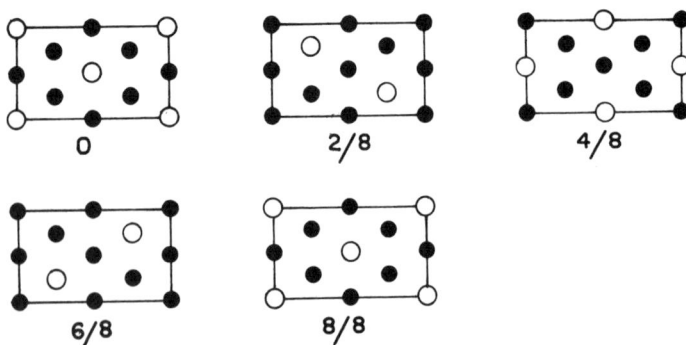

Fig. W-13. Schematic indication of the arrangement of vacancies in monoclinic 4c pyrrhotite. Solid circles indicate Fe atoms occupying octahedral sites in the NiAs-type structure, the larger open circles indicate a vacancy. Sulfur atoms are not shown.

THE NICKEL SULFIDES (Cf. *Ch. 3, p. PR-38; Ch. 5, p. CS-77*)

NIckel forms a large group of sulfides which provide an illustration of the structural chemistry of transition metals. Some of these minerals assume simple sulfide structure types. Vaesite, NiS_2, has the pyrite structure type, polydymite, Ni_3S_4, assumes the spinel structure type, while the high-temperature form of NiS has the NiAs type arrangement.

The synthetic compound Co_8S_9 (Linqvist *et al*, 1936; Geller, 1962), and pentlandite, $(Ni,Fe)_9S_8$ (Linqvist *et al*, 1936; Pearson and Buerger, 1956; Rajamani and Prewitt, 1973) have a more complicated face-centered cubic structure. Pure Fe and Ni analogues of Co_9S_8 do not exist. The unit cell contains 32 sulfur atoms located in two independent positions to provide a 2x2x2 ccp array. The 36 metal atoms per cell are distributed in 4 octahedral interstices and 32 tetrahedral sites. The tetrahedrally coordinated metals are clustered in cubic groups of eight, thus forming a small subunit of anti-fluorite structure which is identical to that depicted previously in Fig. W-2a. It has been suggested that Ni and Fe atoms order on alternate corners of the cubelet of tetrahedral sites, but this could not be confirmed by X-ray analysis. The metal-metal separation between tetrahedral sites is 2.5Å, comparable to the interatomic separation in metallic Ni or Fe. Each tetrahedrally coordinated metal thus has three orthogonal bonds to neighboring metals. The structure of an argentian pentlandite, $(Fe,Ni)_8AgS_8$ (Hall and Stewart, 1973) has been refined, and the octahedral sites were found to be occupied exclusively by the Ag atoms.

Millerite, the low-temperature form of NiS, has a rhombohedral structure in which both the metal atom and sulfur atom have 5-fold coordination. A projection of the structure along *c* (Grice and Ferguson, 1974*; Rajamani and Prewitt, 1974) is presented in Fig. W-14b. Triangles of S atoms are arranged at lattice points of the rhombohedral cell to form trigonal-prismatic columns along *c*. The Ni atoms are located in triangles about the sides of these prisms and acquire a distorted square-pyramidal coordination. The Ni-Ni separation in the triangle (2.534Å) is again equal to the bond distance in metallic Ni (2.492Å). Metal-metal bonds thus serve to stabilize the cluster. The square pyramidal coordination of S about Ni and the "split" location of the sixth neighbor of octahedral configuration (in the present case, the Ni neighbors) is reminiscent of the 7-coordination of S found about Pb and Bi in many sulfosalts. In the latter cases, however, the metal atom is invariably displaced from the base of the pyramid in a direction *away* from the apex of the pyramid, while in millerite the displacement is in the opposite direction.

Heazlewoodite, Ni_3S_2, is a rhombohedral structure based upon a body-centered cubic arrangement of S. The Ni atoms are tetrahedrally coordinated by S with four Ni atoms as next nearest neighbors.

---

*Grice, J.D., and R.B. Ferguson (1974) Crystal structure refinement of millerite (β-NiS). *Canadian Mineral*. 12, 248-252.

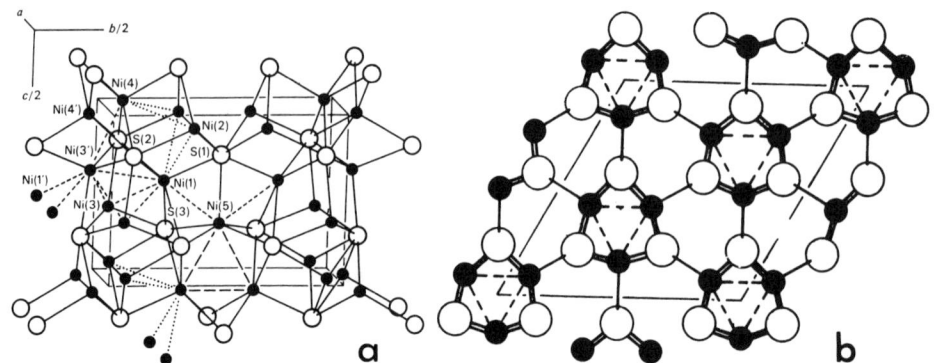

Fig. W-14. (a) The structure of α-Ni$_7$S$_6$ (after Fleet, 1972) (b) projection of the
structure of millerite, NiS, along c (solid circles represent Ni;
dotted lines indicate metal-metal bonds).

Fleet (1972) has recently determined the structure of the synthetic compound
α-Ni$_7$S$_6$. The structure contains three independent S atoms and five independent Ni
atoms. A portion of the structure is depicted in Fig. W-14a. All positions (except
that of one S atom) are only partially occupied. Four of the Ni positions have a
square-pyramidal coordination, and one is in tetrahedral coordination. Each Ni site
has at least one Ni neighbor at a distance equal to the Ni-Ni separation in Ni metal,
and several additional Ni neighbors at distances a few tenths of an Å longer.
Because of the statistical nature of the occupancy of each site, the precise nearest
neighbor configuration of a given Ni atom cannot be specified. Fleet found evidence
for partial longer-range ordering in the structure.

The coordination polyhedra in pentlandite, millerite and heazlewoodite are
depicted in Fig. W-15. It should perhaps again be emphasized that all of these
structures contain clusters of metal atoms stabilized by metal-metal bonds.

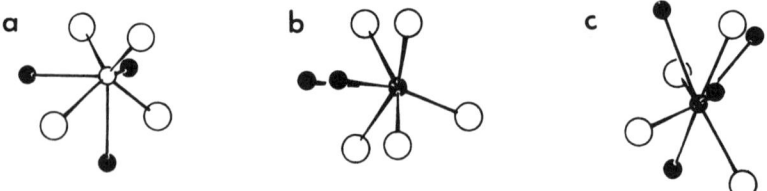

Fig. W-15. The coordination of Ni in (a) pentlandite, (Ni,Fe)$_9$S$_8$, (b) millerite,
NiS and (c) heazlewoodite, Ni$_3$S$_2$.

SILVER AND COPPER SULFIDES (Cf. *Ch. 5, p. 58*)

Copper in sulfides may assume either a tetrahedral or triangular coordination.
Silver has been observed to assume 2-fold coordination, triangular coordination
or a distorted [1+3] flat tetrahedral coordination. The copper and silver sulfides

stand apart from other sulfides in that they frequently display rapid phase transformation at very low temperatures.

The unusual structure of covellite, CuS, has already been described. Chalcocite, Cu_2S, displays an extremely complicated monoclinic structure which is pseudohexagonal (Evans, 1971) and a cell which contains 48 Cu_2S. The structure is based on hcp sulfur atoms. One-third of the Cu atoms occupy triangular interstices within the close-packed S layers to form trigonal sheets similar to those observed in covellite. The remaining Cu atoms occupy triangular sites *between* the close-packed S layers. At 104°C chalcocite transforms rapidly to a hexagonal form whose unit cell corresponds to that of an hcp S array. The electrical conductivity and the diffusion coefficient for Cu in this phase are quite large. The structure (Buerger and Wuensch, 1963; Sadanaga *et al*, 1965[*]) contains a "smeared" distribution of mobile Cu atoms distributed between triangular, tetrahedral, and 2-coordinated sites. At about 350° a further phase transformation takes place to a cubic phase of broad stoichiometry known as digenite (Morimoto and Kullerud, 1963). Anilite, Cu_7S_4, is orthorhombic (pseudocubic) and transforms readily upon grinding to a metastable form of digenite. A small deficiency of Cu in chalcocite results in a new phase, $Cu_{1.96}S$, djurleite, which would appear to have a structure closely related to that of chalcocite.

The distribution of metal atoms among sites in the ccp sulfur atom array in the polymorphs of digenite is statistical (Morimoto and Kullerud, 1963). Similar disorder occurs among the tetrahedral sites in the ccp array in bornite, Cu_5FeS_4 (Morimoto, 1964). The metal atom locations were described by placing atoms in an equipoint of higher order which scattered fractional atoms about the ideal tetrahedral position. The true metal atom distribution may be "smeared" as in high-chalcocite.

Stromeyerite, CuAgS (Frueh, 1955), has a structure related to chalcocite. The mineral is orthorhombic and has a unit cell which corresponds to the orthohexagonal cell of a hcp array of sulfur atoms (Fig. W-16). The Cu atoms occupy triangular interstices in the close-packed layers as in covellite and chalcocite. The Ag atoms have 2-fold coordination and occupy sites between trigonal CuS sheets such that the layers are linked by a zig-zag AgS chain. The structure of other Cu-Ag sulfides (jalpaite, $Ag_{1.55}Cu_{0.45}S$; mckinstryite, $Ag_{1.2}Cu_{0.8}S$) are unknown.

Silver sulfide, Ag_2S, has a rapid phase transformation at 173° from acanthite (α), a monoclinic structure (Frueh,

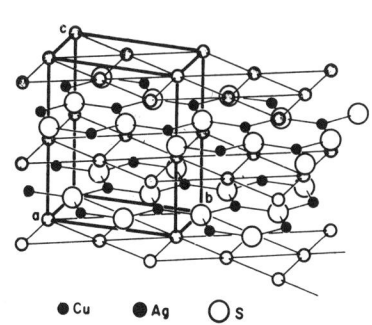

● Cu   ● Ag   ○ S

Fig. W-16.   The crystal structure of stromeyerite, CuAgS.

------

[*]Sadanaga, R., M. Ohmasa, and N. Morimoto (1965). *Mineral. J. Japan* 4, 275.

1958, 1961; Sadanaga and Sueno, 1967), to a cubic form. Acanthite has an arrangement of S atoms which is approximately bcc. Two types of Ag atoms occur in the structure, one having 2-fold coordination, the second forming a flattened trigonal pyramid. The high-temperature modification has a bcc arrangement of S and a statistical distribution of Ag among three different types of interstices (Lowenhaupt and Smith, 1974).

ROCKSALT DERIVATIVES

Galena, PbS, has the rocksalt structure type. Sulfosalts which contain Pb almost invariably are composite structures built of domains with structure closely related to the rocksalt structure type. (Some illustrations are provided in subsequent sections). Several additional sulfosalts are derivatives formed from rocksalt by substitution and distortion. Although all of the metal atoms occupy nominally octahedral sites, only three or four of these sulfur atoms may be bonded to an atom such as Ag or a Group V metal. The displacement of the sulfur atoms from their ideal positions may exceed 1Å.

Miargyrite, $AgSbS_2$ (Knowles, 1964) is monoclinic with $[a,b,c] = [21\bar{1}$ / 0, 1/2, 1/2 / 1, -3/2, 3/2] $[a_1,a_2,a_3]$ NaCl. The Sb atom has but three nearest neighbors; of two independent Ag atoms, one has [3 + 1 + 2] coordination, the other [3 + 3]. Two polymorphic As analogues of miargyrite are known: smithite, $AgAsS_2$ (Hellner and Burzlaff, 1964), is monoclinic with $[a,b,c] = [3, -1/2, 1/2$ / 011 / 02$\bar{2}$] $[a_1,a_2,a_3]$ NaCl. The Ag atoms have [3 + 1 + 2] and [2 + 2 + 2] coordination, while the As atom again forms the characteristic trigonal pyramid with three nearest S neighbors. Trenchmannite, a polymorph of smithite (Matsumoto and Nowacki, 1969), is rhombohedral and has a = 1/2 [14$\bar{3}$] NaCl and c = [111] NaCl, but certain metal and sulfur positions in the ideal rocksalt array are unoccupied. Both smithite and trenchmannite contain a three-membered As_3S_6 ring built of AsS_3 trigonal pyramids.

Marrite, $PbAgAsS_3$ (Wuensch and Nowacki, 1967), and freieslebenite, $PbAgSbS_3$ (Ito and Nowacki, 1974), are isomorphous monoclinic rocksalt derivatives which have $[a,b,c,] = [110$ / -3/2, 3/2, 0 / 001] $[a_1,a_2,a_3]$ NaCl. The Ag atom has a [3 + 1 + 2] distorted tetrahedral coordination; the Group V metal atom forms the usual trigonal pyramid with three of the six available S neighbors.

GROUP V METAL SULFIDES AND RELATED SULFOSALTS

The Group V metal sulfides form ring, chain, or layer structures as a result of the limited number of bonding electrons available from the metal. In the case of As, three nearest S neighbors, together with the metal, form a trigonal bipyramid. This nearest-neighbor configuration is also found for Bi, but fourth and fifth sulfur atoms are present at somewhat larger distances. The resulting polyhedron is almost always a [1 + 2 + 2] square pyramid, with the metal atom displaced slightly from the plane of the base. A few exceptions are known. The Bi

atom in parkerite, $Ni_3Bi_2S_2$ (Fleet, 1973), is involved in metal-metal bonds.
The coordination of the Bi atom in wittichenite, Cu_3BiS_3 (Matzat, 1972; Kocman and
Nuffield, 1973), is trigonal pyramidal but has three S neighbors at much larger
distances which complete a trigonal prismatic coordination. A few Cu-Bi sulfides
have Bi in octahedral coordination, as discussed below. Antimony is found in both
trigonal (e.g., tetrahedrite) and square pyramidal coordination (the case in most
sulfosalts). Both configurations sometimes occur simultaneously in the same
structure (e.g., stibnite, Sb_2S_3). The dimensions of the Group V metal polyhedra
and the effect of linkage on bond distances in these groups has been reviewed by
Takeuchi and Sadanaga (1969).

The Group V Sulfides

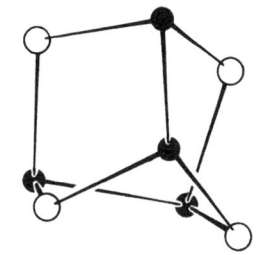

Realgar, AsS, contains a cage-like As_4S_4 molecule
(Ito et al, 1952; Mullen and Nowacki, 1972). Each
of four independent As atoms is bonded to one As and
two S atoms; each sulfur is coordinated by two As
atoms (Fig. W-17). Orpiment, As_2S_3, and stibnite,
Sb_2S_3, form very different structures. In orpiment
(Morimoto, 1954; Mullen and Nowacki, 1972) AsS_3
trigonal pyramids share corners to form six-
membered rings which build a corrugated layer
(Fig. W-18a). The structure of stibnite (Hofmann,

Fig. W-17. The As_4S_4 molecule in realgar. (Solid circles represent As).

1933; Scavnicar, 1960; Baylerr and Nowacki, 1972) is more complex. Two independent Sb atoms are present in the structure. The first has three-fold coordination
and forms a trigonal SbS_3 pyramid. The second has square-pyramidal coordination

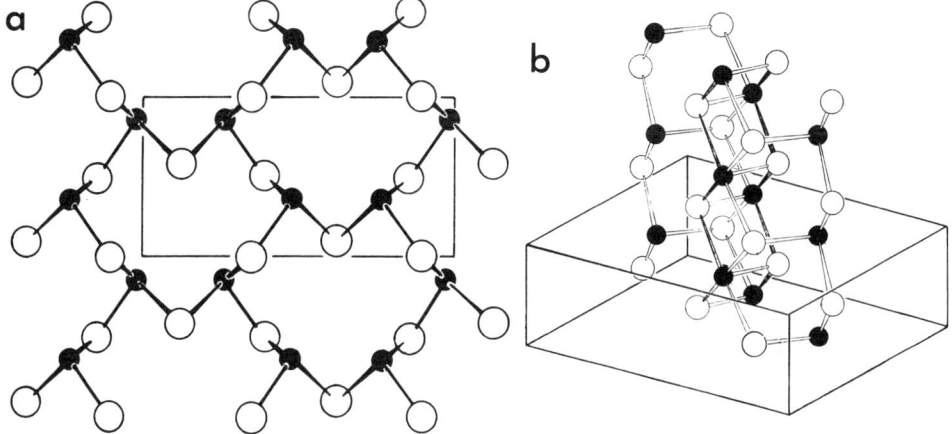

Fig. W-18. (a) Projection along b of one sheet in the layer structure of orpiment, As_2S_3 (b) one of the quadruple chains in the structure of
stibnite, Sb_2S_3. The long bond in the square pyramidal group is
shaded.

with a variation in bond lengths which is typical of this configuration: 2.455, 2.678[2], and 3.373Å[2], where the numbers in brackets indicate multiplicity. The 5-coordinated Sb's share an apical edge to build a double chain. The 3-coordinated Sb's share a corner at either extreme of the double chain to build a two-layer slab of rocksalt structure extended indefinitely along [110]. An idealized projection of this ribbon is presented in Fig. W-19b. Bismuthinite, Bi_2S_3, is isostructural with stibnite (Hofmann, 1933; Kupčík and Vesela-Novakova, 1970).

Getchellite, $AsSbS_3$, is a recently discovered monoclinic mineral with a composition intermediate between stibnite and orpiment. The structure (Guillermo and Wuensch, 1973) consists of thick layers of trigonal pyramids containing As and Sb in disorder. The linkage of pyramids within layers is open and meandering. The simplest structural units are irregular 8-membered rings which are arranged above one another and oriented normal to the plane of the layers. By contrast, orpiment has its 6-membered rings parallel to the plane of sheets which are essentially 2-dimensional.

The Group V chalcogenides all readily form glasses which are of interest because of their semi-conducting and infrared-transmitting properties (Pearson, 1964).

Stibnite Derivatives

Several structures of sulfosalts are simple derivatives of the stibnite structure type. The oxysulfide kermesite, Sb_2S_2O (Kupčík, 1967), contains a pair of five-coordinated Sb atoms which form a double chain similar to the central portion of the stibnite ribbon. Two additional Sb's share a S atom with this ribbon but have three oxygen atoms as additional neighbors. The additional oxygen bond links the quadruple chains into layers.

Aikinite, $CuPbBiS_3$ (Wickman, 1953; Ohmasa and Nowacki, 1970; Kohatsu and Wuensch, 1971), is a stuffed derivative of the stibnite-type structure of bismuthinite. A Pb atom substitutes for the interior five-coordinated Bi in the stibnite chain. The Cu atom is added to a tetrahedral interstice in the structure. The lattice constants and atomic positions in aikinite remain remarkably similar to those of bismuthinite in spite of these substitutions. Seligmannite, $CuPbAsS_3$, and bournonite, $CuPbSbS_3$, the As and Sb analogues of aikinite, are not derivatives of the stibnite structure type, but yet have strikingly similar structures in projection along a direction corresponding to that of the stibnite ribbon (Hellner and Leineweber, 1956; Leineweber, 1956; Takéuchi and Haga, 1969; Edenharter and Nowacki, 1970). Their cell geometries are also similar, with $[a,b,c]$ of seligmannite and bournonite approximately equal to $[-1/2, 0, 1/2 / 1/2, 0, -1/2 / 0\bar{2}0]$ $[a,b,c]$ of aikinite. Stibnite-like regions and an interstitial Cu in a position analogous to that in aikinite may be defined in projection. The structure differs, however, in two important respects: (1) The Pb and Group V metal alternate along the chain in all four metal positions, rather than being relegated to the exterior

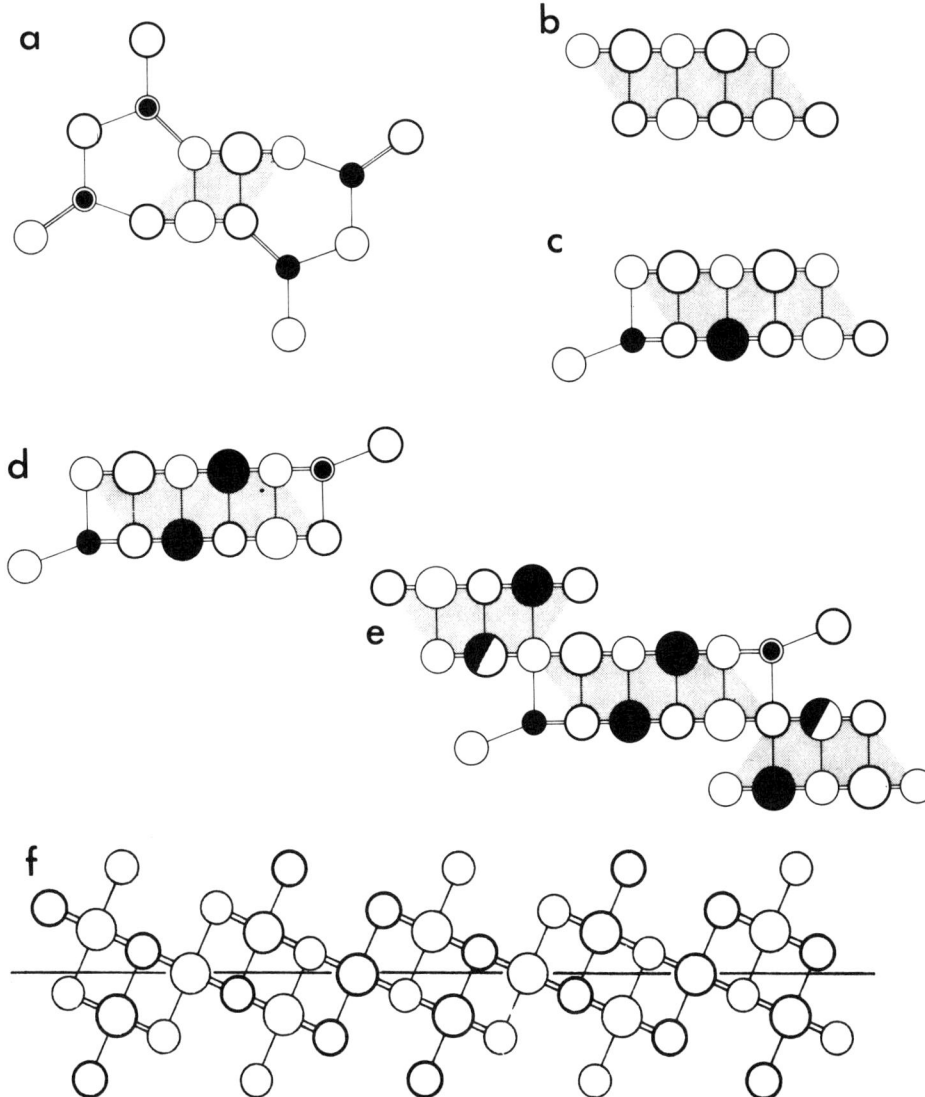

Fig. W-19: Idealized representations of chains found in Bi sulfosalts of high Bi content. Heavy circles and light circles represent atoms located at 1/4 and 3/4, respectively, along a 4Å lattice constant. Large shaded circles represent Pb, large open circles Bi, small open circles S, and the smallest solid circle Cu. (a) Double chain and associated Cu tetrahedra found in emplectite, $CuBiS_2$. The double chain also occurs in $Cu_4Bi_4S_9$, $CuBi_5S_8$, and $Cu_3Bi_5S_9$. (b) The quadruple stibnite-type chain of bismuthinite, Bi_2S_3. (c) The Pb-containing chain in gladite, $PbCuBi_5S_9$. (d) The quadruple chain in aikinite, $PbCuBiS_3$. (e) The 10-membered chain in nuffieldite, $Pb_2Cu(Pb,Bi)Bi_2S_7$, which displays a central unit identical to the aikinite ribbon. (f) The continuous sheet of Bi octahedra found in $CuBi_5S_8$ and $Cu_3Bi_5S_9$.

and interior sites, respectively (this is the feature which results in the doubling of the periodicity of the ribbon). (2) Metal atoms on opposing surfaces of the ribbon occur opposite one another, rather than being staggered by one-half of the period of the stibnite ribbon. This similarity between the aikinite and seligmannite or bournonite structures thus occurs only in projection. The two structure types are quite distinct.

Welin (1966) has recently described three minerals (gladite, $PbCuBi_5S_9$; hammarite, $Pb_2Cu_2Bi_4S_9$; and $Pb_3Cu_3Bi_7S_{15}$) which are superstructures based upon aikinite (or bismuthinite) and which have one lattice constant equal to 3, 3, and 5 times, respectively, that of the parent structure. These structures are presumably ordered superstructures in which only a fraction of the 5-coordinated Bi sites and tetrahedral interstices in bismuthinite are occupied by Pb and Cu, respectively. The crystal structure of gladite has been determined (Kohatsu and Wuensch, 1973) and confirms this relationship. In spite of the similarity between the structures of aikinite and bismuthinite, however, the structure of gladite does not consist of a mixture of bismuthinite (Bi_4S_6) and aikinite ($Cu_2Pb_2Bi_2S_6$) ribbons. It instead consists of bismuthinite ribbons plus a new type of $CuPbBi_3S_6$ chain. Moore (1967) has proposed a system for classification of these derivatives as well as an extensive series of hypothetical members of the family.

Related Bi Sulfosalts

The periodicity of the stibnite-like chain of bismuthinite, the edge of a CuS_4 tetrahedron, and the edge of a PbS_6 octahedron are all close to 4Å. These units might therefore be joined to form a large variety of complex structures in a crystal chemistry which is essentially two-dimensional. With but rare exceptions (e.g., wittichenite, Cu_3BiS_3), all Bi sulfides indeed display one lattice constant equal to this periodicity. Therefore, while the majority of the Bi sulfides are not stibnite derivative, the stibnite-like quadruple ribbon and a double chain of 5-coordinated metals (which forms the kernel of the stibnite ribbon) both figure prominently in the crystal chemistry of the Bi sulfides.

The Cu-Bi sulfides form a family of phases with closely related structures. Emplectite, $CuBiS_2$ (Hofmann, 1933; Kupčik, 1965; Jenkins, 1969) contains double chains linked by a pair of Cu tetrahedra. Their arrangement is shown schematically in Fig. W-19a. (The structure of cuprobismutite, a polymorph of emplectite, is unknown.) The synthetic compound $Cu_4Bi_4S_9$ (Ozawa and Takéuchi, 1972) also contains the double chain but connected by a different Cu-S network. Unlike emplectite, the network contains Cu in both tetrahedral and triangular coordination and a sulfur-sulfur bond; this region bears a similarity to the structure of covellite. Two additional synthetic compounds, $CuBi_5S_8$ (Ohmasa and Nowacki, 1973), and $Cu_{2+x}Bi_{6-x}S_9 \approx Cu_3Bi_5S_9$ (Ohmasa, 1973), have virtually identical structures. Double chains are linked by a single Cu tetrahedron in the former, and by a pair of partially occupied tetrahedra in the latter -- to form a layer. The layer alternates with a second sort of layer built of octahedrally-coordinated Bi,

(Fig. W-19f).

Both a single and a double chain occur in berthierite, $FeSb_2S_2$ (Buerger and Hahn, 1955), linked by a chain of Fe octahedra sharing edges. Galenobismutite, $PbBi_2S_4$ (Wickman, 1951; Iitaka and Nowacki, 1962) has a somewhat similar but distinct structure.

Certain trends are becoming clear in the crystal chemistry of the Pb-Cu-Bi sulfides. If the Pb content of the mineral is high, the structure assumed is a "composite" structure based upon PbS-like domains. Lillianite, $Pb_3Bi_2S_6$ (Takagi and Takéuchi, 1972), and cosalite, $Pb_2Bi_2S_5$ (Weitz and Hellner, 1960), provide examples. If a mineral contains no Pb, or if the ratio of Bi and Cu to Pb is high, the structure contains chain-like units (i.e., layers limited in thickness to two atomic slabs, and of finite width). When the Cu content is high, a network of Cu tetrahedra or triangles link Bi-containing chains together, and form an important feature of the structure. This occurs in emplectite ($CuBiS_2$), hodrushite ($PbCu_4Bi_5S_{11}$ -- Kupcik and Makovicky, 1968), and $Cu_4Bi_5S_9$. On the other hand, if the Cu content is relatively low, the copper atoms play a passive role in the structure and occupy interstices between bismuthinite or emplectite-like chains. This occurs in $CuBi_5S_8$ (Ohmassa and Nowacki, 1973), aikinite, ($PbCuBiS_3$), and gladite ($PbCuBi_5S_9$ -- Kohatsu and Wuensch, 1973). Nuffieldite, $Pb_2Cu(Pb,Bi)Bi_2S_7$ (Kohatsu and Wuensch, 1973), contains a complex 10-membered ribbon (Fig. W-19e). It is of interest to note that the central portion of this chain is identical to that found in aikinite. The 4-membered aikinite ribbon is flanked by a pair of 3-membered ribbons which contain Pb, Bi, and a site containing Pb and Bi in disorder. Idealized representations of the ribbon-like units described above are presented in Fig. W-19f.

THE LEAD-ANTIMONY SULFOSALTS

At least 17 intermediate minerals occur in the system $PbS-Sb_2S$. Table W-2 lists the mineral species which are known in order of decreasing Pb content. Many contain As in solid solution, and indication of this is given when an end-member containing primarily Sb has not yet been reported. Structures are known for very few of these minerals. A substructure determination for boulangerite (Born and Hellner, 1960) shows the mineral to be a composite structure based upon extended slabs of stibnite structure. The substructure of zinkenite (Takeda and Horiuchi, 1971) has similar features. The majority of the Pb-Sb sulfosalts have acicular habits, and are elongated parallel to a 4Å or, more commonly, an 8Å lattice constant. On the basis of the substructure determinations available for boulangerite and zinkenite, and the structures known for related antimony sulfosalts such as jamesonite, $FePb_4Sb_6S_{16}$ (Niizeki and Buerger, 1957b), or meneghinite, $CuPb_{13}Sb_7S_{24}$ (Euler and Hellner, 1960), the majority of these minerals would appear to have structures based upon stibnite or PbS-like chains. Members of the plagionite group (Table W-2) are exceptions. These minerals have tabular

Table W-2. The Lead-Antimony Sulfosalts, $Pb_x Sb_y S$.

(Solid solution is indicated when a pure Sb end-member has not yet been found. A composition followed by (?) is the most satisfactory representation of an analysis and has not been confirmed by a structure determination.)

x	y	Mineral	Formula	a(Å)	b(Å)	c(Å)	β(°)	Space Group	Reference
.565	.304	geocronite	$Pb_{13}(Sb,As)_8 S_{23}$ (?)	9.0	31.9	8.5	118	$P2_1/m$	Born & Hellner (1960)**
.454	.363	boulangerite	$Pb_5 Sb_4 S_{11}$	21.52	23.46	8.07	100.8	$P2_1/a$	Jambor (1967b)**
.444	.370	sterryite	$Pb_{12}(Sb,As)_{10} S_{27}$ (?)	28.4	42.6	8.201		Pbam*, Pba2	Jambor (1967b)**
.429	.381	semseyite***	$Pb_9 Sb_8 S_{21}$	13.603	11.935	24.452	106.046	C2/c	Kohatsu & Wuensch (1974)
.425	.550	sorbyite	$Pb_{17}(Sb,As)_{22} S_{40}$	44.9	8.28	26.4	113.42	C2,Cm,C2/m	Jambor (1967b)**
.415	.390	madocite	$Pb_{17} Sb_{16} S_{41}$ (?)	27.2	34.1	8.12		Pba2*, Pbam	Jambor (1967a)**
.400	.400	veenite	$Pb_2(Sb,As)_2 S_5$	8.44	26.2	7.90		$P2_1 cn$*, Pmcn	Jambor (1967a)**
.379	.414	dadsonite	$Pb_{11} Sb_{12} S_{29}$	19.05	4.11	17.33	96.33	$P2,Pm,P2/m$	Jambor (1969a)**
.372	.419	playfairite	$Pb_{16} Sb_{18} S_{43}$ (?)	45.4	8.29	21.3	92.5	$P2_1$*, $P2_1/m$	Jambor (1967b)**
.368	.421	heteromorphite***	$Pb_7 Sb_8 S_{19}$	13.60	11.93	21.22	90.83	C2/c	Jambor (1969b)**, Kohatsu & Wuensch (1974)
.361	.426	launayite	$Pb_{22} Sb_{26} S_{61}$ (?)	42.6	8.04	32.3	102.08	C2*,Cm,C2/m	Jambor (1967b)**
.294	.491	plagionite***	$Pb_5 Sb_8 S_{17}$	13.4857	11.8656	19.9834	107.168	C2/c	Cho & Wuensch (1970,1974)
.280	.480	robinsonite	$Pb_7 Sb_{12} S_{25}$ (?)	16.51 α 96.7	17.62 β 96.4	3.97 γ 91.2		$P1,P\bar{1}$	Berry et al (1952)**
.273	.485	guettardite	$Pb_9(Sb,As)_{16} S_{33}$ (?)	20.0	7.94	8.72	101.58	$P2_1/a$	Jambor (1967b)**
.250	.500	twinnite	$Pb(Sb,As)_2 S_4$	19.6	7.99	8.60		$P\bar{1}$?	Jambor (1967b)**
.222	.519	zinkenite	$Pb_6 Sb_{14} S_{27}$	44.06		8.60		$P6_3, P6_3/m$	Takeda & Horiuchi (1971)*
.200	.553	fülöppite***	$Pb_3 Sb_8 S_{15}$	13.441	11.726	16.930	94.71	C2/c	Nuffield (1974)

 * Substructure determination only.   ** Lattice constants only.

*** Member of the plagionite group, a homologous series $Pb_{3+2n} Sb_8 S_{15+2n}$ in which c increases with n.

habit and appear to constitute an homologous series, $Pb_{3+2n}Sb_8S_{15+2n}$ ($n = 0$ to 3) in which two lattice constants remain invarient, while the third increases with n. Structures have recently been determined for fuloppite, $n = 0$ (Nuffield, 1974), plagionite, $n = 1$ (Cho and Wuensch, 1970; 1974) and semseyite, $n = 3$ (Kohatsu and Wuensch, 1974). The series of structures again consist of slabs of rocksalt structure, two atoms in thickness, but oriented parallel to {112}, a cleavage plane in the structure. Successive members differ only in the width of the rocksalt unit. Structure with n odd may be derived from those with n even through removal of a Pb-S pair from the central portion of the slab (where the coordination of all atoms is pseudo-octahedral) followed by shear of the structure by 1/4 a, and collapse of (001) to fill the void. The unknown structure of heteromorphite has been predicted by means of this model (Kohatsu and Wuensch, 1974).

THE LEAD-ARSENIC SULFOSALTS

The system $PbS-As_2S_3$ displays almost as many intermediate minerals (Table W-3) as does its antimonian counterpart. Interestingly, with but few exceptions (*e.g.*, jordanite-geocronite, dufrenoysite-veenite, scleroclase-twinnite), there is little correspondence of composition or structure between minerals of the two systems. Unlike the Sb sulfosalts, however, detailed structures are known for the majority of the lead arseno-sulfides, and interesting variations in structure are observed as the Pb content of the series decreases.

Gratonite, $Pb_9As_5S_{15}$ (Rosch, 1963; Ribar and Nowacki, 1969), the mineral of highest Pb content, is a rhombohedral rocksalt superstructure which has $c = $ [111] PbS. Jordanite (Wuensch and Nowacki, 1966; Ito and Nowacki, 1974) is mono clinic (pseudohexagonal) and has a structure which consists of two rocksalt-like slabs, five octahedra in thickness rotated 180° about [111] of PbS relative to one another.

Minerals with higher As content break into regions of distorted rocksalt-like structure joined by slabs of 9-coordinated Pb. The rathite group (Table W-3) constitutes a subfamily of this series of minerals in which 32 metal atoms and 40 S atoms are contained in a cell of essentially constant geometry. The crystal structure of dufrenoysite, $Pb_2As_2S_5$ (Marumo and Nowacki, 1967b; Ribar *et al*, 1969), is a representative of this subgroup (Fig. W-20). One type of Pb atom is 9-coordinated by S in the form of a trigonal prism, with three additional neighbors along normals to the prism faces. The triangular bases of the prisms are indicated in Fig. W-20 by the dotted lines. These groups are linked by sharing of prism edges to form a slab. Alternating with these slabs are layers of structure which are described in the literature as layers of distorted rocksalt structure. As shown in Fig. W-20, however, the structure of the second type of region closely resembles a collection of stibnite-like ribbons. The difference between members of the rathite group lies in the metal content of these ribbons: Pb_4As_7 in dufrenoysite; $(Pb,Tl)_2AgAs_9$ in rathite I (Marumo and Nowacki, 1965); Pb_2As_{10} in rathite III (LeBihan, 1962); Pb_3As_9 in rathite Ia (LeBihan, 1962).

Table W-3. The Lead-Arsenic Sulfosalts, Pb_xAs_yS.

x	y	Mineral	Formula	a(Å)	Cell Parameters b(Å)	c(Å)	β(°)	Space Group	Reference
.600	.267	gratonite	$Pb_9As_4S_{15}$	17.758		7.807		R3m	Rösch (1963), Ribár & Nowacki (1969)
.565	.304	jordanite	$Pb_{14}As_6S_{23}$	8.918	31.899	8.462	117.79	$P2_1/m$	Wuensch & Nowacki (1966)*, Ito & Nowacki (1974)
.400	.400	dufrenoysite**	$Pb_2As_2S_5$	7.90	25.74	8.37	90.35	$P2_1$	Marumo & Nowacki (1967b), Ribár et al (1969)
.350	.450	rathite Ia**	$Pb_7As_9S_{20}$	7.91	25.80	8.43	90	$P2_1$	LeBihan (1962)
.322	.456	rathite II**	$Pb_{18.5}As_{25}S_{56}$	8.371	70.49	7.914	90.13	$P2_1$	LeBihan (1962), Engel & Nowacki (1970)
.306	.473	acentric baumhauerite	$Pb_{11}As_{17}S_{36}$	22.80 α 90.05	8.36 β 97.26	7.89 γ 89.92		P1	Engel & Nowacki (1969)
.300	.500	rathite I**	$(Pb,Tl)_3As_5S_{10}$	25.16	7.94	8.47	100.47	$P2_1/a$	LeBihan (1962), Marumo & Nowacki (1965)
.300	.500	rathite III**	$Pb_3As_5S_{10}$	24.52	7.91	8.43	90.0	$P2_1$	LeBihan (1962)
.278	.500	centric baumhauerite	$Pb_5As_9S_{18}$	22.78 α 90.0	8.33 β 97.4	7.90 γ 90.0		$P\bar{1}$	LeBihan (1962)
.250	.500	scleroclase	$PbAs_2S_4$	19.62	7.89	4.19	90.0	$P2_1/n$	Iitaka & Nowacki (1961)*
.222	.556	hutchinsonite	$(Pb,Tl)_2As_5S_9$	10.81	35.36	8.61		Pbca	Takéuchi et al (1965)

* Substructure determination only. ** Member of the rathite group; 32 metal and 40 S atoms contained within a cell of essentially constant geometry.

Disordered crystals are known in which the layering may be mixed (Marumo and Nowacki, 1967b).

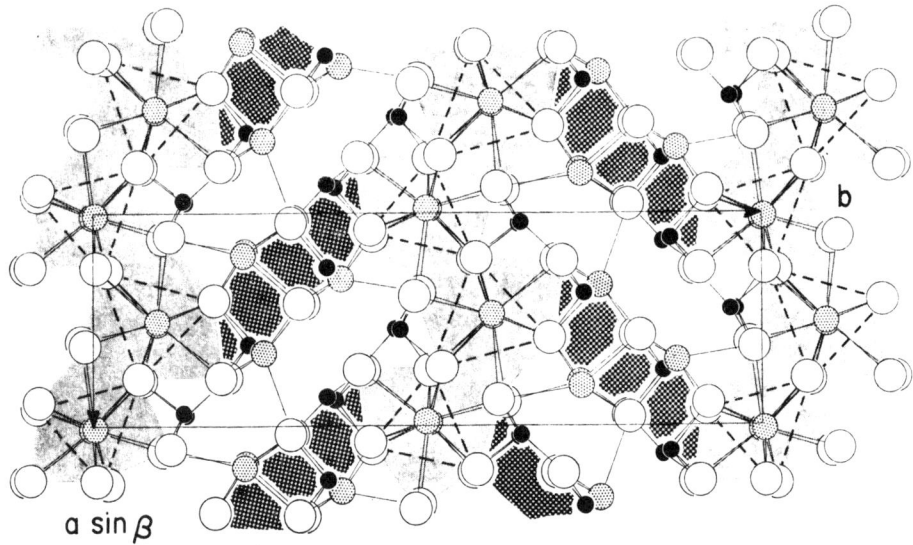

Fig. W-20. Projection along *c* of the crystal structure of dufrenoysite, $Pb_2As_2S_5$, a representative member of the rathite group. Layers of 9-coordinated Pb (light shading) alternate with ribbons of stibnite-like structure (dark shading). The composition and/or thickness of the latter region differs in related Pb-As sulfo-salts. (Shaded circles represent Pb, solid circles As, open circles S atoms).

In minerals of yet higher As content, the thickness as well as the compo-sition of the stibnite-like layer varies. Minerals of this sort include rathite II and baumhauerite (LeBihan, 1962; Engel and Nowacki, 1969), and scleroclase (Iitaka and Nowacki, 1961). Hutchinsonite, $(Pb,Tl)_2As_5S_9$, the mineral with highest As content, also contains PbS-like slabs (Takeuchi *et al*, 1965). The layer of 9-coordinated Pb, however, is replaced by a slab which contains laterally-joined spiral chains of AsS_3 pyramids. Similar chains have been found in lorandite, $TlAsS_2$ (Zemann and Zemann, 1959; Knowles, 1965; Fleet, 1973).

ADDITIONAL SULFOSALT STRUCTURES

Crystal structures are known for several additional sulfosalt minerals which have not been mentioned in preceding sections because of their uniqueness. Sulvanite, Cu_3VS_4 (Pauling and Hultgren, 1933; Trojer, 1966), for example, is isometric. Both the Cu and V atoms are tetrahedrally coordinated by S, yet the structure is not related to other tetrahedral structures. The sulfur atom is coordinated by three Cu atoms in locations close to those for tetrahedral con-figuration. The fourth S neighbor, V, is found to be located in the *inverse* of the direction corresponding to tetrahedral coordination. Pauling (1965) has offered an explanation for this curious configuration.

Pyrargyrite, Ag_3SbS_3, and proustite, Ag_3AsS_3, are isomorphic structures containing Group V metal pyramids stacked along their axes of symmetry and linked by a trigonal spiral formed of Ag atoms having a 2-fold coordination by S (Harker, 1936; Engel and Nowacki, 1966). Both minerals have polymorphs -- pyrostilpnite, Ag_3SbS_3 (Kutoglu, 1968), and xanthoconite (Engel and Nowacki, 1966) -- whose crystal structures are known. Stephanite, Ag_5SbS_4 (Ribar and Nowacki, 1970), another sulfosalt with a large fraction of Ag, contains isolated SbS_3 pyramids. Samsonite, $MnAg_4Sb_2S_4$ (Hruskova and Synecek, 1969) is the only sulfosalt which contains Mn. Its structure contains isolated SbS_3 pyramids; the Mn atom is octahedrally coordinated by S and each sulfur atom has a distorted tetrahedral coordination of Mn, Sb, and two Ag atoms.

Vrbaite, $Hg_3Tl_4As_8Sb_2S_{20}$ (Ohmasa and Nowacki, 1971) provides an interesting example of As-Sb ordering in a sulfosalt. Additional Tl sulfosalt structures are provided by the rare minerals hatchite, $PbTlAgAs_2S_5$ (Marumo and Nowacki, 1964), and its Cu-analogue, wallisite (Takéuchi et al, 1968).

Ch. 3

ELECTRON INTERACTIONS AND CHEMICAL BONDING IN SULFIDES

C. T. Prewitt and *V. Rajamani*

INTRODUCTION

The understanding of electronic processes and chemical bonding in sulfides has increased considerably in recent years. Much of the work that has been done is theoretical in nature and has been presented in a form which is generally unpalatable to geologists or mineralogists who want to understand something about current sulfide research, but who have neither the time nor the inclination to become specialists in theoretical chemistry. The problem is, however, that sulfide chemistry is complicated, and one must know much more than is given in most mineralogical textbooks in order to understand sulfides and to read and understand current literature in such journals as *The American Mineralogist*. A substantial boost was given to theoretical mineralogy by Burns' (1970) book, *Mineralogical Applications of Crystal Field Theory*, but only a small portion of this book is devoted to the molecular orbital theory necessary for sulfides. A number of good inorganic chemistry texts are available, but most chemists are neither interested in nor aware of the extensive worldof mineral chemistry, and, consequently, mineralogists must sift through a lot of material to find what they want. In this Short Course we face difficulties which arise from the fact that we have selected a topic which could be the subject of several undergraduate and graduate college courses but which must be presented and digested in one afternoon. This is an extremely difficult task and, inevitably, important aspects of the subject must be omitted or glossed over. The notes have been prepared to give a quick review of the electronic structures of atoms important in sulfide chemistry. Also, included is an abbreviated review of group theory because group theory is essential for the understanding of bonding theories, and many mineralogists and geologists have never been exposed to group theory concepts. The treatments will be too elementary for some and too brief for others. Hopefully, these notes and the Short Course will inspire many to dig deeper into the subject.

The coverage of bonding in sulfides begins with an introduction to crystal field theory which is actually more useful for oxides than for sulfides. The more applicable molecular orbital and band theories are developed more fully and are illustrated with examples from current sulfide research. Alternative ways of looking at sulfide structures are considered through a somewhat more empirical covalence/interatomic distance approach and, finally, different types of bonding are contrasted by describing electronic structures in the pyrite, thiospinel, and pentlandite types of structures.

ELECTRONIC STRUCTURES OF THE ELEMENTS

Over the years mineralogists have been able to rationalize many features of silicate crystal structures and chemical phenomena such as solid solutions with relatively simple conceptual ideas of how atoms interact. Because silicate structures are at least partly ionic in character, the electrostatic interactions are important and much use has been made of Pauling's Rules, ionic radii, radius ratios, ionic charge, and even electronegativity. However, the situation is different with sulfides and useful theory is more complicated. For an intelligent approach to the understanding of sulfide chemistry, it is necessary to know about the electronic structures of the atoms and how these structures interact in the real sulfide structures. Most readers will have been exposed to quantum chemistry to at least some degree, but it is felt that a short review will be helpful for those who do not use quantum chemistry concepts in their everyday work. Therefore, this section is devoted to a discussion of the basic ideas of atomic structure. There are many books which may be consulted for more detailed information about atomic structure than is given here. Some of those used in the preparation of these notes are Gray (1965), Orgel (1966), Cotton and Wilkinson (1966), Krebs (1968), and Dickerson, Gray, and Haight (1974).

Wave Equations

The structure of an atom consists of a nucleus surrounded by one or more electrons which behave in certain ways, depending on the nature of the nucleus, the number of electrons in the atom, and the environment of the atom itself. This behavior of the electrons affects many of the chemical and physical properties in which we are interested and results from the wave properties of the electron. Information about the nature of the electron can be obtained by solving a *wave equation*, i.e., the Schrödinger equation.

$$\frac{\partial^2 \psi}{\partial x^2} + \frac{\partial^2 \psi}{\partial y^2} + \frac{\partial^2 \psi}{\partial z^2} + \frac{8\pi^2 m}{h} \left[E - V_{(x,y,z)} \right] \psi_{(x,y,z)} = 0$$

The square of the amplitude $|\psi_{(x,y,z)}|^2$ is the probability density of an electron at (x,y,z). E and V are kinetic and potential energy functions, m is the mass of the electron, and h is Planck's constant. Solutions of this equation are restricted by boundary conditions which state that $|\psi|^2$ must be continuous, single-valued, and finite everywhere. Furthermore, the Pauli exclusion principle states that no two electrons in a system can give identical solutions to the Schrödinger equation. The mathematics of the solution of the wave equation for the hydrogen atom are complex, but straightforward, and, as a result, we can get a good idea of how an atom is constructed. However, this picture for the hydrogen atom is only an approximation of the situation in many-electron atoms because additional electrons interact with one another as well as with the nucleus. The calculations of wave functions for many-electron atoms are dependent on various approximations and assumptions about the behavior of electrons and are beyond the scope of these notes.

Quantum Numbers and the Hydrogen Atom

For a complete description of the state of an electron in an atom, four quantum numbers, n, ℓ, m_ℓ, and s are required. These result from the solution of the Schrödinger equation for the hydrogen atom and are necessary for describing the electronic structures of atoms. Application of the Pauli exclusion principle means that no two electrons in an atom have four identical quantum numbers. The quantum numbers are:

n = principal quantum number,

ℓ = azimuthal or orbital angular momentum quantum number,

m_ℓ = orbital orientation (magnetic) quantum number,

s = spin moment quantum number.

Table PR-1 gives the possible combinations for the quantum numbers through $n = 4$. In the hydrogen atom each electron occupies an *orbital* which is described by a set of unique quantum numbers. The principal quantum number, n, designates the shell K, L, M,..., etc., in which the electron is found; the smaller the number, the lower

TABLE PR-1. Possible Quantum Numbers for the Hydrogen Atom through n = 4.

Shell	n	ℓ	m_ℓ	s	Designation	Sum of possible states
K	1	0		±1/2	1s	2
L	2	0	0	±1/2	2s	2
		1	-1,0,1	±1/2	2p	6
M	3	0	0	±1/2	3s	2
		1	-1,0,1	±1/2	3p	6
		2	-2,...,2	±1/2	3d	10
N	4	0	0	±1/2	4s	2
		1	-1,0,1	±1/2	4p	6
		2	-2,...,2	±L/2	4d	10
		3	-3,...,3	±1/2	4f	14

(or more negative) the energy of the electron. The quantum number ℓ represents the classical angular momentum of the electron, although this is not physically defined in wave mechanics because the electron is not considered as a particle with definite position and velocity. The magnitude of the orbital angular momentum is $\ell(\ell+1)h/2\pi$ where h is Planck's constant. Azimuthal quantum numbers are generally designated by s, p, d, f,..., instead of 1, 2, 3, 4,.... Thus, when we speak of s electrons or d orbitals, the azimuthal quantum number is specified.

The third quantum number, m_ℓ, is called the *orbital orientation* quantum number because it tells something about the orientation of the orbital in space. It is also called the magnetic or axial quantum number.

Electrons act as though they have a magnetic moment which results from the electron spinning around its axis. The spin moment quantum number (s = +1/2, or -1/2) is assigned to indicate the direction or orientation of this spin. Each orbital can have no more than two electrons. When an orbital contains two electrons, they must be of opposite spin.

The orbitals for the hydrogen atom, illustrated in Fig. PR-1, are differentiated by unique combinations of n, ℓ, m_ℓ and the square of the function $\psi_{n\ell m_\ell}$ is the probability function of finding the electron somewhere in space. The orbitals for s electrons are spherical in shape and the electron density decreases with distance from the nucleus. The electron distributions are sperical in all s orbitals although nodes exist in the s orbitals with $n > 1$. These nodes are points where the wave functions change sign and where the radial distribution function goes to zero.

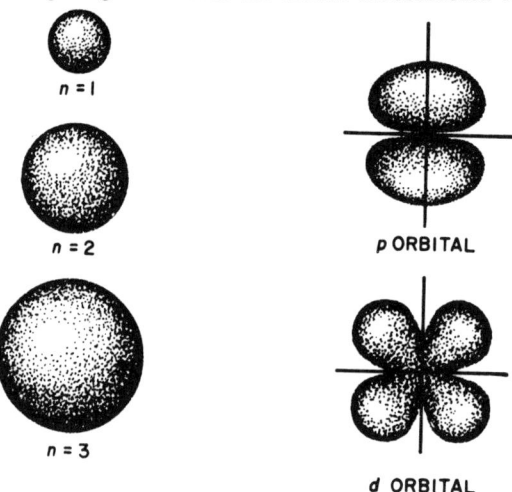

$n = 1$

$n = 2$

$n = 3$

ns ORBITALS

p ORBITAL

d ORBITAL

Fig. PR-1. Hydrogen orbitals for different combinations of n and ℓ.

The next orbitals to appear are the three $3p$ orbitals, corresponding to m_ℓ = -1, 0, 1. Each orbital is cylindrically symmetrical around one cartesian axis, x, y, or z, and consists of two lobes of high electron density separated by a nodal plane of zero density. The sign of the wave function (although not of the electron density) is positive in one lobe and negative in the other.

It should be noted that orbital diagrams such as those shown in Fig. PR-1 contain most, but not all, of the electron density. Exactly what percentage of the electron density they contain is unimportant (perhaps 90%), but what is important is that they represent *concentrations* of electron density and that as n increases the electron clouds become more diffuse and the highest density regions are further from the nucleus.

The Periodic System

As was stated earlier, exact solution of the Schrödinger equations is not possible for many-electron atoms and the hydrogen orbitals are only approximations of the orbitals in the heavier atoms. Nevertheless, the quantum numbers are the same and a reasonably good picture of the periodic table can be obtained by seeing how the hydrogen orbitals would be filled with electrons, starting with those of lowest energy. Table PR-2 can be used as a guide to this progression. In general, energy increases down the table. The $3s$, $3p$, and $3d$ orbitals in the hydrogen atom have the same energy but differ in their radial extent from the nucleus. There are, for example, two $1s$ electrons in the K shell, two $2s$ and six $2p$ electrons in the L shell, and so on. When p, d, and f orbitals of equal energy are being filled, electron spins remain unpaired, if possible. This is known as *Hund's Rule* and will be especially important when d orbitals are discussed in a later section.

Filling of the d Orbitals of the Transition Metals

The transition metals, characterized by the presence of unpaired d electrons, are very important in sulfide chemistry. In these notes, we will consider only the first-row transition elements and ions, i.e. Sc → Cu, in detail.

Table PR-3 shows the common valence states of the atoms in the first transition

TABLE PR-2. Alternative Arrangement of Quantum States

$1s^2$
$\quad 2s^2 \quad 2p^6$
$\qquad\quad 3s^2 \quad 3p^6 \quad 3d^{10}$
$\qquad\qquad\quad 4s^2 \quad 4p^6 \quad 4d^{10} \quad 4f^{14}$
$\qquad\qquad\qquad\quad 5s^2 \quad 5p^6 \quad 5d^{10} \quad 5f^{14}$
$\qquad\qquad\qquad\qquad\quad 6s^2 \quad 6p^6 \quad 6d^{10} \quad 6f^{14}$
$\qquad\qquad\qquad\qquad\qquad\quad 7s^2 \quad 7p^6 \quad 7d^{10}$
$\qquad\qquad\qquad\qquad\qquad\qquad\quad 8s^2 \quad 8p^6$

TABLE PR-3. The First-Row Transition Elements and Ions[*]

	d^0	d^1	d^2	d^3	d^4	d^5	d^6	d^7	d^8	d^9	d^{10}
0+		Sc	Ti	V		Cr/Mn	<u>Fe</u>	<u>Co</u>	<u>Ni</u>		Cu/Zn
1+							<u>Mn</u>			Ni	<u>Cu</u>
2+			Ti	V	Cr	Mn	Fe	Co	Ni	Cu	<u>Zn</u>
3+	Sc	Ti	V	<u>Cr</u>	<u>Mn</u>	<u>Fe</u>	<u>Co</u>	<u>Ni</u>	Cu		<u>Ga</u>
4+	<u>Ti</u>	<u>V</u>	<u>Cr</u>	<u>Mn</u>	Fe	Co	<u>Ni</u>				<u>Ge</u>
5+	<u>V</u>	<u>Cr</u>	Mn								<u>As</u>
6+	<u>Cr</u>	Mn	Fe								<u>Se</u>
7+	<u>Mn</u>										<u>Br</u>

[*]*Oxidation states corresponding to the underlined entries in this table have been reported for minerals.*

series and of the atoms at either end of the series. Except for Cu^+, the lowest oxidation states for the transition-metal ions in minerals is 2+, with both s electrons lost. The highest oxidation states attainable are equal to the number of *unpaired* electrons in the $3d$ and $4s$ orbitals. This is why the maximum oxidation number increases from 3+ in Sc (one d and two s electrons) to 7+ in Mn (five d and two s electrons). There are several irregularities in the filling of the five d orbitals which are a result of small decreases in energy when the orbitals are exactly half-filled or completely occupied. For example, Cu has a $d^{10}s^1$ configuration, and it is relatively easy to remove the s electron, and thus the Cu^+ ion is observed rather frequently. However, Zn has two easily removed s electrons and the 2+ state is the common one. For the intermediate ions, the d^5 configuration is favored and this explains why Cr^0 does not exist in the d^4 configuration.

GROUP THEORY

Basic Principles

Most people with a geological background have taken mineralogy courses which provided some information about the principles of symmetry and their application to problems in mineralogy and crystallography. However, relatively few have had much exposure to group theory and its use in the study of the structural chemistry of mineral systems. Therefore, this section attempts to provide some of the background in group theory necessary to understand current research papers on sulfide chemistry.

For those familiar with the crystallographic point groups and space groups, group theory is a systematic approach to the ways in which symmetry elements can be combined to generate the point and space groups and it also provides a mechanism for working with chemical phenomena affected by crystal symmetry. Elementary group theory is not difficult to understand and use. However, it can be very complicated, and advanced treatment is beyond the scope of these notes. There are many books available on group theory which are of interest to the mineralogist, particularly Cotton (1971), Hall (1969), McWeeny (1963), Cracknell (1968), and Hollingsworth (1967).

Although group theory can be applied to all sorts of groups, mathematical and otherwise, the treatment here will be confined to *crystallographic groups*, i.e., combinations of the familiar symmetry elements (mirror planes, rotation axes, inversion centers, translations) which make up the crystallographic point and space groups. Consider first the 32 point groups. Each point group is a unique combination of symmetry elements which result when a motif or pattern is symmetrically arranged about a central point. The pattern can have two or three dimensions. Associated with each symmetry *element* is a symmetry *operation* which transforms one unit of the pattern into its symmetrical equivalent. A basic principle of group theory is that the combination of two symmetry operations results in a third operation which is also a member of the group. We say that the set of symmetry operations is *closed* with respect to a law of combination, called *multiplication*.

In order to provide examples of the application of group theory, it is necessary to use symbols to represent symmetry elements and symmetry operations. The International symbols used by crystallographers to designate point groups are generally not used in group theory and spectroscopy. Instead, symbols based on the Schoenflies notation are used with some additions.

Basically, the following definitions apply:

C_n n-fold rotation axis, n = 2, 3, 4, 6

C_{nh} as above with a horizontal mirror plane perpendicular to C_n

C_{nv} n-fold rotation axis with vertical mirror planes parallel to C_n

D_n n-fold rotation axis with n 2-fold axes normal to D_n

D_{nh} as above with a perpendicular mirror plane

D_{nd} n-fold rotation axis with diagonal mirror planes parallel to and half-way between twofold axes

S_n n-fold rotoreflection axes

T,O special symbols for tetrahedral and octahedral point groups, respectively

σ_h plane of symmetry perpendicular to the principal rotation axis

σ_v plane of symmetry containing the principal rotation axis

σ_d plane of symmetry containing the principal rotation axis and bisecting the angle between two twofold axes perpendicular to the principal rotation axis.

The best way to illustrate group properties is with an example. Point group 4mm in the International notation or C_{4v} in the Schoenflies notation is shown in Fig. PR-2. A motif (represented by a comma) in the general position (a position not on a symmetry element of the point group) has coordinates x,y,z and is related to the other seven commas by the symmetry operations of the point group. There are eight symmetry operations represented by

E the identity operation (xyz → xyz)

C_4 90° rotation around z (xyz → $\bar{y}$xz)

$\bar{C}_4$ or $C_4{}^3$ -90° or 270° rotation around z (xyz → y$\bar{x}$z)

C_2 180° rotation around z (xyz → $\bar{x}\bar{y}$z)

σ_x reflection across mirror plane ⊥ to x (xyz → $\bar{x}$yz)

σ_y reflection across mirror plane ⊥ to y (xyz → x$\bar{y}$z)

σ_2 reflection across diagonal mirror plane (xyz → yxz)

σ_1 reflection across diagonal mirror plane (xyz → $\bar{y}\bar{x}$z)

Using the law of combination mentioned earlier, we can construct a *group multiplica-table* which shows that if two symmetry operations in turn operate on a motif, the result is equivalent to the effect of a third symmetry operation.

C_{4v}	E	C_4	$\bar{C}_4$	C_2	σ_x	σ_y	σ_1	σ_2
E	E	C_4	$\bar{C}_4$	C_2	σ_x	σ_y	σ_1	σ_2
C_4	C_4	C_2	E	$\bar{C}_4$	σ_1	σ_2	σ_y	σ_x
$\bar{C}_4$	$\bar{C}_4$	E	C_2	C_4	σ_2	σ_1	σ_x	σ_y
C_2	C_2	$\bar{C}_4$	C_4	E	σ_y	σ_x	σ_2	σ_1
σ_x	σ_x	σ_2	σ_1	σ_y	E	C_2	$\bar{C}_4$	C_4
σ_y	σ_y	σ_1	σ_2	σ_x	C_2	E	C_4	$\bar{C}_4$
σ_1	σ_1	σ_x	σ_y	σ_2	C_4	$\bar{C}_4$	E	C_2
σ_2	σ_2	σ_y	σ_x	σ_1	$\bar{C}_4$	C_4	C_2	E

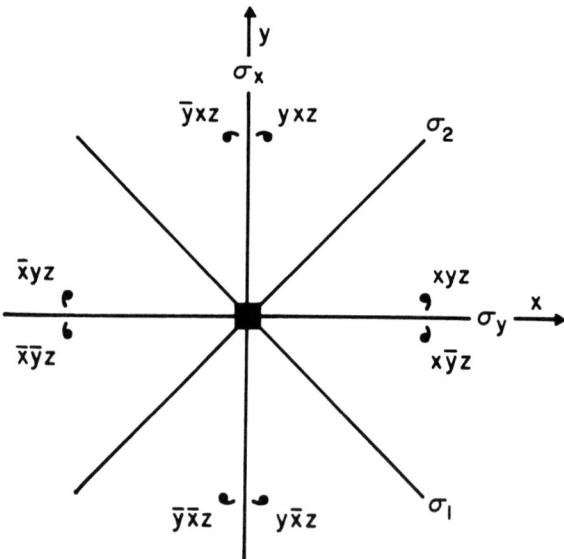

Fig. PR-2. Point group C_{4v} = 4mm. The commas represent the motif which is repeated by the operations of the point group.

For example, the relation $\sigma_x \cdot C_4 = \sigma_2$ transforms xyz to yxz. The convention adopted here is the usual one; that is, the operation designated by a column heading is left-multipled by the row heading. In the C_{4v} example this means that if a point at xyz is transformed by C_4 and then σ_x, this is equivalent to a reflection across σ_2. Note that this combination is not the same as $C_4 \cdot \sigma_x = \sigma_1$.

This leads us to the four postulates of group theory:

1. A set of symmetry operations is closed with respect to a law of combination called multiplication; i.e., the result of each combination is also a member of the group.
2. Group multiplication is associative, i.e., $(C_2 \cdot C_4) \cdot \sigma_x = C_2 \cdot (C_4 \cdot \sigma_x)$.
3. There exists an identity operation E such that $E \cdot C_4 = C_4 \cdot E = C_4$.
4. To each operation there corresponds a unique reciprocal or inverse operator such that $C_4^{-1} \cdot C_4 = C_4 \cdot C_4^{-1} = E$, where $C_4^{-1} \equiv \bar{C}^4 = a$ $90°$ rotation in the negative direction.

If a set of symmetry operations conforms to these postulates, then they constitute a group. A group multiplication table can be generated for each point group and each table contains all the information necessary for the study of the associated group. The advantage of applying group theory to physical problems is that all properties based on the same group multiplication table can be treated by the same mathematical formalism.

Several properties of the group should be apparent to anyone familiar with elementary crystallography. It was noted earlier that in C_{4v} there are eight points in the general position and eight symmetry elements. This equivalence is always present, and the number of symmetry elements is called the *order* of the group. A *subgroup* contains a subset of the elements of the group which themselves also form a group. For example, point group C_4 is a subgroup of C_{4v} and is of order 4. Another property is that the symmetry operations can be represented by matrices. The C_4 operation is given by

$$\begin{bmatrix} x_2 \\ y_2 \\ z_2 \end{bmatrix} = \begin{bmatrix} 0 & \bar{1} & 0 \\ 1 & 0 & 0 \\ 0 & 0 & 1 \end{bmatrix} \begin{bmatrix} x_1 \\ y_1 \\ z_1 \end{bmatrix}$$

where the 3 x 3 matrix is the symmetry operation applied to the point x_1, y_1, z_1. A different matrix is associated with each symmetry operation, and for C_{4v} the matrices are simple. However, for point groups containing three- and six-fold axes, the operations are more complicated because of the non-orthogonal axial systems. The combinations of symmetry matrices through matrix multiplication follow exactly the same rules as the combination for C_{4v} given above.

Classes of Symmetry Operations

In general, whenever sets of symmetry elements are related by a symmetry operation of another element of the group, they are said to form a *class*. In C_{4v}, σ_x and σ_y are related by C_4 and, therefore, are in the same class. Similarly, C_4 and $\bar{C}_4$ are related by σ_y. E by itself forms a class and the complete list of classes in C_{4v} is E, $2C_4$, C_2, $2\sigma_v$, $2\sigma_d$. Mathematically, the elements belonging to a class may be found by performing *similarity transformations*, but it is not necessary to go into these here.

Representations of Groups

The *representation* of a symmetry group is characterized by the way in which a set of basis vectors are transformed by the symmetry operations of the group. Referring again to C_{4v}, the matrices for the symmetry operations transform the basis vectors as follows:

$$E: \begin{bmatrix} x \\ y \\ z \end{bmatrix} = \begin{bmatrix} 1 & 0 & 0 \\ 0 & 1 & 0 \\ 0 & 0 & 1 \end{bmatrix} \begin{bmatrix} x \\ y \\ z \end{bmatrix} \qquad \sigma_x: \begin{bmatrix} \bar{x} \\ y \\ z \end{bmatrix} = \begin{bmatrix} \bar{1} & 0 & 0 \\ 0 & 1 & 0 \\ 0 & 0 & 1 \end{bmatrix} \begin{bmatrix} x \\ y \\ z \end{bmatrix}$$

$$C_4: \begin{bmatrix} \bar{x} \\ y \\ z \end{bmatrix} = \begin{bmatrix} 0 & \bar{1} & 0 \\ 1 & 0 & 0 \\ 0 & 0 & 1 \end{bmatrix} \begin{bmatrix} x \\ y \\ z \end{bmatrix} \qquad \sigma_y: \begin{bmatrix} x \\ \bar{y} \\ z \end{bmatrix} = \begin{bmatrix} 1 & 0 & 0 \\ 0 & \bar{1} & 0 \\ 0 & 0 & 1 \end{bmatrix} \begin{bmatrix} x \\ y \\ z \end{bmatrix}$$

$$\bar{C}_4: \begin{bmatrix} y \\ \bar{x} \\ z \end{bmatrix} = \begin{bmatrix} 0 & 1 & 0 \\ \bar{1} & 0 & 0 \\ 0 & 0 & 1 \end{bmatrix} \begin{bmatrix} x \\ y \\ z \end{bmatrix} \qquad \sigma_1: \begin{bmatrix} \bar{y} \\ \bar{x} \\ z \end{bmatrix} = \begin{bmatrix} 0 & \bar{1} & 0 \\ \bar{1} & 0 & 0 \\ 0 & 0 & 1 \end{bmatrix} \begin{bmatrix} x \\ y \\ z \end{bmatrix}$$

$$C_2: \begin{bmatrix} \bar{x} \\ \bar{y} \\ z \end{bmatrix} = \begin{bmatrix} \bar{1} & 0 & 0 \\ 0 & \bar{1} & 0 \\ 0 & 0 & 1 \end{bmatrix} \begin{bmatrix} x \\ y \\ z \end{bmatrix} \qquad \sigma_2: \begin{bmatrix} y \\ x \\ z \end{bmatrix} = \begin{bmatrix} 0 & 1 & 0 \\ 1 & 0 & 0 \\ 0 & 0 & 1 \end{bmatrix} \begin{bmatrix} x \\ y \\ z \end{bmatrix}$$

The basis vectors, xyz, are oriented as in Fig. PR-2. The matrices multiply according to the group multiplication table and are the representation of the group using these basis vectors. In group theory, the sum of the diagonal terms of a matrix is called the *trace* or *character* of the matrix. The character is

$$\chi = \sum_j a_{jj}.$$

It can be shown that if the characters are taken for the representation, they are independent of the basis vectors selected for the reference frame.

We can now begin to construct a character table using first the basis vectors xyz. Note that the z vector is not affected by any of the transformations; hence, it can be treated separately from the x and y vectors.

E	$2C_4$	C_2	$2\sigma_v$	$2\sigma_d$	
1	1	1	1	1	z
2	0	-2	0	0	x,y

Note that in this table, the symmetry operations have been grouped in classes and that σ_v represents σ_x and σ_y, and σ_d represents σ_1 and σ_2. When the rows and columns containing z are blocked out, there remain eight 2 x 2 matrices operating only on x and y. It is important to note that the p orbitals of Fig. PR-3 transform just as do the xyz basis vectors.

Rotations, R_x, R_y, and R_z, around x, y, and z can also be used as bases for the representation. Using the same kind of development as for the vectors, transformation matrices for the rotations are

$$
E: \quad
\begin{bmatrix} R_x \\ R_y \\ R_x \end{bmatrix}
=
\begin{bmatrix} 1 & 0 & 0 \\ 0 & 1 & 0 \\ 0 & 0 & 1 \end{bmatrix}
\begin{bmatrix} R_x \\ R_y \\ R_z \end{bmatrix}
\qquad
\sigma_x: \quad
\begin{bmatrix} R_x \\ -R_y \\ -R_z \end{bmatrix}
=
\begin{bmatrix} 1 & 0 & 0 \\ 0 & \bar{1} & 0 \\ 0 & 0 & \bar{1} \end{bmatrix}
\begin{bmatrix} R_x \\ R_y \\ R_z \end{bmatrix}
$$

$$
C_4: \quad
\begin{bmatrix} R_y \\ R_x \\ R_z \end{bmatrix}
=
\begin{bmatrix} 0 & 1 & 0 \\ 1 & 0 & 0 \\ 0 & 0 & 1 \end{bmatrix}
\begin{bmatrix} R_x \\ R_y \\ R_z \end{bmatrix}
\qquad
\sigma_y: \quad
\begin{bmatrix} -R_x \\ R_y \\ -R_z \end{bmatrix}
=
\begin{bmatrix} \bar{1} & 0 & 0 \\ 0 & 1 & 0 \\ 0 & 0 & \bar{1} \end{bmatrix}
\begin{bmatrix} R_x \\ R_y \\ R_z \end{bmatrix}
$$

$$
\bar{C}_4: \quad
\begin{bmatrix} -R_y \\ R_x \\ R_z \end{bmatrix}
=
\begin{bmatrix} 0 & \bar{1} & 0 \\ 1 & 0 & 0 \\ 0 & 0 & 1 \end{bmatrix}
\begin{bmatrix} R_x \\ R_y \\ R_z \end{bmatrix}
\qquad
\sigma_1: \quad
\begin{bmatrix} R_x \\ R_y \\ -R_z \end{bmatrix}
=
\begin{bmatrix} 0 & 1 & 0 \\ 1 & 0 & 0 \\ 0 & 0 & \bar{1} \end{bmatrix}
\begin{bmatrix} R_x \\ R_y \\ R_z \end{bmatrix}
$$

$$
C_2: \quad
\begin{bmatrix} -R_x \\ -R_y \\ R_z \end{bmatrix}
=
\begin{bmatrix} \bar{1} & 0 & 0 \\ 0 & \bar{1} & 0 \\ 0 & 0 & 1 \end{bmatrix}
\begin{bmatrix} R_x \\ R_y \\ R_z \end{bmatrix}
\qquad
\sigma_2: \quad
\begin{bmatrix} -R_y \\ -R_x \\ -R_z \end{bmatrix}
=
\begin{bmatrix} 0 & \bar{1} & 0 \\ \bar{1} & 0 & 0 \\ 0 & 0 & \bar{1} \end{bmatrix}
\begin{bmatrix} R_x \\ R_y \\ R_z \end{bmatrix}
$$

Again the matrices can be blocked out because operations mix R_x and R_y but operations involving R_z are independent. The character table can be expanded as

E	$2C_4$	C_2	$2\sigma_v$	$2\sigma_d$	
1	1	1	1	1	z
1	1	1	-1	-1	R_z
2	0	-2	0	0	$(x,y)(R_x,R_y)$

The (x,y) and (R_x,R_y) labels are listed in parentheses to indicate that they are mixed in this representation. There are other bases which must be considered because the d orbitals which we want to study later cannot be characterized by vectors and rotations alone. These additional bases are the squares and binary products of coordinates, i.e., x^2, y^2, xy, yz, xz, $x^2 + y^2$, and $x^2 - y^2$. The analogy to the labeling of the d orbitals is obvious and we will see that the d orbitals of Fig. PR-3 can be used as bases for additional representations. For example, the $d_{x^2-y^2}$ orbital is compatible with the representation:

E	$2C_4$	C_2	$2\sigma_v$	$2\sigma_d$
1	-1	1	1	-1

PR-11

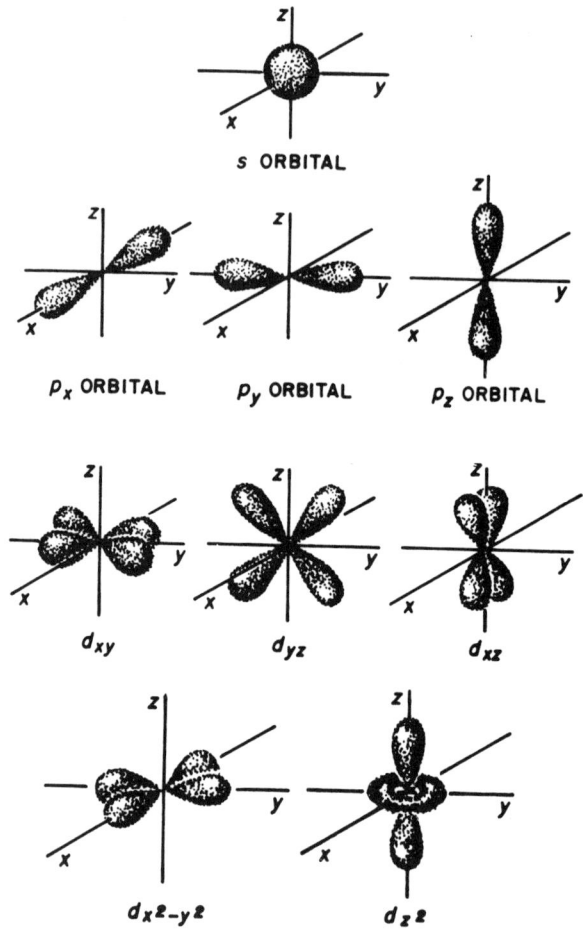

Fig. PR-3. The s, p, and d orbitals.

The reader should verify this and how the other d orbitals transform. The complete character table is:

C_{4v}	E	$2C_4$	C_2	$2\sigma_v$	$2\sigma_d$		
A_1	1	1	1	1	1	z	$x^2 + y^2$, z^2
A_2	1	1	1	-1	-1	R_z	
B_1	1	-1	1	1	-1		$x^2 - y^2$
B_2	1	-1	1	-1	1		xy
E	2	0	-2	0	0	$(x,y)(R_x,R_y)$	(xz,yz)

The only additional base not included in the d orbitals is $x^2 + y^2$ which could represent an s orbital or ellipsoid cylindrically symmetric around the z axis. The symbols to the left of the character table are Mulliken symbols used to label each irreducible representation. This system is based on the characters for each symmetry operation. A and B are used for χ_E = 1, E for χ_E = 2, and $\bar{I}$ for χ_E = 3. A and B are differentiated as to whether the representation is symmetric or anti-symmetric to the principal symmetry axis (χ_{C_4} = +1 or -1). The subscripts for A and B are 1 for χ_{σ_x} = +1 and 2 for χ_{σ_x} = -1. Modifiers used for other point groups are single (') and double ('') primes for planes perpendicular to the principal rotation axis and subscripts g (*gerade* = even) and u (*ungerade* = odd) for symmetry and antisymmetry with respect to an inversion center.

Thus far we have spoken only of irreducible representations. It is possible to select other bases which lead to *reducible* representations. One aspect of the character table is that the character is related to the number of basis vectors, rotations, or binary products left unchanged by the symmetry operations.

Orbitals and Energy Levels

Now that some of the basic ideas of group theory have been presented, it is possible to show how group theory can be applied to chemical problems. The two most important point groups for sulfides are the tetrahedral T_d and the octahedral O_h groups. Character tables for these groups are given below:

T_d	E	$8C_3$	$3C_2$	$6S_4$	$6\sigma_d$		
A_1	1	1	1	1	1		$x^2 + y^2 + z^2$
A_2	1	1	1	-1	-1		
E	2	-1	2	0	0		$(2z^2 - x^2 - y^2, x^2 - y^2)$
T_1	3	0	-1	1	-1	(R_x, R_y, R_z)	
T_2	3	0	-1	-1	1	(x,y,z)	(xy, xz, yz)

O_h	E	$8C_3$	$6C_2$	$6C_4$	$3C_2$	i	$6S_4$	$8S_6$	$3\sigma_h$	$6\sigma_d$		
A_{1g}	1	1	1	1	1	1	1	1	1	1		$x^2+y^2+z^2$
A_{2g}	1	1	-1	-1	1	1	-1	1	1	-1		
E_g	2	-1	0	0	2	2	0	-1	2	0		$(2x^2-x^2-y^2, x^2-y^2)$
T_{1g}	3	0	-1	1	-1	3	1	0	-1	-1	(R_x, R_y, R_z)	
T_{2g}	3	0	1	-1	-1	3	-1	0	-1	1		(xz, yz, xy)
A_{1u}	1	1	1	1	1	-1	-1	-1	-1	-1		
A_{2u}	1	1	-1	-1	1	-1	1	-1	-1	1		
E_u	2	-1	0	0	2	-2	0	1	-2	0		
T_{1u}	3	0	-1	1	-1	-3	-1	0	1	1	(x,y,z)	
T_{2u}	3	0	1	-1	-1	-3	1	0	1	-1		

CHEMICAL BONDING IN SULFIDES

We have previously discussed how orbitals on a free atom are filled with electrons, and we have also given some of the background in group theory necessary to discuss how electrons behave when atoms are combined into molecules or chemical compounds. This subject is an enormously complicated one when all types of chemical combinations are considered. In this treatment, we will concentrate on providing information which will be most easily assimilated and applied by the reader to bonding in sulfides. As the general mineralogical community becomes more knowledgable and sophisticated about the theoretical aspects of sulfide crystal chemistry, items discussed here will be taken for granted and more rigorous explanations of bonding will be required.

Crystal Field Theory

The chemistry of transition metals which is largely characterized by the incompletely filled d or f shells of the metal atoms has become an extremely important and useful subject in the earth sciences for two reasons: (1) These elements together constitute 40 weight percent of the Earth, and (2) they exhibit various physical and chemical properties which are markedly different from those of non-transition elements and, therefore, enable us to understand the various physico-chemical processes taking place in the Earth. One of the several theoretical models proposed to explain the properties of transition metal compounds is based on Crystal Field Theory (CFT) which deals with interactions between transition metal ions and surrounding anions (ligands) in these compounds. Bethe in 1929 developed CFT by considering anions (ligands) as negative point charges (i.e., he assumed that bonding is purely ionic). Although this assumption is not entirely realistic (because an anion has definite structure and finite size), this theory generally does give results which are qualitatively correct. Because of its conceptual simplicity, this theory has been extensively applied to understand and solve several mineralogical and geochemical problems involving silicate minerals (Burns, 1970). The basic tenets of the CFT and its modifications (as applied to sulfides) are summarized in the following pages. Because the first row transition elements (from $3d^1$ to $3d^9$, i.e., $Ti^{3+} - Cu^{2+}$) are by far the most important elements in many geochemical processes, the discussion is restricted to these elements.

The five $3d$ orbitals are designated as d_{xy}, d_{yz}, d_{xz}, $d_{x^2-y^2}$, and d_{z^2}. Two of the orbitals $d_{x^2-y^2}$ and d_{z^2} have lobes directed along cartesian axes, as shown in Fig. PR-3 and are designated as the e_g group. The other three orbitals, d_{xy}, d_{yz}, d_{xz}, called t_{2g}, have lobes projecting between these axes. All five $3d$ orbitals are "degenerate" (have the same energy) when a transition metal ion is isolated and free from surrounding ligands and, therefore, an electron has equal probability of occupying any of the d orbitals. Let us consider a transition metal ion in a regular octahedral site of a mineral structure. The presence of six ligands (usually identical) located on the cartesian axes results in a nonspherical electrostatic field on the ion which results in the removal of the degeneracy of

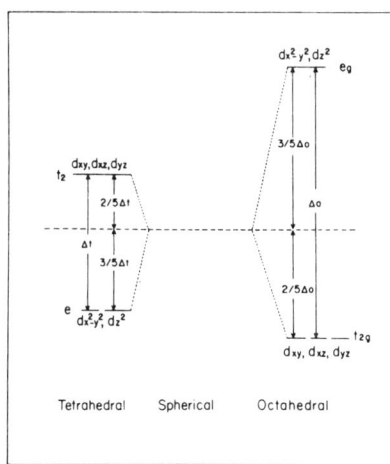

dx²,y²,dz² eg

dxy,dxz,dyz
t₂
2/5Δt
Δt
3/5Δt
e
dx²y², dz²

3/5Δo

Δo

2/5Δo

t₂g
dxy, dxz, dyz

Tetrahedral   Spherical   Octahedral

Fig. PR-4. Crystal field-splitting
of d orbitals in tetrahedral and
octahedral environments.

the five (originally) degerate d orbitals.
It should be noted that the e_g group orbitals
($d_{x^2-y^2}$ and d_{z^2}) are oriented in one way
pointing towards ligands and the t_{2g} group
orbitals (d_{xy}, d_{yz} and d_{xz}) are all oriented
in another way between ligands. It is easy
to see that the e_g group orbitals would
undergo greater electrostatic repulsion due
to the negatively charged ligands (and,
therefore, orbitals will have higher potential energy) than the t_{2g} group orbitals.
In other words, the former group of orbitals
are less stable (lie higher on an energy
level diagram) than the latter group. This
is illustrated in the right-hand side of
Fig. PR-4. This induced energy separation
between the two groups of orbitals is
called *crystal field splitting* (designated
as Δ_o for an octahedral site). For a given
row of transition elements, this quantity Δ_o
is a function of (a) valencies of the cations--the higher the valency, the greater
the splitting, (b) nature of the coordinated ligands, for example, increasing Δ_o
is observed in the sequence $I^- > Br^- > Cl^- > F > OH^- > S > 0$, and (c) metal-ligand
interatomic distances--the smaller the distance the greater the magnitude of Δ_o,
because $\Delta \sim K/R^5$, where K is a constant and R is the metal-ligand interatomic distance. Note: a small change in the interatomic distance will cause a significant
change in Δ_o.

The splitting of the energy of d orbitals obeys a "center of gravity" rule
and, therefore, t_{2g} orbitals are lowered by $2/5\Delta_o$ below and e_g orbitals are raised
by $3/5\Delta_o$ above the center (referred to as the *baricenter*). Thus, each electron in
a t_{2g} orbital stabilizes the ion by $2/5\Delta_o$ whereas each electron in an e_g orbital
destabilizes the ion by $3/5\Delta_o$. The resultant stabilization energy is called the
crystal field stabilization energy (designated as CFSE). The distribution of d
electrons in the two groups of orbitals is controlled by two opposing forces--the
intra-atomic exchange stabilization (Hund's rule) and the magnitude of crystal field
splitting, Δ_o(crystal field). These two factors lead to high-spin and low-spin
electronic configurations, respectively, in certain cations having 4, 5, 6, and 7 d
electrons (Cr^{2+}, Mn^{3+}; Mn^{2+}, Fe^{3+}; Fe^{2+}, Co^{3+}; Co^{2+} and Ni^{3+}). If the amount of
energy gained by the crystal field (i.e., by adding extra electrons in the t_{2g}
orbitals, which would otherwise occupy e_g orbitals) outweighs the energy required
to pair these electrons in the t_{2g} orbitals, then electrons will pair in these
orbitals rather than be distributed over all the orbitals. To illustrate this by
way of an example, consider Co^{2+} which has seven d electrons. When the magnitude of

Δ_o is small, as in many silicates, five electrons would be distributed in the five d orbitals and the other two electrons would go into the low-lying t_{2g} orbitals to gain maximum stabilization. This electronic configuration, $t_{2g}^5 e_g^2$, is known as high-spin cobalt or $^{VI}Co^{2+}$H.S. However, if a ligand can cause a high Δ_o so that the energy gained by pairing one more electron in the t_{2g} orbitals outweighs the energy required to pair that electron, then Co^{2+} assumes a different electronic configuration, $t_{2g}^6 e_g^1$, and this is known as low-spin cobalt or $^{VI}Co^{2+}$L.S. It should be noted here that for cations having less than four or greater than seven d electrons (i.e., d^1, d^2, d^3, d^8 and d^9; e.g., Ti^{3+}, V^{3+}, Cr^{3+}, Ni^{2+} and Cu^{2+}) there exists only one electronic configuration (i.e., only one spin-state) regardless of the magnitude of crystal field splitting. Electronic configurations and CFSE for the first row transition metal cations in octahedral coordination with ligands are listed in Table PR-4.

Table PR-4. Electronic Configurations and CFSE of Transition Metal Ions.

Number of 3d electrons	Ions	Octahedral						Tetrahedral		
		High-spin			Low-spin					
		t_{2g}	e_g	CFSE	t_{2g}	e_g	CFSE	e	t_2	CFSE
1	Ti^{3+}	1	0	$2/5\Delta_o$	–	–	–	1	0	$3/5\Delta_t$
2	Ti^{2+} V^{3+}	2	0	4/5	–	–	–	2	0	6/5
3	V^{2+} Cr^{3+}	3	0	6/5	–	–	–	2	1	4/5
4	Cr^{2+} Mn^{3+}	3	1	3/5	4	0	8/5	2	2	2/5
5	Mn^{2+} Fe^{3+}	3	2	0	5	0	10/5	2	3	0
6	Fe^{2+} Co^{3+}	4	2	2/5	6	0	12/5	3	3	3/5
7	Co^{2+} Ni^{3+}	5	2	4/5	6	1	9/5	4	3	6/5
8	Ni^{2+} --	6	2	6/5	6	2	6/5	4	4	4/5
9	Cu^{2+} --	6	3	3/5	6	3	3/5	4	5	2/5

In the splitting of d orbitals due to the presence of ligands, we assumed earlier that the octahedral site in which a cation is located is regular. However, such "regular" octahedral sites are rare in many ferromagnesian silicates and usually the sites are distorted (i.e., symmetry is lower than O_h). This distortion causes a further resolution of d orbitals whereby certain cations gain additional stabilization in distorted octahedral sites. For example, the M(1) octahedral site in the olivine structure is tetragonally distorted, i.e., one axis is elongated relative to the other two. If the elongated direction is taken as the z axis, then $d_{x^2-y^2}$ (with respect to d_{z^2}) and d_{xy} (with respect to d_{xz} and d_{yz}) orbitals are de-stabilized as shown in Fig. PR-5. If we consider divalent high-spin cobalt (d^7) in that site, it will gain additional stabilization by $2/3\alpha$ (total stabilization energy = $4/5\Delta_o + 2/3\alpha$) because the sixth and seventh electrons will go into the lower

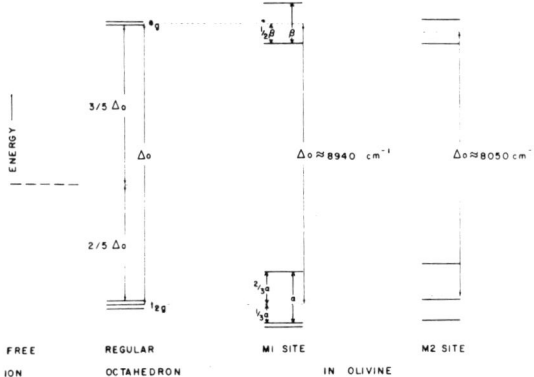

Fig. PR-5. Crystal field splitting of the M(1) and M(2) sites in olivine.

two of the resolved t_{2g} orbitals. Although it is claimed that such distortions
increase the stabilization energy (Burns, 1970), it should be noted that distortions
might increase the mean cation-ligand distance R which will decrease the magnitude
of Δ_o which, in turn, will decrease the CFSE. Sometimes distortions are caused by
certain cations with four and seven d electrons (e.g., Mn^{3+}, low-spin Co^{2+}, and
Ni^{3+}). Such spontaneous distortions are due to what is known as the *Jahn-Teller*
effect. Jahn Teller distortion arises "if one of the d orbitals is completely empty
or completely filled while another of equal energy is half filled, the environment
about the transition metal ion is predicted to distort spontaneously to a different
geometry in which a more stable electronic configuration is attained by making the
occupied orbital lower in energy" (Burns, 1970, p. 20).

So far we have seen the effects of an octahedral crystal field on transition
metal cations. Although tetrahedral sites are relatively less abundant in mineral
structures, it is interesting to consider the effect of a tetrahedral crystal field
on cations (Note: tetrahedral sites are important in sulfide and oxide structures
and also in silicate magmas). For a regular tetrahedral site, ligands are considered
as occupying the alternate vertices of a cube and, therefore, t_{2g} orbitals (those
projecting between Cartesian axes) would undergo greater electrostatic repulsion than
the e_g orbitals. Because a tetrahedron lacks a center of symmetry, these two groups
are designated as t_2 and e. The relative energies of the two groups of orbitals are
shown in Fig. PR-4. The magnitude of the tetrahedral crystal field splitting (re-
ferred to as Δ_t) is much smaller than Δ_o (usually $\Delta_t = -4/9\Delta_o$). Because of this
small crystal field splitting, the low-spin electronic configuration is unknown in
tetrahedral coordination. An electron in an e orbital is stabilized by $3/5\Delta_t$ and an
electron in a t_2 orbital is destabilized by $2/5\Delta_t$. Electronic configurations and
CFSE for the first row transition elements in tetrahedral coordination with ligands

are included in Table PR-5. The difference between the octahedral and tetrahedral
CFSE for a given cation is termed octahedral site preference energy (OSPE), which is
a useful quantity in predicting the distribution of these cations in structures containing both octahedral and tetrahedral sites. OSPE of transition metal cations
in the spinel structure, as given by McClure (1957) and Dunitz and Orgel (1957) are
listed in Table PR-5. It should be noted that divalent nickel in octahedral coordination and divalent cobalt in tetrahedral coordination gain maximum CFSE among the
common divalent transition metal ions. This aspect appears to be one of the
important differences in the crystal chemical properties of nickel and cobalt.

Table PR-5. Crystal Field Stabilization Energies for Transition Metal Ions in
Octahedral and Tetrahedral Coordination Sites in Oxide Structures.

Number of 3d electrons	Ions	Octahedral CFSE kcal mole^{-1}	Tetrahedral CFSE kcal mole^{-1}	Octahedral site preference energy OSPE
1	Ti^{3+}	20.9	14.0	6.9
2	V^{3+}	38.3	25.5	12.8
3	Cr^{3+}	53.7	16.0	37.6
4	Mn^{3+}	32.4	9.6	22.8
5	Mn^{2+} Fe^{3+}	0	0	0
6	Fe^{2+}	11.9	7.9	4.0
	Co^{3+}	45.0	26.0	19.0
7	Co^{2+}	22.2 (17.1)	14.8	7.4 (2.3)
8	Ni^{2+}	29.2	8.6	20.6
9	Cu^{2+}	21.6	6.4	15.2

Sources of data: Dunitz and Orgel (1957); values in parentheses
are from McClure (1957).

Molecular Orbital Theory

As mentioned earlier, the assumption that anions (ligands) are negative point
charges (i.e., bonding is purely ionic) is not a realistic one, particularly as the
degree of ionicity of the bond decreases. Therefore, in highly covalent structures
where there is extensive overlapping of metal and ligand orbitals, application of
crystal field theory to explain the physico-chemical properties of these compounds
is not strictly valid. A more satisfactory theory which considers both the overlap
of orbitals as well as electrostatic interactions is termed molecular orbital theory
or MO theory. (This theory is particularly useful for sulfide minerals because of
the more covalent nature of metal-sulfur bonds relative to metal-oxygen bonds).
The overlap of orbitals, to a large extent, depends on the symmetry and spatial
properties of metal and ligand orbitals.

When the many-electron atom was discussed previously, electrons were added successively to orbitals of higher and higher energy. The same approach can be taken with molecules or combinations of atoms; electrons can be added to molecular orbitals, no more than two to an orbital, and the resulting number of orbitals is just the sum of the orbitals on the respective atoms.

First, consider two hydrogen atoms (Fig. PR-6). When the atoms are far apart there is no interaction but, as they are brought closer together, the positive charges of the nuclei repel one another as do the two negatively-charged electron clouds. However, if the attraction of a nucleus of one atom to the electron cloud of the other is strong enough, the electron density between the atoms begins to increase, the energy decreases and a molecule is formed. The final internuclear distance is a compromise between the attraction mentioned above and the repulsion between the nuclei. The increased electron density between nuclei is the overlap of the $1s$ atomic orbitals. If the electron density of atom a is represented by $|\psi_{1s}^a|^2$, a molecular orbital can be constructed by a linear combination of atomic orbitals and the resulting electron density is represented by $|\psi_{1s}^a + \psi_{1s}^b|^2$. This is called a *bonding orbital* and is shown diagrammatically in Fig. PR-7. If instead of adding the atomic orbitals together, they are subtracted, the molecular orbital

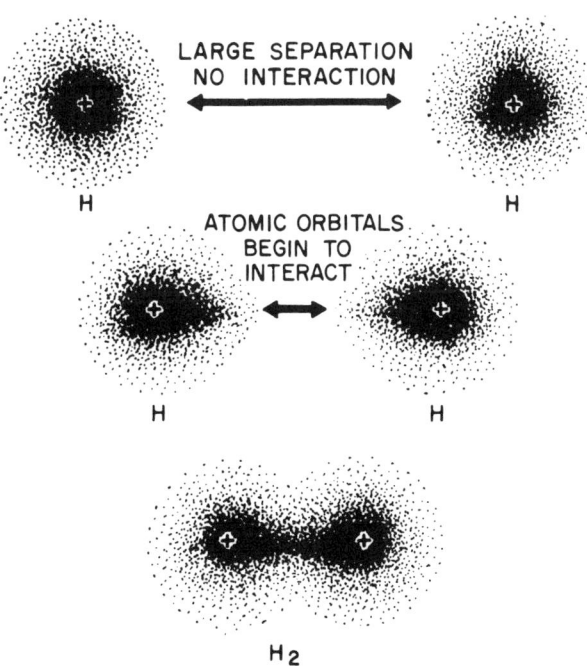

LARGE SEPARATION
NO INTERACTION

H   H

ATOMIC ORBITALS
BEGIN TO
INTERACT

H   H

H₂

Fig. PR-6. Formation of a hydrogen molecule from two hydrogen atoms.

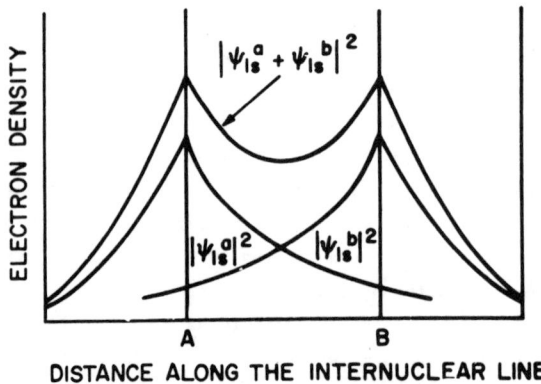

Fig. PR-7. Electron density in the hydrogen molecule resulting from the linear combination of two atomic orbitals of hydrogen.

is represented by $\left| \psi_{1s}^{a} - \psi_{1s}^{b} \right|^2$ and the electron density falls to zero halfway between the nuclei. The electrons are concentrated outside of the internuclear region which tends to decrease the attraction between nuclei. This type of orbital is called an *antibonding orbital*. Figure PR-8 shows an energy-level diagram for the H_2 molecular orbital. The two $1s$ orbitals on the atoms become σ^b bonding and σ^* antibonding molecular orbitals. The σ indicates that the orbital is invariant to rotation around its axis.

This simple picture of the hydrogen molecule helps to understand more complicated combinations. Let us consider, for example, a transition metal ion in a regular octahedral site surrounded by six identical ligands. Two of the five $3d$

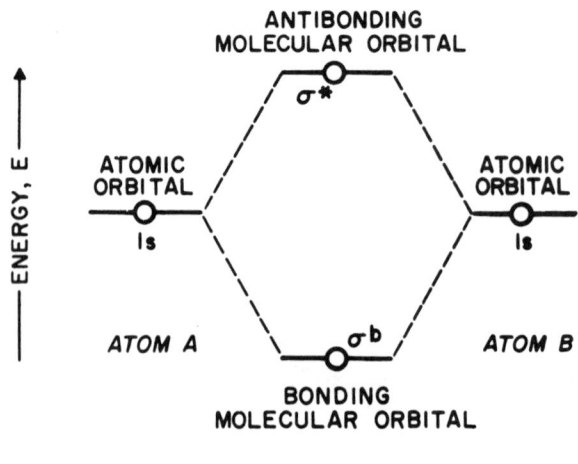

Fig. PR-8. Energy-level diagram for the hydrogen molecule.

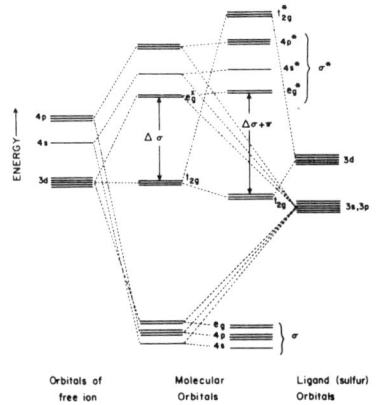

Fig. PR-9. Energy-level diagrams for molecular orbitals of a transition-metal sulfide.

orbitals (e_g group) which are directed towards the ligands, one $4s$ orbital and three $4p$ orbitals combine to form six hybrid molecular orbitals (usually referred to as d^2sp^3) which are again directed towards the six ligands. When sufficient overlap takes place between these six metal orbitals and ligand orbitals (usually s and p orbitals), the orbitals become mixed to form two sets of new orbitals-- bonding orbitals (σ), and antibonding orbitals (σ^*) (Fig. PR-9). Bonding orbitals which are completely filled with electrons have the characteristics of the lower energy atomic orbitals and are more stable than the individual atomic orbitals as can be seen from Fig. PR-9. Excess electrons that are needed to completely fill the bonding orbitals will go into the antibonding orbitals which have the characteristic of the higher energy atomic orbitals and are less stable than the individual atomic orbitals. Usually, the bonding orbitals (σ) have the ligand character and the antibonding orbitals (σ^*) have the character of the metal atoms. It should be noted here that   three of the five $3d$ metal orbitals (t_{2g}) (those projecting between the ligands) do not participate in the σ-bonding (direct overlap). These orbitals are called nonbonding (with respect to σ-bonding in octahedral coordination) orbitals. Because of σ-bonding in which only the e_g group of metal $3d$ orbitals participate to become less stable antibonding orbitals, the two groups of $3d$ orbitals are separated by an energy gap ($\Delta_{cov-\sigma}$) which bears resemblance to Δ_o of crystal field theory. The nonbonding metal t_{2g} orbitals may overlap with suitable ligand orbitals (for example $3d$ orbitals of sulfur ligands) to either side of the direct cation-anion vector to form what is known as a π-bond. Depending upon the relative stabilities of metal t_{2g} orbitals and ligand π-orbitals, the t_{2g} orbitals are stabilized or destabilized as in σ-bonding. For example, the $3d$ orbitals of the sulfur ligand are vacant and less stable than the metal t_{2g} orbitals. Overlapping of these orbitals (i.e., π-bonding) leads to stabilization of the metal t_{2g} orbitals (lowered in energy) as shown in Fig. PR-9. Thus, π-bonding in sulfides causes an increase in the energy gap between the two groups of metal $3d$ orbitals. π-bond formation and the consequent increase in Δ_o as well as in bond strength are believed to be responsible for the sulfophile nature of the transition elements. Further, the presence of cations in low-spin states (e.g., iron in pyrite and cobalt in several sulfide minerals containing octahedral sites) is also attributed to π-bond formation.

It is worthwhile to mention here that the spin state of transition metal ions is a function of the type of ligand, pressure, and temperature because these parameters control the magnitude of crystal field splitting (Δ_o). Table PR-6 shows the frequently observed relation between the type of ligands and the spin-state of Mn, Fe, Co, and Ni. Trivalent Co and Ni seem to occur as low-spin ions even in oxides. Temperature and pressure are expected to have opposing influences on the interatomic distances (Note: crystal field splitting, which determines the electronic configuration, is inversely proportional to the fifth power of the interatomic distances). Increasing temperature would favor a high-spin state because of increasing interatomic distances and thermal motion of atoms. This was found to be the case in the compounds $LaCoO_3$ and $GdCoO_3$ where low-spin Co^{3+} goes to high-spin Co^{3+} at $\sim450°C$. Similarly, it has been reported that the spin-state Co^{3+} in Co_2O_3 is pressure dependent, i.e. Co_2O_3 (high-spin) is a stable phase at atmospheric pressure whereas Co_2O_3 (low-spin) is the high-pressure phase (Chenevas, Joubert, and Marezio, 1971).

Table PR-6. Relation between the Spin-State of Cations and the Type of Ligands.

Ligand	Mn	Fe	Co	Ni^{3+}
F	HS	HS	HS	HS
O	HS	HS	HS/LS	LS
S	HS	HS	LS	LS
S_2	HS	LS	LS	LS

Band Theory

Although ligand field theory has been used with reasonable success to understand the aspects of bonding in sulfide minerals, several physical properties exhibited by these sulfides cannot be explained by this theory. For example, most transition metal sulfides have a metallic appearance, and a few of them even show metallic conduction and temperature independent paramagnetism (Pauli-paramagnetism). These physical properties clearly indicate that the assumption that the valence electrons are localized on cation sites is not strictly applicable to these sulfides. In recent years band theory of crystalline solids has been used to explain the bonding and physical properties of transition metal sulfides. For a complete description of band theory as it applies to sulfides the reader is referred to the papers by Goodenough (1967) and Jellinek (1970). Briefly, the $4s$ and $4p$ orbitals of the metal atom along with e_g or t_2 group of $3d$ orbitals (depending on whether the metal atom is located in an octahedral or tetrahedral ligand field) overlap with $3s$ and $3p$ orbitals of sulfur leading to the formation of a valence band (consisting of bonding orbitals) and a conduction band (consisting of antibonding orbitals).

A band is described as consisting of closely spaced, allowed electronic energy levels in a crystalline solid. The valence band is mainly due to $3p$ and $3s$ orbitals of sulfur and is usually filled with electrons. The conduction band, which is usually empty, is due to $4p$ and $4s$ orbitals of the transition metal. The valence and the conduction bands are separated by an energy gap, the magnitude of which depends on the degree of covalent mixing of the metal and sulfur s and p orbitals, which, in turn, depend on the difference in electronegativity of the cation and the anion. Although it was stated earlier that the valence band is mainly due to sulfur, this is strictly true only in the extreme ionic model. In a more general case, the character of the valence band is also determined by the electronegativity difference-- the smaller the difference, the greater would be the contribution by the cation.

As mentioned in the discussion of molecular orbital theory, the transition metal $3d$ orbitals, the e_g orbitals in an octahedral ligand field and the t_2 orbitals in a tetrahedral ligand field, will also overlap with the sulfur orbitals to form narrow bands. Usually these d orbitals are localized on the metal and lie below the conduction band (Fig. PR-10). If the bottom of the conduction band and the highest filled $3d$ orbitals are separated by an energy gap, then the compound will be a semiconductor. If the energy of the highest filled d level falls near or on the bottom of the conduction band, then part of the electrons go from the d orbital to the conduction band and the compound will exhibit metallic properties as in $Co_{1-x}S$ and CoS_2. In some cases, the $3d$ orbitals of neighboring cations may interact directly or through anions to form broad d bands. The $3d$ electrons are no longer localized to any specific cation sites and become collective, i.e., completely delocalized in the d bands. These d bands (particularly those of antibonding orbitals, σ^*, with respect to cation-anion bonding) are usually incompletely filled and these compounds exhibit metallic conduction and temperature independent paramagnetism, as for example in Co_9S_8, NiS, Ni_3S_2, and $Fe_{1+x}S$. In these crystal structures these d-d interactions are manifested by short metal-metal distances.

ASPECTS OF METAL-METAL BONDING IN Fe, Co, AND Ni SULFIDES

In addition to forming several isotypic sulfides, Fe, Co, and Ni each form specific structures with sulfur which appear to be characteristic of each metal (in particular, Ni has several such structures--see Ch. 5). These include mackinawite ($Fe_{1+x}S$), cobalt pentlandite (Co_9S_8), millerite (NiS), heazlewoodite (Ni_3S_2) and godlevskite (Ni_7S_6). One important aspect of these minerals is their metallic nature including Pauli-paramagnetism (Hulliger, 1968). Structural reasons for their metallic properties and the reason for the formation of specific structures by a specific metal are not clear. Recent structural studies on these minerals revealed that their structures contain short metal-metal distances suggesting the presence of metallic bonds (Taylor and Finger, 1969; Geller, 1962; Rajamani and Prewitt, 1973, 1974; Fleet, 1972). Although these metal-metal bonds could account for the metallic properties, the reasons for the formation of such metal-metal bonds and their influence on the cation-anion bonding are not understood. The above authors suggested that these phases are stabilized by metal-metal bonds which apparently

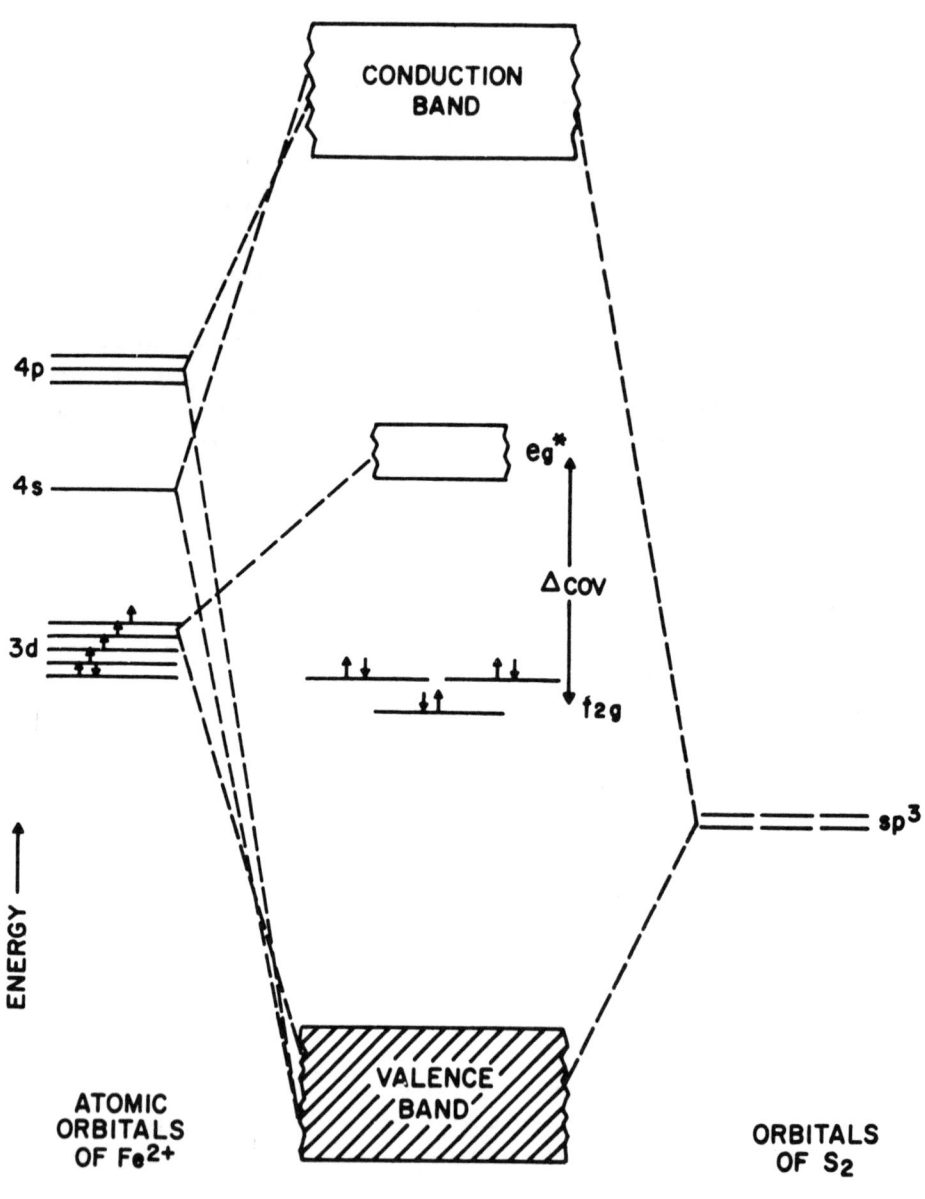

Fig. PR-10. Energy-level diagram showing the valence and conduction bands of FeS_2.

control the solid-solution behavior of these metals in these structures. The
observed metal-metal bond distances and the number of bonds in each structure are
listed in Table PR-7. It is useful to understand the relation between the stoichiometry of a phase, the metal-sulfur coordination, and the number of metal-metal
bonds, so that predictions can be made on the structure and properties of the compounds of a given transition metal ion, which is one of the aims of crystal chemistry. To this end, Pearson's general valence rule for valence compounds which show

Table PR-7. Metal-Metal Coordination in Selected Sulfides.

Name	Composition	M-S Coordination	M-M Coordination	M-M Distance, Å
Mackinawite	$Fe_{1+x}S$	4	4	2.602
Co-Pentlandite	Co_9S_8	6, 4	3	2.505
Pentlandite	$(Fe,Ni,Co)_9S_8$	6, 4	3	2.531
Millerite	NiS	5	2	2.534
Heazlewoodite	Ni_3S_2	4	4?	2.49
αNi_7S_6	Ni_7S_6	5, 4	2?	2.492

semiconductivity may be used. This rule has been used to classify structures containing a chalcogen as the anion (Takéuchi, 1970). It should be noted here that the
Pearson's valence rule could not be used to classify structures containing metal-
metal bonds (because it took into account only valence electrons) and, therefore,
these structures remain "unclassified" (Pearson, 1972). This rule was based on the
idea that the valence shell of anion (in the present case sulfur) contains eight s
and p electrons and is expressed by the relation:

$$\frac{n_a + n_c + b_a - b_c}{N_a} = 8,$$

where n_a is the number of valence electrons on the anions, n_c is the number on the
cations less any unshared valence electrons, b_a is the number of electrons involved
in forming anion-anion bonds, b_c is the number of electrons forming cation-cation
bonds and N_a is the number of anions; all of these numbers are calculated per formula
unit of the compound. For example, in NiS_2, $n_a = 2 \times 6 = 12$, $n_c = 2$, $b_a = 2$, $b_c =$
0 and $N_a = 2$. Thus $(12 + 2 + 2 - 0)/2 = 8$. NiS_2 with the pyrite structure is a
semiconductor (Hulliger, 1968). In hexagonal NiS with the NiAs structure, $n_a =$
6, $n_c = 2$, $b_a = b_c = 0$ and $N_a = 1$, $(6 + 2 + 0)/1 = 8$. Although Pearson's valence
rule is obeyed, the phase is observed to be metallic above $263°K$ (Trahan et al,
1971). Further, even in other "normal" sulfides such as CoS, Co_3S_4, Ni_3S_4, and
CoS_2 (which obey Pearson's rule) metallic conductivity has been observed
(Hulliger, 1968). It is believed that the metallic conductivity in these sulfides

is due to the delocalization of cations' d electrons in the antibonding (σ^* with respect to M-S bonding) d orbitals. Thus, for these transition-metal sulfides we may have to consider additional conditions for d electrons which are involved either directly or indirectly in metal-metal interactions. Available structural evidence suggests that unpaired d electrons in the antibonding d orbitals could be involved in the metal-metal bonding. Then n_c in Pearson's equation may be defined as the number of valence plus any unpaired d electrons in the antibonding d orbitals. For example, in millerite, $n_a = 6$, $n_c = 4$, $b_a = 0$, $b_c = 2$ and $N_a = 1$, $(6 + 4 + 0 - 2)/1 = 8$. If cation-cation bonds and anion-anion bonds are considered as electron pair bonds (Pearson, 1972; Hulliger, 1968), then

$$C_a = b_a/N_a \text{ and } C_c = b_c/N_c$$

where C_a is the number anion-anion bonds per anion and C_c is the number of cation-cation bonds per cation and N_c = number of cations in the formula unit.

Pearson's equation can be modified as

$$\frac{n_a + n_c + C_a N_a - C_c N_c}{N_a} = 8; \quad C_c = \frac{n_a + n_c + C_a N_a - 8N_a}{N_c}$$

Applying this equation to

1) Mackinawite ($Fe_{1+x}S$): $C_c = (6 + 6 + 0 - 8) = 4$

2) Cobalt pentlandite (Co_9S_8): $C_c = \dfrac{(8 \times 6) + (8 \times 5) + 0 - (8 \times 8)}{8} = \dfrac{88 - 64}{8} = 3$
   (Note: only tetrahedral cobalt atoms are taken into account because these are the ones involved in direct metal-metal interactions.)

3) Millerite (NiS): $C_c = \dfrac{(6 + 4 + 0 - 8)}{1} = 2$

4) Heazlewoodite (Ni_3S_2): $C_c = \dfrac{(2 \times 6) + (3 \times 6) + 0 - 16}{3} = 14/3 = 2.67$

From Table PR-7 it is evident that, except in heazlewoodite, the above equation seems to apply for all structures containing short metal-metal distances. From X-ray powder diffraction data it was suggested that in the structure of heazlewoodite (rhombohedral) each Ni is coordinated to four other Ni atoms in addition to four sulfur atoms (Hulliger, 1968). In view of this discrepancy, the structure of heazlewoodite needs careful study using single-crystal diffraction techniques. Nevertheless, the possibility exists that the discrepancy is due to the assumption that cation-cation bonds are essentially electron-pair bonds, an assumption which may not be strictly correct.

It is worthwhile mentioning here that metallic properties in a sulfide does not necessarily imply the existence of direct metal-metal bonds and vice-versa. Indirect, weak metal-metal interactions through anion σ or π bonds could cause broadening of d levels into bands into which d electrons are delocalized. Partial occupancy of such d bands could cause metallic properties and Pauli-paramagnetism, as in Co_9S_8. Factors which influence the formation of direct metal-metal bonds are not completely understood.

The physical properties of sulfides, especially the magnetic and electrical properties, have been intensely studied in recent years. Apart from their industrial applications, these properties of sulfides could be profitably used in mineral exploration and also are important in geomagnetism and geoelectricity. Indirectly, these properties enable us to understand certain aspects of bonding in particular sulfides. An attempt is made in the following pages to explain the physical properties, including magnetic and electrical properties and reflectivity of transition-metal sulfides with the help of band theory.

On the basis of magnetic properties, sulfides can be classified as diamagnetic, paramagnetic, and Pauli-paramagnetic. The magnetic moment of an atom is determined by the spin of electrons and by their orbital motion. Usually in transition metal compounds the orbital contribution to the magnetic moment of an atom is negligible and, therefore, only the electronic spin contribution is important. It is well known that the magnetic moments of ions are due to unpaired electrons in the outermost shells. Thus, ions which have completely filled outer shells will not have magnetic moments. In transition metals, because $4s$ electrons are involved in bonding with sulfur and become part of the valence band electrons, $3d$ electrons alone are responsible for their magnetic moments. $3d$ orbitals will also overlap sulfur $3s$ and $3p$ orbitals but only to a much smaller extent than $4s$ and $4p$ orbitals. Therefore, in many cases, the $3d$ orbitals remain essentially localized on the metal, although the five-fold degeneracy is removed because of the presence of a ligand field (referred earlier as nonbonding and antibonding orbitals) Each $3d$ orbital is split into an α level and a β level corresponding to two different electron spins (Fig. PR-11). In an octahedral ligand field the t_{2g} α level has the lowest energy (potential energy) and e_g β level has the highest energy. If the energy of the t_{2g} β level is higher than the energy of the e_g α level, the d electrons would be distributed in t_{2g} α and e_g α levels before pairing takes place in the t_{2g} β level. The result is a high-spin electronic configuration for the ion. Because the high-spin state has the maximum number of unpaired electrons for a given transition metal ion, the sulfides containing high-spin ions will be paramagnetic with high magnetic susceptibility. An example is MnS_2 (Fig. PR-11). However, if the energy of t_{2g} β levels is lower than the energy of e_g α levels, the electrons will pair in t_{2g} orbitals before occupying e_g orbitals. This electronic configuration becomes low-spin with a minimum of unpaired electrons and consequently low susceptibility values (Fig. PR-12).

In the majority of the first-row transition-metal sulfides, cations commonly exist in the high-spin state and, therefore, these sulfides are paramagnetic with positive values of magnetic susceptibility. Coupling of the magnetic moments of the adjacent cations takes place either directly or by cation-anion-cation exchange and this leads to ordering of magnetic moments, especially at low temperatures. This exchange coupling usually results in antiferromagnetism (e.g., $Fe_{1-x}S, Co_{1-x}S, Ni_{1-x}S$, NiS_2,etc.) or ferrimagnetism (monoclinic Fe_7S_8, $FeCr_2S_4$,etc.) in many paramagnetic sulfides. Only a very few sulfides exhibit ferromagnetism (e.g., CoS_2 and $CuCr_2S_4$). Sulfides containing transition metal ions with d^0 and d^{10} or spin paired d^6 (as Fe^{2+} in pyrite) have no net spin moment and thus are diamagnetic. In certain sulfides the $3d$ orbitals of neighboring cations interact to form d-bands. This effect becomes particularly significant at low temperatures and in structures where the cation coordination number is small, as in pentlandite $(Fe,Co,Ni)_9S_8$, millerite (NiS), heazlewoodite (Ni_3S_2), and others. The d electrons in the resultant d-bands

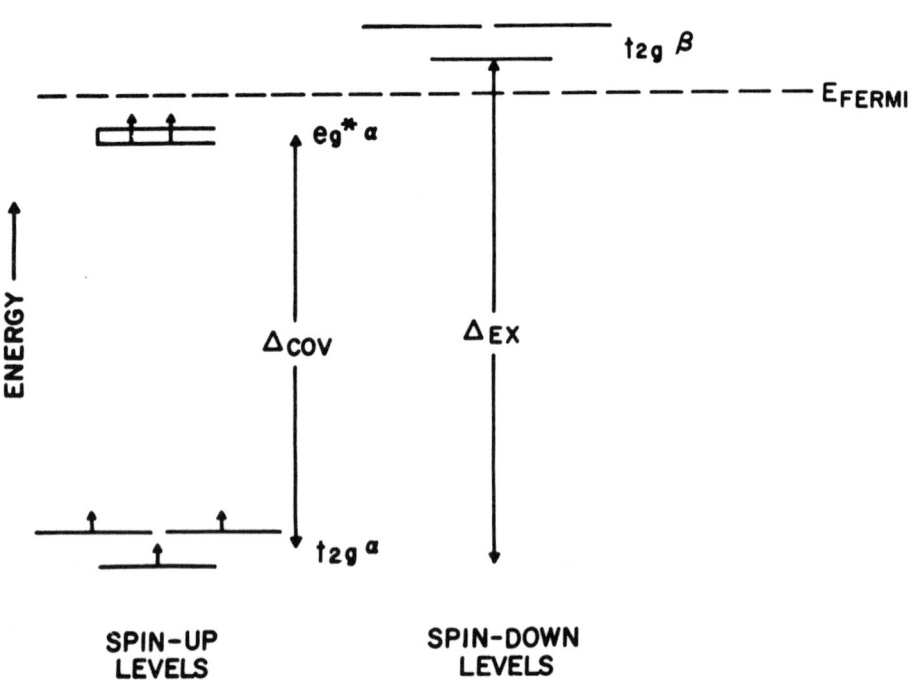

Fig. PR-11. Schematic energy-level diagram for MnS_2 d orbitals (Bither *et al*, 1968). Mn^{2+} is in the high-spin electronic configuration.

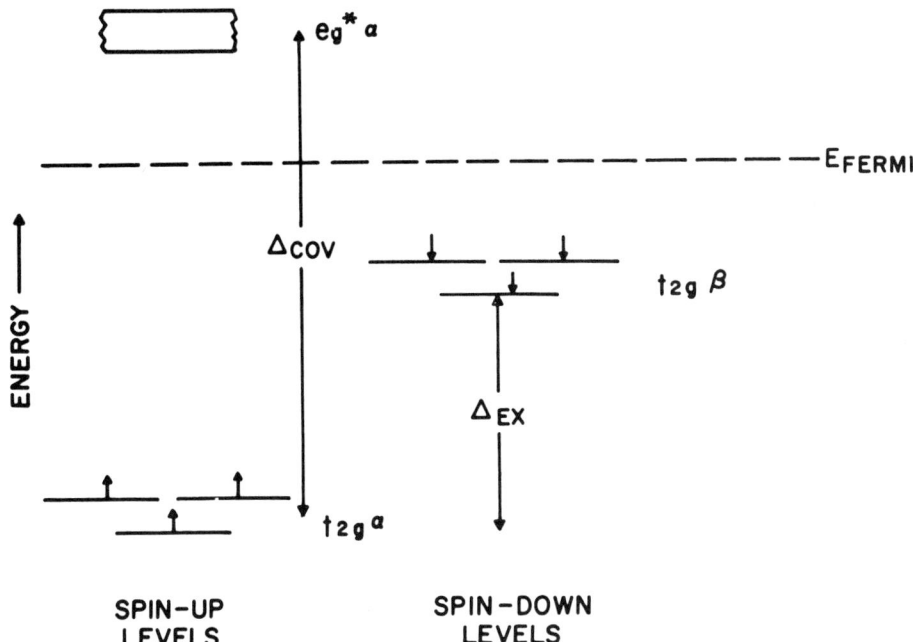

Fig. PR-12. Schematic energy-level diagram for FeS_2 d orbitals. Fe^{2+} is in the low-spin electronic configuration.

are completely delocalized. If the d-bands are incompletely filled, then the compounds exhibit temperature–independent paramagnetism (Pauli-paramagnetism with very small susceptibility values). Thus, measurement of magnetic properties would be very useful in understanding the nature of bonding in sulfides and also the electronic configuration of cations present in them.

In many transition metal sulfides, the localized and split $3d$ orbitals fall between the filled valence band and the empty conduction band. The energy gap between the bottom of the conduction band and the highest partly filled d level is usually small. At moderately high temperatures thermal energy could cause promotion of a small number of d electrons from d orbitals (usually σ^* antibonding) to the conduction band. In an applied electric field the movement of promoted electrons in the conduction band causes electrical conductivity. Because of the smaller number of current carriers, the conductivity in such sulfides is usually small, increasing with temperature as the probability of electron promotion rises. These sulfides are semiconductors, e.g., hexagonal $Fe_{1-x}S$, FeS_2, NiS_2, etc.

Several transition metal sulfides exhibit metallic conduction (with low resistivity which increases with increasing temperature as opposed to semiconductors in which resistivity decreases with increasing temperature). The metallic conductivity in these sulfides arises in different ways:

(1) It is possible that the highest filled d level overlaps with the empty conduction band and hence some of the d electrons will be transferred to the conduction band in which they are delocalized causing metallic conduction when an electric field is applied. Such compounds are called n-type metallic conductors, e.g., CoS_2 and $Co_{1-x}S$.

(2) If the energy of the d orbital of a cation M^{m+} falls below the top of the valence band and if the energy of the d orbital of the same cation but in a different oxidation state $M^{(m-1)+}$ also falls below the top of the valence band, then the cation will be reduced by removal of an electron from the valence band. Because of this, holes are created within the valence band and metallic conduction results due to the movement of electrons within the (incompletely) filled valence band. Such compounds are called p-type metallic conductors and this type of conductivity is usually exhibited by sulfides containing the transition metal ion in which the energy of the d orbital is low because of a high effective nuclear charge, e.g. Cu^{2+} and Cu^{1+}. It should be pointed out here that in both n-type and p-type metallic conductors the d orbitals are localized on cations.

(3) A third type of metallic conduction arises because of the interaction of the d orbitals of adjacent cations either directly or through intervening sulfur. This interaction leads to the formation of d bands and, therefore, the d electrons are no longer localized to any specific metal atom. More commonly these d bands are incompletely filled, and movement of electrons within the band also causes metallic conduction. Such sulfides also exhibit Pauli-paramagnetism, e.g., Co_3S_4, Co_9S_8, NiS, Ni_3S_2, and others.

Jellinek (1970) divided the transition metal sulfides into four main classes on the basis of their magnetic and electrical properties. These include:

(1) Semiconductors with paramagnetism or diamagnetism.

(2) Metallic conductors of the p-type with paramagnetism.

(3) Metallic conductors of the n-type with paramagnetism.

(4) Metallic conductors with temperature-independent paramagnetism (Pauli-paramagnetism) or diamagnetism.

Reflectivity

Reflectivity is one of the important physical properties by which sulfide minerals are commonly identified and is defined as the ratio of intensity of reflected light (I) to that of incident light (I_o), $R = I/I_o$. Reflectivity is related to two important optical constants such as the refractive index of the medium (n) and absorption coefficient (k) by Fresnel's equation:

$$R = \frac{(n - 1)^2 + k^2}{(n + 1)^2 + k^2}$$

When $R = 1$, reflectivity is 100 percent. This equation shows that minerals with large values of n and k will have high reflectivity. The absorption coefficient is related to a parameter, n_{eff}, representing effective number of free electrons per molecule. Band theory of solids enables one to understand the significance of n_{eff} in determining the reflectivity of sulfide minerals (Burns and Vaughan, 1970).

To illustrate the relation between reflectivity and bonding, let us first consider a simple case such as sodium metal which has the electronic configuration $1s^2 \, 2s^2 \, 2s^6 \, 3s^1$. In a crystal of sodium metal, the $3s$ orbital of each Na atom overlaps with the $3s$ orbitals of neighboring Na atoms to form the valence band. Because each Na atom has only one $3s$ electron, the valence band thus formed will be exactly half-filled at absolute zero. (Note: each orbital can have two electrons with opposite spins). The position of the electron potential energy on the energy scale at a particular temperature is known as the Fermi level above which the orbitals are empty. Because the valence band is only half-filled, the electrons within the band can be easily excited into overlying empty levels within the band by absorbing electromagnetic radiation. The amount of radiation absorbed depends on the number of available energy levels above the Fermi level. When the excited electrons return to the ground state, light is re-emitted, causing high reflectivity for Na metal. It must be emphasized here that metallic lustre and reflectivity in any compound are due to the availability of empty energy levels above the Fermi level and delocalization of excited electrons in these energy levels.

In sulfides the situation is somewhat complex. We have seen earlier the formation of valence and conduction bands by the overlap of $4s$ and $4p$ orbitals of transition metal atoms and $3s$ and $3p$ orbitals of sulfur. The antibonding $3d$ orbitals are either localized on cations or form narrow d bands depending upon the magnitude of covalent mixing of $3d$ orbitals with the sulfur orbitals. In most of the transition-metal sulfides, the valence band is thought to be completely filled. Thus, in stoichiometric sulfides, electronic transitions within the valence band could not account for their reflectivity. However, electronic transitions could occur between the two groups of $3d$ orbitals and/or between the $3d$ orbitals and the conduction band. Because in many sulfides the $3d$ orbitals are localized on cations, only transitions between $3d$ orbitals and the conduction band could account for their metallic appearance and reflectivity. When there is extensive covalent bonding to cause the formation of broad d bands where electrons could be effectively delocalized, then electronic transitions between the two groups of $3d$ orbitals could give rise to high reflectivity. For example, in FeS_2, pyrite the valence band is completely filled and the six $3d$ electrons occupy the nonbonding t_{2g} orbitals. The antibonding e_g^* orbitals are empty and form d bands because of extensive covalent bonding between iron and sulfur (Burns and Vaughan, 1970). Electrons are excited into the e_g^* band from the t_{2g} orbitals by absorbing electromagnetic radiation and become effectively free electrons accounting for the metallic lustre and high reflectivity of pyrite. In the pyrite series, the reflectivity decreases in the sequence $FeS_2 > CoS_2 > NiS_2 > CuS_2$ (51.6:34:27:17, respectively). Because Co^{2+}(L.S.), Ni^{2+}, and Cu^{2+} have seven, eight and nine $3d$ electrons, respectively, the e_g^* bands will have one, two and three $3d$ electrons in CoS_2, NiS_2 and CuS_2. Therefore, the number of available unoccupied energy levels above the Fermi level decreases in the above sequence. It was observed that the values of n_{eff} in these sulfides are roughly proportional to the number of energy levels available to t_{2g} electrons excited into the e_g^* band (Burns and Vaughan, 1970). Therefore, substitution of Fe by Co and Ni in pyrite would decrease its reflectivity. Thus, compositional variations in sulfide minerals would affect markedly their reflectivity values. It is important to mention here that increasing covalent character of the metal-chalcogen bond causes an enhancement of electron delocalization which, in turn, increases reflectivity. For example, in the isotypic Cu-dichalcogenides CuS_2, $CuSe_2$, and $CuTe_2$ the reflectivity increases in the sequence $CuS_2 < CuSe_2 < CuTe_2$. This sequence is in the order of increasing size and polarizability of anions.

INTERATOMIC DISTANCES AND COVALENCY IN SULFIDES

When working with silicate and other oxide crystal structures, the mineralogist is able to depend on average interatomic distances for a particular ion in a particular coordination being reasonably uniform from one crystal structure to another. This feature of oxide structures enabled Shannon and Prewitt (1968) to compile a set of effective ionic radii which, when the radius for a cation coordinated by N oxygens is added to the radius of oxygen coordinated by M cations, usually give an interatomic distance within $\sim 0.01\overset{o}{A}$ of that observed. This has proved to be extremely useful in a variety of applications in oxide crystal chemistry. Shannon and Prewitt's work showed that there is a linear relation between cell volumes and radius cubed (r^3) of the varying ion in a series of isotypic compounds. However, such relationships are not found in sulfide structures and, in fact, are not found with other anions in the periodic table except possibly for fluorine. The reason for our inability to use the simple ionic radius concept with sulfides is that the bonds in sulfides are not ionic. One might say that they are not completely ionic in silicates either, so what's the difference? The answer to this question is complex, and we are not yet completely sure about all parts of it. The first part is that Shannon and Prewitt (and others) did find that small inconsistencies exist between observed and calculated (from radii) bond distances in oxides and fluorides. Second, the Shannon and Prewitt radii table was derived from more than 700 refined oxide and fluoride crystal structures and because oxygen and fluorine are similar in both electronegativity and size, the amount of covalence in a given bond, say Si-O, does not prevent giving a radius to Si^{4+} which will "work" in most silicate tetrahedra. Third, the outer electrons in sulfur are $3s^2 3p^4$, their orbitals are more diffuse and extend farther from the atom than those of oxygen, and the d orbitals are close enough in energy to be involved in bond formation.

What does this mean for sulfides? First, Shannon (1971) showed that many of the inconsistencies observed for oxides and sulfides could be explained by taking the electronegativity of a third atom into account, e.g., Co in $CoGeO_3$. The effect of a third atom A in compound $A_p B_q X_r$ ($A_x B_y X_z$) is to increase the interatomic distance of the B-X bond, when the electronegativity of A is greater and to decrease it when the electronegativity is less. The magnitude of this effect seems to increase with the covalence of the B-X bond and hence we see greater variations in sulfides than we do in oxides. Second, it might be possible to derive a set of radii exclusively for use with sulfides, but there are not nearly enough good structure data to do so even if there are no other problems. One other problem that does exist, but which has not been well documented, is the extensive metal-metal bonding that we find in sulfides. We know that metal-metal interactions will affect the cation-anion distances, even in oxides. However, not enough is yet known to determine a correction factor for metal interaction. Third, the type of bonding in a sulfide may change depending on the types of cations present, on stoichiometry, or even on the stereochemistry of the compound.

Recently Gamble (1974) and Shannon and Vincent (1974) published papers giving important data and discussing relationships between interatomic distances, ionic/covalent ratios, and magnetic properties in halides and chalcogenides. Gamble compares radius ratios of IVb, Vb, and VIb transition metal sulfides with ionic character and bond lengths. One result is that catatom-anatom distances are proportionally shorter than calculated from ionic radii in compounds such as Ti, V, Hf, Zr, Nb, and Ta chalcogenides when the covalence is greater. That is, the more covalent it is, the shorter the bond. Another interesting result is that the catatoms in these structures are coordinated by either a trigonal prism or an octahedron of sulfur atoms, depending on the (ionic) radius ratio (r^+/r^-). Figure PR-13 shows the sharp division between the two types of coordination. Gamble (1974) discusses the reasons for the two types of coordination: these involve band structure calculations and do not concern us here. What is important is that the published ionic radii *can* be used to investigate sulfide crystal chemistry, but only if covalence effects are taken into account either in the model being investigated or in corrections to the radii.

Shannon and Vincent (1974) in a somewhat similar, but more comprehensive study of different types of structures, investigated the effects of covalence on cell volumes of isotypic compounds. We have long been concerned with the radii of Ni^{2+} and Mg^{2+}. In some structures Ni^{2+} seems to be larger than Mg^{2+}, and in others the reverse is true. A few cell

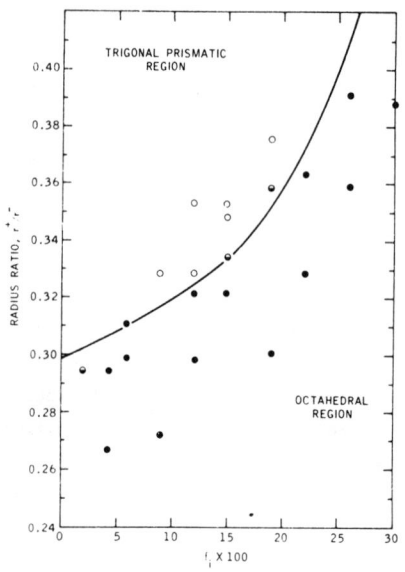

Fig. PR-13. Raduis ratio r^+/r^- vs. the fractional ionic character f_i of the metal-chalcogen bond in Group IVb, Vb, and VIb transition-metal dichalcogenides (after Gamble, 1974).

volumes ($\overset{o}{A}^3$) from Shannon and Vincent (1974) are given below for illustration. We assume that the larger volume of a pair of isotypic compounds represents a larger radius for the substituting atom.

NiO	72.4	NiF$_2$	33.4	NiI$_2$	88.3
MgO	74.7	MgF$_2$	32.6	MgI$_2$	102.1
	Fe$_2$SiO$_4$	307.9	Fe$_2$GeS$_4$	530.7
	Mg$_2$SiO$_4$	290.0	Mg$_2$GeS$_4$	570.3

The reversals in volume which appear to be a function of covalence of the bonds are dramatic. In the first three pairs the Ni-X bonds are more covalent than the Mg-X bonds in each pair and the shortening of Ni-X becomes greater as the electronegativity difference increases. For the latter two pairs, the Fe-X bonds are more covalent and shorter relative to Mg-X as the covalency increases. Shannon and Vincent (1974) provide many more examples and devise a *covalency contraction* parameter to represent the magnitude of the covalency effect.

APPLICATION OF BONDING THEORIES TO SPECIFIC SULFIDES

To illustrate how molecular orbital and band theories are used to help understand bonding and solid-solution behavior of transition metal ions in sulfide minerals, three specific sulfides have been chosen which include (1) pyrite in which there is no direct or indirect metal-metal interaction, (2) thiospinels in which there are indirect metal-metal interactions through sulfur intermediaries, and (3) the pentlandite series in which there are both direct and indirect metal-metal interactions or bonding.

<u>Pyrite</u> *(cf. Ch. 2, pp. W23-W25)*

The structure of pyrite is based on a face-centered cubic array of ions with the NaCl-type structure. The cations are located on Na positions. The anions are present as S_2 units, and the center of each S-S bond is located on a Cl position. Each sulfur atom is bonded to three metal atoms and one sulfur atom (in the S_2 unit) in the form of a distorted tetrahedron. Each cation is coordinated to six sulfur atoms and the sulfur octahedron is compressed along the trigonal axis and shares corners with neighboring octahedra. The metal-sulfur distance in pyrite, for example, is 2.26Å, which is significantly shorter than the theoretical distance 2.62Å obtained by adding the radii of $^{VI}Fe^{2+}$ and $^{IV}S^{2-}$. This indicates that in the pyrite structure Fe-S bonding is essentially covalent. The atomic arrangement and the interatomic distances in the structure led several investigators to assume hybridization of sulfur $3s$ and $3p$ orbitals to form sp^3 hybrid orbitals (Bither *et al*, 1968; Burns and Vaughan, 1970; Goodenough, 1972; Brostigan and Kjekshus, 1970). One of the four hybrid orbitals is involved in bonding with another sulfur ion and the other three are involved with three metal ions. The above authors also assumed that $4s$, $4p$ and the e_g group of $3d$ orbitals (those that are directed towards sulfur ions) of the cations form hybrid orbitals of the type d^2sp^3. Covalent mixing of the hybrid orbitals of both anions and cations give rise to a set of bonding molecular orbitals (σ orbitals: $1\sigma_{S-S} + 6\sigma_{M-S}$, totalling seven) and another set of antibonding orbitals (σ^*). Because sulfur is more electronegative than the transition metals (Fe, Co, Ni, etc.), the bonding (σ) orbitals are primarily anionic (Fig. PR-9, and are completely filled with 14 electrons (12 from the S_2 unit and 2 from the metal ion); the antibonding orbitals are primarily cationic and are usually empty. The three t_{2g} orbitals are not involved in σ bonding with sulfur ions and, therefore, remain nonbonding. However, Burns and Vaughan (1970) considered that these t_{2g} orbitals are involved in π-bonding with sulfur $3d$ orbitals, although proof for the existence of π bonding is by no means clear. It should be noted here that the t_{2g} orbitals are no longer degenerate in the pyrite structure because of the trigonal distortion of the sulfur octahedron. For FeS_2 the six $3d$ electrons of divalent $Fe(d^6)$ occupy the three t_{2g} orbitals with their spins paired. Therefore, iron in pyrite is a diamagnetic semiconductor. The covalent mixing of e_g orbitals of Fe^{2+} with the sp^3 hybrid orbitals of sulfur results in the destabilization of these orbitals which form a narrow antibonding (σ^*) band. Any metallic conductivity in pyrite-type compounds could be attributed to partial filling of this e^* band. Because all the six $3d$ electrons of Fe^{2+} are present in the t_{2g} orbitals in pyrite, the e_g^* band is empty. In the series FeS_2, CoS_2, NiS_2, and CuS_2, the e_g^* band is filled with 0, 1, 2, and 3 electrons, respectively. Increasing the electron population in the e_g^* band results in increasing bond distances because of greater electronic repulaion and, therefore, a reduction in the covalent mixing of e_g orbitals with sulfur orbitals. Consequently, all parameters also increase in the series, in the order listed in Table PR-8. Furthermore, electrical and magnetic properties change with the increasing population of electrons in the e_g^* band. For example, FeS_2 is a diamagnetic semiconductor, CoS_2 is a ferro-magnetic metallic conductor.

TABLE PR-8. Interatomic Distances and Cell Parameters in Transition Metal Disulfides

	FeS_2	CoS_2	NiS_2	CuS_2
M^{2+} - S distance (Å)	2.26	2.34	2.40	--
Cell edge, a (Å)	5.42	5.53	5.69	5.79
Number of electrons in the e_g^* band	0	1	2	3

It is evident from the foregoing discussion that substitution of Fe^{2+} by Co^{2+} and Ni^{2+} in pyrite will increase the cell parameter of pyrite. This obser-vation is contrary to what we would expect on the basis of ionic radii considera-tions alone. Furthermore, the limits of solid solution in the ternary system FeS_2 - CoS_2 - NiS_2 can also be explained by the bonding scheme discussed above. Figure PR-14 shows that the solid solution between FeS_2 and NiS_2 is not complete. Perhaps this could be due to the large difference in the interatomic distances between Fe-S (2.26Å) and Ni-S (2.40Å) in the pyrite structure. The Co-S distance (2.32Å) is intermediate between Fe-S and Ni-S and, therefore, could account for complete solid solution with both FeS_2 and NiS_2 (Nickel, 1970).

Thiospinels

Fig. PR-14. Solid solution in the system FeS_2-CoS_2-NiS_2 at 700°C (after Nickel, 1970).

Common sulfide minerals with the spinel structure include carrolite ($CuCo_2S_4$), linnaeite (Co_3S_4), siegenite [$(Co,Ni)_3S_4$], polydymite (Ni_3S_4), viola-rite [$(FeNi_2)S_4$], greigite (Fe_3S_4) and daubreelite [$(Fe,Mn,Zn)Cr_2S_4$]. The thio-spinel structure is based on the cubic close packing of sulfur atoms. The cations occupy one half of the available octahedral and one eighth of the tetra-hedral interstices in the unit cell which contains 8 AB_2S_4 (where A stands for diva-lent cations and B stands for trivalent cations). When the tetrahedral site is occupied by one kind of cation (i.e., by a divalent cation) and the octahedral sites by another kind (i.e., by trivalent cations) the spinel is referred to as normal. Whereas, when the octahedral sites are occupied by A and one B cations and the tetrahedral site by another B, the spinel type is designated as inverse. Each sulfur atom in the structure is coordinated to three octahedral cations and to one tetrahedral cation. Although there is no indication of the presence of direct metal-metal interaction in the structure, indirect interactions are believed to exist through sulfur intermediaries between octahedral cations and between octahedral and tetrahedral cations (Vaughan et al, 1971).

Goodenough (1969) and Vaughan et al (1971) explained the bonding aspects in thiospinel using molecular orbital and band theories. To illustrate this, let us take a common thiospinel, Co_3S_4, which is a normal spinel whose octahedral sites are occupied by trivalent cobalt and tetrahedral site by divalent cobalt. As in pyrite, the sulfur ions are assumed to form four sp^3 hybrid orbitals. The $4s$ and $4p$ orbitals of octahedral trivalent cobalt (d^6) along with the e_g group of $3d$ orbitals mix with the sp^3 hybrid orbitals of sulfur to form a set of stable bonding orbitals, σ_B (which are essentially anionic) and another set of unstable antibonding orbitals, σ_B^* (which are essentially cationic). The six $3d$ electrons of trivalent cobalt occupy the three t_{2g} nonbonding $3d$ orbitals with their spins paired (i.e., low-spin state). Therefore, the antibonding e_g^* band in Co_3S_4 will be empty. In the tetrahedral site, the $4s$, $4p$ and t_2 group of $3d$ orbitals of divalent cobalt (d^7) mix with sulfur orbitals to form bonding (σ_A) and antibonding (σ_A^*) orbitals. The two e group $3d$ orbitals remain nonbonding and are completely filled with four electrons with their spins paired. The other three electrons of $^{IV}Co^{2+}$ occupy the three t_2^* group of antibonding orbitals with their spins unpaired. The bonding orbitals in both cases (σ_B and σ_A) form part of the valence band and are completely filled with electrons. The antibonding σ_B^* and σ_A^* orbitals form the conduction band and are usually empty. The antibonding e_g^* group of octahedral Co^{3+} and t_2^* group of orbitals of tetrahedral Co^{2+} form narrow bands and fall below the main conduction band. Interactions between the octahedral and tetra-hedral cobalt ions result in the coalescence of these e_g^* and t_2^* narrow bands to form a single, broad, partially filled d band in which three $3d$ electrons are completely delocalized (Fig. PR-15). Therefore, Co_3S_4 exhibits metallic and temperature-independent paramagnetic (Pauliparamagnetic) properties. However, the two antibonding bands of $3d$ orbitals may or may not overlap, depending upon the type of cations occupying the two crystallographic sites and also upon the extent of metal-sulfur covalent bonding in each site. If there is an energy gap between the e_g^* and t_2^* bands as, for example, in $FeCr_2S_4$, then the d electrons tend to be localized on the cation and exchange magnetic interactions result in ferro-, antiferro-, and ferrimagnetism and the sulfide becomes a semiconductor.

The various physical properties exhibited by the thiospinel group could be explained by the bonding scheme outlined above (Vaughan et al, 1971). In the linnaeite-polydymite series, substitution of Co by Ni results in an increase in the number of $3d$ electrons in the antibonding d band. In polydymite, an inverse spinel with a formal cation distribution $Ni^{3+}[Ni^{2+},Ni^{3+}]S_8$, there will be six $3d$ electrons in the d band as opposed to only three electrons in Co_3S_4. Therefore, substitution of Co by Ni in the linnaeite-siegenite-polydymite solid-solution series is believed to increase the cell parameter. Furthermore, it is stated that "nickel substituting for cobalt in Co_3S_4 does not fundamentally alter the band structure" and, therefore, "the existence of this solid solution series is readily explained" (Vaughan *et al*, 1971, p. 374). In the case of the Fe-thiospinel, greigite (Fe_3S_4), trivalent high-spin iron occupies the tetrahedral and one of the two octahedral sites, whereas divalent high-spin iron occupies the other octahedral site. The presence of high-spin iron in the structure results in negligible overlap between the e_g^* and t_2^* antibonding d bands, and, therefore, greigite is semiconducting. Because of the difference in the spin states of Fe and Co in the two structures, the solid solution between these two members is very limited (Fig. PR-16). Similarly the lack of complete solid solution between violarite ($FeNi_2S_4$) in which one of the octahedral sites is occupied by divalent low spin iron, and greigite is again explained as due to the difference in the spin states of iron in the two structures.

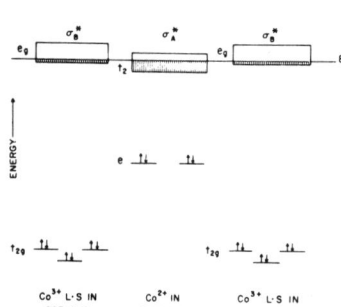

Fig. PR-15. Schematic energy level diagram for the $3d$ orbitals in Co_3S_4 (after Goodenough, 1969).

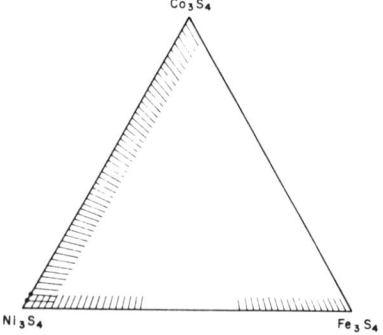

Fig. PR-16. Solid solution in the thiospinel system Fe_3S_4-Co_3S_4-Ni_3S_4. (The diagram is schematic, indicating observed natural compositions.)

Pentlandites

The composition of natural pentlandites is given by the general formula M_9S_8 where M stands for Fe, Co, and Ni. Occasionally pentlandites contain substantial amounts (up to 10 wt.%) of Ag. The structure of pentlandite is based on the pseudo-cubic closest packing of sulfur atoms. In the unit cell of pentlandite, there are four formula units of M_9S_8. The cations occupy one-eighth of the available octahedral and one-half of the tetrahedral interstices (see Ch. 1). The S_4 tetrahedra are trigonally distorted. The sulfur atoms are located on two equipoints in the unit cell 8(c) (coordinated by four cations) and 24 (e) (coordinated by five cations). The bonding can be explained with the help of the Co_9S_8 structure. The interatomic distances in Co_9S_8 (also in natural pentlandites) and magnetic and electrical properties suggest extensive metal-sulfur covalent bonding and metal-metal interactions in the structure. The $4s$, $4p$, and t_2 group of $3d$ orbitals of tetrahedral Co overlap with $3s$ and $3p$ orbitals of S to form bonding (σ) and antibonding (σ^*) molecular orbitals. The nonbonding e group of $3d$ orbitals are filled with four electrons with their spins paired. The remaining three d electrons of Co occupy the antibonding t_2^* band. Similarly, mixing of octahedral Co and S orbitals results in the formation of a narrow band of e_g^* orbitals of Co. The nonbonding t_{2g} orbitals will be filled with six electrons with their spins paired and the remaining one d electron (assuming Co is divalent) will occupy the e_g^* band. Although it is assumed here that Co in the pentlandite structure is divalent, the assumption is not strictly correct when we consider the structural formula, Co_9S_8. If S can be considered as divalent (S^{2-}), the structural formula gives the apparent valency of Co in the structure as 1.78 (the apparent valency, μ', of a cation in a compound of the type $M_m^{\mu'}X_\chi^{x'}$ is given by the expression $m\mu' = \chi x'$. For example in Co_9S_8, $9 \times \mu' = 8 \times 2$. Therefore, $\mu' = 16/9 \simeq 1.78$). The excess $4s$ and $4p$ electrons of Co could be located in the conduction band which should otherwise be empty) causing metallic conductivity.

In the pentlandite structure other effects, especially metal-metal inter-actions, complicate the picture. The interatomic distances indicate that each tetrahedral Co is coordinated to three other tetrahedral Co atoms and this coor-dination leads to the formation of a cube cluster of eight Co atoms, located at the corners of a small cube within the unit cell (Fig. PR-17). The three unpaired electrons in the t_2^* band of each tetrahedral Co are involved in bonding with three other Co's in the cluster. The bonding orbitals (with respect to metal-metal bonding) form a broad band which is completely filled with electrons ($\sigma_{M(T)}$ in

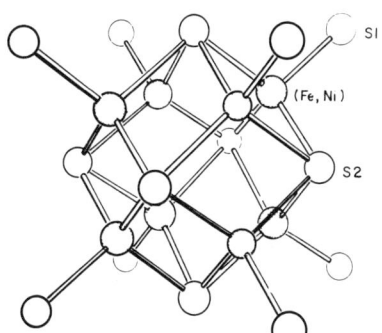

Fig. PR-17. Part of the pentlandite structure showing the "cube cluster" of tetrahedral metal atoms which are coor-dinated to one sulfur (S1) and three sulfur (S2) atoms.

Fig. PR-18) and the antibonding orbitals, $\sigma^*_{M(T)}$, remain empty. Therefore, it has been considered that the number of $3d$ electrons in the metallic cube cluster is fixed to a constant value of 56 electrons (Rajamani and Prewitt, 1973).

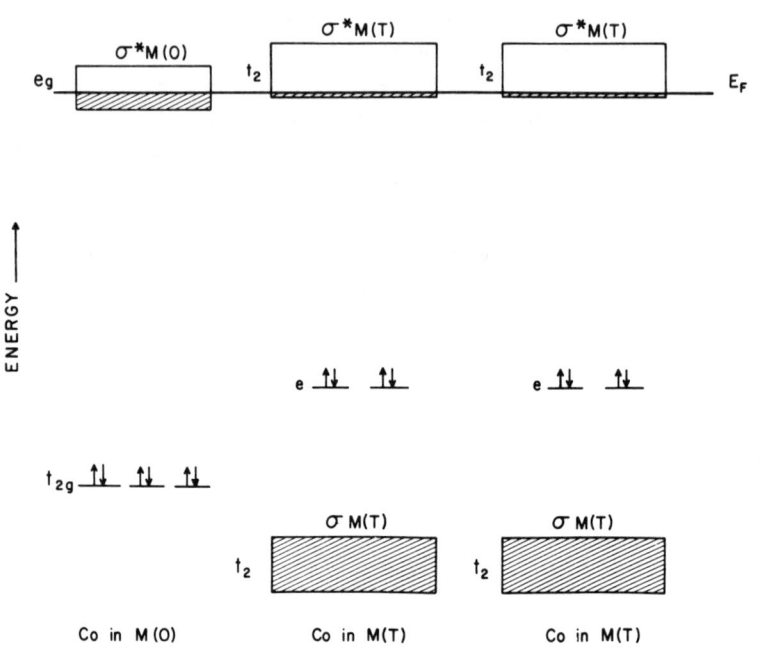

Fig. PR-18. Schematic energy level diagram for the $3d$ orbitals in Co_9S_8.

In addition to the formation of cube cluster, the structure also permits metal-metal interactions between the cube clusters of tetrahedral Co and between tetra-hedral and octahedral Co atoms through S intermediaries. All these interactions could lead to the coalescence of e_g^* and $\sigma^*_{M(T)}$ bands to form a single, broad, partially-filled d band as shown in Fig. PR-18. Therefore, Co_9S_8 exhibits metal-lic conductivity and Pauli-paramagnetism.

The presence of the cube-cluster of metal atoms in the pentlandite structure has an important effect on the chemistry of this mineral. Of the three possible end members, Fe_9S_8, Co_9S_8, and Ni_9S_8, only Co_9S_8 was observed to form a stable homogeneous phase in the system Fe-Co-Ni-S (Knop and Ibrahim, 1961). More recemtly, the synthesis of a thin film of cubic iron sulfide with a diffraction pattern similar to that of pentlandite has been described (Nakazawa et al, 1973). It is easy to visualize that because Ni^{2+} has four electrons in the t_2^* band, it cannot conveniently form three metal-metal bonds, and even if formed, the anti-bonding $\sigma^*_{M(T)}$ will have an unpaired electron, which might destabilize the phase. Knop and Ibrahim (1961) also observed that the individual substitution of Co by Fe and Ni in the structure is limited, whereas when they substitute simultaneously

PR-40

for Co the solid solution was found to be complete (Fig. PR-19). Natural pent-landites also have a restricted range in composition as shown in Fig. PR-19.

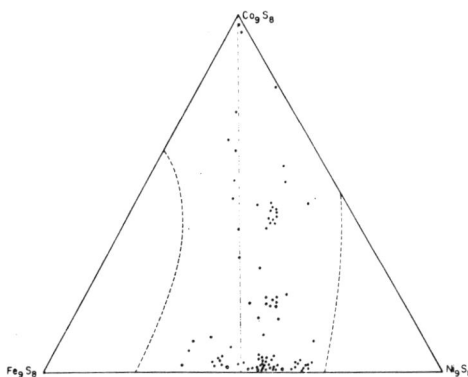

Fig. PR-19. Natural pentlandite compositions (Rajamani and Prewitt, 1973). Dashed lines represent the estimated solid-solution limits out-lined by Knop and Ibrahim (1961) in The M_9S_8 section of the system Fe-Co-Ni-S. Dashed-dot line repre-sents the compositions for which Fe:Ni = 1:1.

A possible reason for this peculiar chemistry is as follows: the cube cluster in the structure fixed the number of d electrons at a constant value. Therefore, we could expect that pentlandite compositions fall close to the join $Co_9S_8-Fe_{4.5}Ni_{4.5}S_8$. This is because $[Fe^{2+}(d^6) + Ni^{2+}(d^8)]/2$ is equivalent to $Co^{2+}(d^7)$. Increasing the Ni content in pentlandite over the ratio Ni:Fe $\geq$ 1 will increase the number of d electrons in the unit cell. Ordering of Ni in the octahedral site and creation of cation vacancies in the tetrahedral sites could effectively keep the number of d electrons at a constant value. Similarly, because Fe has only six $3d$ electrons, increasing the Fe content in pentlandite over the ratio Fe:Ni $\geq$ 1 will result in the addition of excess cations in the unoccupied tetrahedral sites to maintain a constant number of electrons, as in an electron compound (where the ratio of elec-trons to atoms is fixed). Therefore, the structural formula for pentlandite could be $[Fe,Ni,Co](Fe,Ni,Co,\square)_8S_8$ where $\square$ represents tetrahedral vacancies in Ni-rich compositions or excess cations in Fe-rich compositions. Thus, the nonstoichiometry of pentlandite is due to metal addition and omission solid solution.

Conclusions

These examples show that aspects of bonding in transition metal sulfides can be described, to a first approximation, by molecular orbital and band theories. Furthermore, these theories offer satisfactory explanations for the various physical and chemical properties exhibited by these sulfides. However, it should be pointed out that no single theory, when applied to all sulfides, gives a com-pletely satisfactory picture.

Ch. 4

S. D. Scott

INTRODUCTION

Mineralogists have two fundamental aims in their experimental syntheses of sulfide minerals. One is to grow single crystals of sufficient size, perfection, and chemical homogeneity for electrical, optical, and x-ray diffraction studies. The other is to determine phase relationships and underlying thermodynamic variables with a view to unravelling natural processes. Of course, the best possible circumstance is to satisfy both objectives simultaneously which recently has become possible through some advances in sulfide research.

This chapter primarily examines the general principles and techniques guiding laboratory investigation of chemical interactions among metallic sulfides. The techniques described are all in current practice and were used to obtain most of the phase equilibria and thermochemical data discussed by Craig and Scott in the following chapter. I will touch only briefly on the specialized technology of growing large single crystals. This is a topic worthy of a Short Course of its own but many of its techniques are unsuited to phase studies on sulfides. For example, large crystals have been grown from melts, by chemical vapor transport, and by sublimation, but these methods operate at temperatures that are too high to be practical for most geochemically-pertinent phase equilibrium studies. Besides, the steep thermal gradients used in vapor transport and sublimation experiments result in growth that is often too far removed from equilibrium to produce homogeneous crystals of multicomponent solid solutions. Another topic that is not covered here is solution chemistry of sulfides which not only has given us important insights into ore-forming processes but also has led to significant achievements in synthesizing sulfide crystals in hydrothermal solutions.

SOME PRINCIPLES OF EXPERIMENTATION

The basic guidelines for studies of sulfide phase equilibria are about the same as other branches of petrology. Indeed, some experimental methods have been borrowed from silicate petrologists, although special problems encountered in sulfides have led to some unique procedures. A significant point is that all such endeavors are petrology as emphasized by Barton (*Sulfide Petrology, 1970*) and by Stanton (*Ore Petrology, 1972*). As such, sulfides are best examined in the context of their encompassing rock systems, be they sulfide or silicate. A first approach requires experimentation to determine stability relations among the phases of interest.

S-1

Appearance-of-Phase Method

In its simplest form, the construction of a phase diagram requires the location of all phase boundaries for the system chosen; that is, the determination of which phases are in equilibrium at a particular temperature (T), pressure (P), and bulk composition (X). Not all such solid phases are necessarily minerals; all or part of the P-T-X conditions under study might lie outside of those found in nature or, conversely, some compounds have been synthesized before they were found as minerals.

The usual way of determining phase relationships is by the "appearance-of-phase" method. Carefully weighed bulk compositions are heated at a particular P and T and the products are examined at the completion of the experiments. If different phase assemblages appear in the products of, say, experiment 1 and experiment 2, then, assuming equilibrium was achieved, one or more phase boundaries must lie between the bulk compositions of experiments 1 and 2. More precise location of a phase boundary can then be determined by successively reducing the incremental change in X from experiment to experiment until the boundary is closely bracketed. Tests of whether the boundary actually represents equilibrium are discussed below.

Occasionally, time can be saved in experiments by calculating from thermo-dynamics the equilibria expected. Craig and Scott discuss thermochemical calculations in the following chapter but, for now, consider the following example from Barton and Skinner (1967, p. 279). Within the ternary system Pb-Ag-S there are two possible arrangements of tie-lines among the phases PbS, Ag_2S, Pb and Ag; these are Ag_2S + Pb and PbS + Ag. Which is the stable assemblage? For the reaction, Ag_2S + Pb $\stackrel{\rightarrow}{\leftarrow}$ PbS + 2Ag, the free energy, ΔG_R°, is negative at all temperatures of mineralogical interest. Therefore, the reaction proceeds spontaneously to the right and PbS + Ag is the equilibrium pair, not Ag_2S + Pb.

In systems where some data are already available, further time can be saved by carefully choosing key univariant or divariant assemblages for study rather than using a "shot gun" approach of examining all bulk compositions at, say, 10 wt. % increments of the components. Natural assemblages can sometimes be used as a guide to which phases or composition intervals could be most profitably studied. For example, pyrite and pyrrhotite are commonly found in terrestrial environments but troilite and iron are not. Therefore, within the context of the Fe-S system, it would be more pertinent mineralogically to concentrate experiments on the former phases. On the other hand, if thermodynamic data such as FeS activities are sought, troilite must also be included (as discussed in the next chapter). Care must be taken not to over-interpret natural assemblages as many contain phases nucleated during retrograde cooling from higher P-T conditions. For example, in the Fe-S system, close examination of ores often reveals monoclinic pyrrhotite + hexagonal pyrrhotite + pyrite. At a fixed confining pressure, this represents an invariant assemblage suggesting that

one phase--monoclinic pyrrhotite as it turns out, nucleated upon cooling without
concomitant readjustment of the other phases.

Thus far our appearance-of-phase experiments have told us only what phases
are compatible. To derive maximum information from a system we also need to
know the compositions of its solid solutions. A case in point is the Fe-As-S
system. Appearance-of-phase experiments by Clark (1960a) demonstrated that
for different T and X conditions arsenopyrite (plus vapor) can be in univariant
equilibrium with (As,S)-liquid + arsenic, arsenic + loellingite, loellingite +
pyrrhotite, pyrrhotite + (As,S)-liquid, pyrite + (As,S)-liquid and pyrite +
pyrrhotite. It was known (Clark, 1960a,b; Morimoto and Clark, 1961) that
arsenopyrite displayed a range in composition approximately from $FeAs_{0.9}S_{1.1}$ to
$FeAs_{1.1}S_{0.9}$ as a function of T, P, and the assemblage in which it occurred. Thus
a complete understanding of the functional relationships of arsenopyrite
composition to P and T would be useful for geobarometry and geothermometry. That
this might be possible was demonstrated by Kretschmar (1973) who found that the
range of arsenopyrite compositions in nature for different assemblages was
compatible with those expected from experimental data.

Equilibrium

Equilibrium has been variously defined as "a state [of rest] from which a
system has no spontaneous tendency to change" (Barton et al, 1963, p. 171) and
"a state [such] that after any slight temporary disturbance of external conditions
it returns rapidly or slowly to the initial state" (Lewis and Randall, 1961, p. 15).
The first definition has given rise to two tests for equilibrium in experimental
studies:

1. The system is held at constant conditions and the products are examined
periodically. If, after initial reaction, no further changes are noted with
time, the system is assumed to have reached equilibrium.

2. The experiment is repeated, perhaps using different starting materials.
If the results are reproducible, equilibrium is assumed.
Neither of these tests is completely reliable. Kinetics of a reaction may be
such as to preclude any possibility of change in the starting materials or else
change may occur in some phases but not in others. For example, consider the
assemblage loellingite + pyrite at 300°C. If $FeAs_2$ and FeS_2 are used as starting
materials, no change will be observed in them even after many months of heating.
If, instead, iron and sulfur in the proportion 1:2 and $FeAs_2$ are the starting
materials, some pyrite will form eventually but loellingite will not change. In
both cases, the system is seemingly in a state of rest, but this is only a
consequence of nonreactivity, not of equilibrium.

The best test of equilibrium, following the Lewis and Randall definition, is
to perturb the system in some manner causing change in the phases, and then return

the system to its original conditions. If the phases reattain their original
state, the reaction is said to have been reversed and equilibrium is assumed.

One major cause of incorrect interpretation of experimental results is the
formation of metastable phases and failure to recognize them as such. Metastability
is an equilibrium state and, particularly if the activation barrier between the
metastable and most stable states is large, it will survive the test of reversibility
and go undetected. In this sense, metastability of the type described above for
loellingite + pyrite presents few problems. The real difficulties are encountered
when complete reaction occurs outside the stable phase field for an assemblage.
A case in point involves monoclinic pyrrhotite. Taylor (1970a) was able to
reverse the metastable reaction monoclinic pyrrhotite $\rightleftarrows$ hexagonal pyrrhotite at
292°C in the solid state whereas Kissin (1974), using a hydrothermal technique
to flux the reaction, found that monoclinic pyrrhotite inverted reversibility to
hexagonal pyrrhotite + pyrite at 254°C.

Applications to Nature

Recognition of equilibrium in nature is an even more cantankerous problem but
is no less important if we are to apply experimental results to the interpretation
of natural processes. Barton *et al* (1973) have recognized several types of
mineral associations (Fig. S-1), most of which do not represent bulk equilibrium
but may, nevertheless, provide some useful information. Even when bulk
equilibrium does occur in nature it usually encompasses many more components than
our laboratory experiments. Some components will be in sufficiently low

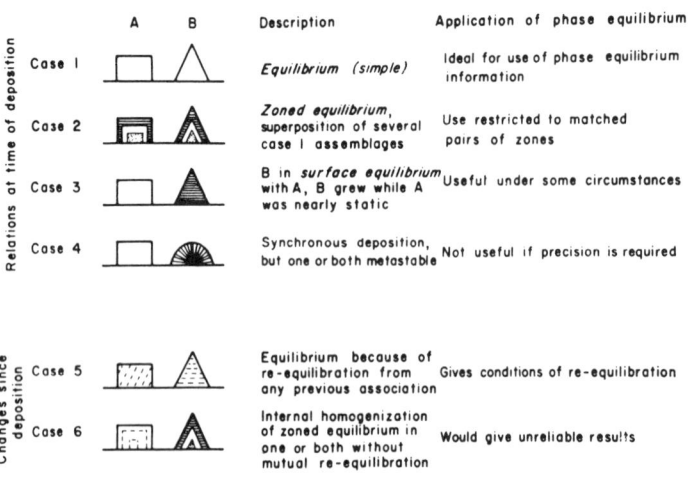

Fig. S-1. Some relations between coexisting minerals. (From Barton *et al*, 1963,
Mineral Soc. Amer. Spec. Pap. 1, 171-185).

concentration that they can be ignored or else corrected for. The latter is possible for those components of a solid solution which obey Raoult's Law in which case their thermodynamic contributions will be proportioned to their concentrations. We may still be left with a large number of components and may be tempted to experiment with all of them simultaneously but by systematically dealing with two, then three, then four, etc. we will eventually gain a better understanding of nature.

Only those sulfides which are sufficiently non-reactive upon cooling in nature to preserve their equilibrium compositions obtained at elevated pressure and temperature are useful geochemical tools in deciphering conditions of formation or subsequent metamorphism of ores. Such "refractory" minerals which react with others only sluggishly are few in number and include pyrite, arseno-pyrite, molybdenite and sphalerite. In his comparison of the reactivity of sulfides in the solid state, Barton (1970) estimated that it takes as long to equilibrate the refractory sulfides at 650°C as it does the nonrefractory pyrrhotites at 250°C and Cu-Fe sulfides at 100°C. For example, experiments by Barton and Toulmin (1966) involving solid-state reaction of sphalerite + hexagonal pyrrhotite failed to reach complete equilibrium after heating at 580°C for more than a year, whereas Yund and Hall (1968) found that troilite + pyrrhotite annealed at 90°C reached equilibrium after only seven days. Thus, as pointed out by Barton and Toulmin (1966), the experimental mineralogist is faced with a problem: on the one hand, refractory sulfides will provide the most useful information on ore-forming processes but, on the other hand, being relatively unreactive in standard, solid-state experiments, their phase relations are difficult to establish in the laboratory within the geologically-important temperature range below 600°C. The next section includes some methods of overcoming this difficulty.

METHODS

Evacuated Silica Tube

Dry reaction in an evacuated silica tube is the time-honored means of synthesizing sulfides, and more sulfide phase equilibria have been determined in this way than any other. Kullerud (1971) described the equipment and procedures in considerable detail, so I will discuss only the main features here.

Fused silica is an excellent container for sulfides. It does not devitrify until approximately 1100°C and does not react with sulfur below this temperature. Its low thermal conductivity enables the tube to be sealed in an oxy-gas flame without disturbing the charge and its low thermal expansion enables the runs to be quenched rapidly. Silica, if thin enough, is somewhat transparent to x-rays so it is possible to follow reactions in a high-temperature x-ray camera by sealing sulfides into a capillary (Moh and Taylor, 1971).

The preparation of a standard evacuated silica-tube experiment is illustrated
by Fig. S-2. An eight-inch length of thoroughly-cleaned tubing is separated into
two tubes by melting the glass in a small, hot oxy-gas flame from a welding
torch (Fig. S-2,A and B). This is accomplished by slowly rotating the tube over
the blue cone of the flame until melting collapses the walls. Pulling the ends
of the tube should be avoided during this process, otherwise the walls will
become too thin. The two ends of the tube can be safely held with your fingers
thanks to the low thermal conductivity of the silica but the bright-white flame
necessitates wearing welder's goggles. When the tubes have been separated they
are chilled by plunging them into cold water and a number is scratched on them
with a diamond scribe for later identification. Next, the starting materials
are carefully weighed into the tube. This is best accomplished by first
weighing the tube empty and again after each addition of pre-weighed starting
components. In this way, the amount of each material actually in the tube is
known despite any losses which might have occurred during loading. The materials
can be placed directly in the bottom of the tube using a narrow piece of creased
weighing paper as a slide or by means of a funnel formed by drawing out a piece
of soft glass. It is common practice to add the sulfur first and gently melt it
in the bottom of the tube in order to avoid losses later during the evacuation
process. After all the materials have been successfully transferred, a short
piece (∿1") of tightly-fitting silica rod is inserted (Fig. S-2,C) to eliminate

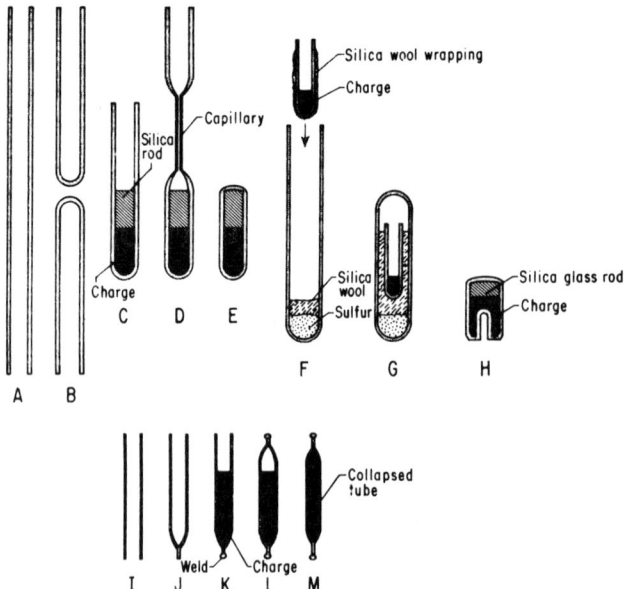

Fig. S-2. Various types of tubes used in sulfide experiments: A-E, simple evacuated
silica tubes with a glass rod; F-G, tube-in-tube; H, DTA tube; I-M, collapsible
precious metal tube. (From Kullerud, 1971, *Research Techniques for High Pressure
and High Temperature* (Ulmer, ed), Fig. 1, p. 291).

much of the vapor space and the tube is ready for evacuating and sealing. This
process is tricky but, with practice, can be done in just a few minutes. First,
the tube is necked down to a capillary just above the filler-rod (Fig. S-2,D),
again being careful to avoid thinning of the walls. A wet strip of cloth
wrapped around the bottom end of the tube will protect the charge and fingers
from the heat. Second, the tube is connected to a vacuum pump and evacuateu
to 0.02mm Hg or better. This should be done in stages by rapidly flipping the
pump switch on and off a few times initially to prevent the fine-grained charge
from being sucked into the vacuum line. It also helps to have a large capacity
vacuum dessicator in the line to act as a "vacuum buffer." Finally, with the
vacuum pump running, the capillary is sealed off in the flame, the top end of
the tube is melted around the filler-rod (Fig. S-1,E), the tube is quenched in
water, and we are ready to start the experiment.

The size of tubing used is dictated in part by the amount of charge to be
reacted, although the larger the tube the more difficult it is to seal. Starting
materials can be prepared in batches of several hundred milligrams in tubes of
6mm I.D. x 8mm O.D. Larger quantities may not homogenize within reasonable lengths
of time. Also, for this reason charges should be as small as possible: 30 mg is
usually sufficient and can be conveniently contained within a tube 3mm I.D. x 4mm
O.D. x 3cm long.

Starting materials are either native elements or sulfides presynthesized from
the native elements by the evacuated silica tube method. Sulfur can be obtained
6N (99.9999%) pure. Most metals are now available 4-5N pure as wire, sheet, sponge,
or powder but many, particularly iron, develop an oxide coating which must be
removed. This is accomplished by heating the powdered metal below its melting
point in a hydrogen gas stream. Water and oxygen can be scrubbed from the
hydrogen by passing it through a dessicant and then over zinc metal at 400°C.
A train for this process is easily set up and will lower f_{O_2} of normal tank
hydrogen to 10^{-46} atm which is well below the metal-metal oxide boundary for
most metals of interest to the experimenter.

The charges which were sealed into evacuated silica tubes are reacted in
furnaces of the type described by Kullerud (1971) and illustrated in Fig. S-3.
The hot-spot of the furnace is spread over several inches by carefully spacing
the resistance windings and inserting a nickel sheath between the inner and outer
alumina tubes. Temperatures can be controlled within ±1°C or better by means of
solid-state, stepless proportional controllers which are available from a large
number of manufactureres in the price range $200-$500 or within ±5°C by means
of an inexpensive variac or a time-proportioning controller. Temperature is
monitored by thermocouples in direct contact with the silica tubes. Chromel-
alumel thermocouples are reliable to 900°C and Pt-Pt + Rh at higher temperatures.
The usual error quoted for quality chromel-alumel thermocouples is ±3/8% which,
at 500°C, is ±2°C.

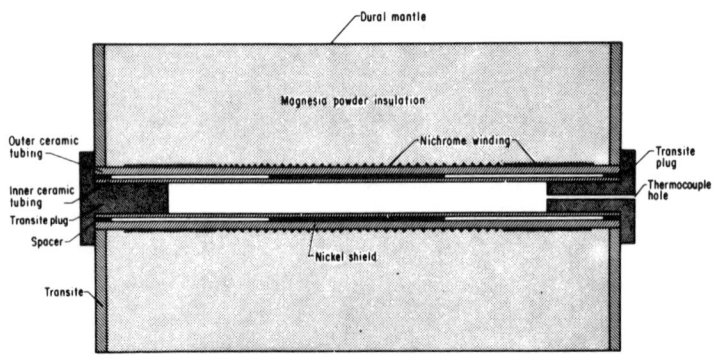

Fig. S-3. Standard tube furnace for evacuated silica tube experiments. (From Kullerud, 1971, *Research Techniques for High Pressure and High Temperature,* (Ulmer, ed), Fig. 3, p. 301).

The vapor pressure of sulfur rises very rapidly above its boiling point at 445°C and will explode thin-walled tubes unless the charge is preheated below 600°C to combine the sulfur and metals before attaining the desired run temperature. A word of caution: because of the danger of explosion <u>always</u> wear a face shield when inserting, inspecting or removing tubes from a furnace. Experiments in which liquid sulfur is one of the phases are particularly problematical. Kullerud (1971) claims that a silica tube with 1.5mm walls and 3mm I.D. can contain the pressure (∿100 bars) of saturated sulfur at 1000°C. However, the safest way to conduct such an experiment is to support the tube with a pressurized gas such as argon in a standard cold-seal pressure vessel.

Reactions in evacuated silica tubes rely on diffusion, either through the vapor phase for the more volatile elements or in the solid phases. Rates of solid state diffusion can be exceedingly slow and it is not unusual for evacuated silica tube experiments to take months to reach equilibrium. Refractory sulfides, in particular, are very difficult to equilibrate within reasonable lengths of time at geologically-significant temperatures. Reactions can sometimes be speeded up by quenching and regrinding the charge periodically or by forcing the grains into closer contact with one another. The latter is accomplished by pelletizing the charge in a hydraulic press before encapsulating in a silica tube or by melting the charge and gradually lowering the temperature to the desired value.

Advantages of the evacuated silica tube method are:

1. The equipment is inexpensive and easy to operate .

2. The system is clean and of precisely known bulk composition.

3. Clear glass tubes permit visual examination of the charge without termination of the experiment.

4. Runs can be quenched rapidly by dropping the tubes into water, an important consideration when dealing with most sulfides.

Disadvantages are:

1. Solid-state diffusion in some sulfides is too slow to attain stable equilibrium within reasonable lengths of time and metastability is frequently encountered.

2. The run products are usually very fine-grained and unsuitable, for example, for single-crystal x-ray diffractometry.

3. Because the silica tube is rigid, confining pressure cannot be varied independently but will always be that of the ever-present vapor phase.

Differential Thermal Analysis (DTA)

When a phase change occurs, heat is either absorbed from or added to the surroundings. By detecting the resulting slight temperature increment we can determine the sign of the enthalpy change (endothermic or exothermic) and the temperature of the phase boundary. In practice a test thermocouple is inserted into a recess in a evacuated silica tube containing the charge (Fig. S-2,H), and a reference thermocouple into a block of inert material such as alumina. Both assemblies are placed a few mm apart in the hot spot of the furnace and the temperature of the reference thermocouple and its difference with respect to the test thermocouple are recorded as the furnace is slowly heated or cooled.

Typical thermograms for pentlandite are shown in Fig. S-4. The small peak at 573°C is caused by the α-β transition in quartz which was added to the pentlandite to calibrate the reference thermocouple. The large peak beginning at 609-610°C represents the breakdown of pentlandite to $(Ni,Fe)_{1-x}S$ and heazlewoodite. The peak beginning at 862-863°C represents the intersection of the solidus.

The chief advantage of DTA is the speed with which phase boundaries can be detected compared with appearance-of-phase experiments. However, DTA is plagued by the same kinetic problems as normal silica tube experiments and, in most cases, its results can be interpreted only in the light of some prior knowledge of the phase relations under investigation. A heat effect by itself does not tell us what phases were consumed or produced.

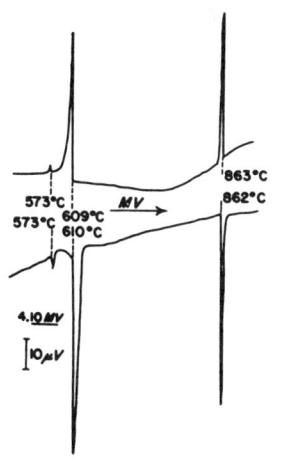

Fig. S-4. Differential thermal analysis curves for synthetic $Fe_{4.5}Ni_{4.5}S_8$ on heating (bottom) and cooling (top). (From Kullerud, 1963, *Canadian Mineral.* ', 358).

Salt Flux

An obvious way of overcoming reaction kinetics in evacuated silica tube experiments is to add a flux or catalyst to the charge which will increase reactivity among the sulfides without shifting their equilibria. Molten salts in which the sulfides are slightly soluble comply with these requirements and have been used for decades in the preparation and coarsening of sulfide phosphors. More recently, Boorman (1967) demonstrated the utility of salt fluxes by using them to extend the sphalerite + pyrite + pyrrhotite equilibrium down to 303°C, 277° below the previous best effort of Barton and Toulmin (1966) using the standard evacuated silica tube method. Kretschmar (1973) found that salt fluxes facilitated reaction in the refractory Fe-As-S system to as low as 300°C.

The lowest temperature at which a molten salt flux will operate is governed by its eutectic composition. Its highest temperature is the boiling point, although fluxes containing NH_4Cl have an upper limit of 350°C where the NH_4Cl dissociates. At present only four binary mixtures have been tested (Table S-1),

Table S-1. Binary Salt Fluxes Used for Sulfide Phase
Equilibrium Experiments (Moh and Taylor, 1971)

Salts	Mol Ratio	Temperature, °C
NaCl - KCl	50:50	>675
KCl - LiCl	42:58	>360
NH_4Cl - LiCl	<50:50	270-350
KCl - $AlCl_3$	34:66	>130

but there is no reason why other mixtures or more components could not be used to extend the operating range and efficiency of the technique. In fact, motivated in part by a need for such information, Smith et al (1974) have compiled a computerized file, ALKHAL, which contains the world's literature on salt systems with an alkali halide component. The file can be drawn on to compute eutectics in multicomponent systems (Smith, 1967). All that remains is the laborious task of testing the computed mixtures for their efficiency in recrystallizing sulfides.

Sample preparation is performed in the same manner as for standard evacuated silica tube experiments with the exception that the salt mixture is thoroughly mixed with the sulfide charge. Special handling is required for fluxes containing LiCl and $AlCl_3$ which are both deliquescent. They must first be dehydrated by passing gaseous HCl through the molten salt, then cooled in a vacuum dessicator and, finally, weighed and mixed with the sulfide charge in a dry glove-box.

Advantages of salt fluxes besides those inherent in the silica tube method lie in their ability to speed up sulfide reactions and to coarsen run products.

Major disadvantages are the special handling and storage required of LiCl- and AlCl$_3$-containing fluxes and inability to produce large single crystals of sulfides except under special conditions (described later).

Hydrothermal Recrystallization

Most of the shortcomings of the previous methods, particularly their inability to produce large single crystals and to permit independent variation of pressure, can be overcome by hydrothermal recrystallization described by Barnes (1971) and by Scott (in preparation) from whom the following account is taken.

The method consists essentially of recrystallizing nutrient sulfides in an appropriate aqueous solution contained in a capsule at elevated temperature and pressure. The aqueous solution in which the sulfides are slightly soluble acts as a flux promoting crystal growth, in some cases at temperatures well below 200°C, and can also be used to control activities of important species in the solution. Figure S-5 illustrates the two procedures that are used: (A) recrystallization over a small temperature gradient, and (B) recrystallization *in situ*. Equilibrium is achieved or at least closely approached in either case because the solution is slowly precipitating all sulfides simultaneously. Therefore, if the crystals are in equilibrium with the solution, they must be in equilibrium with each other.

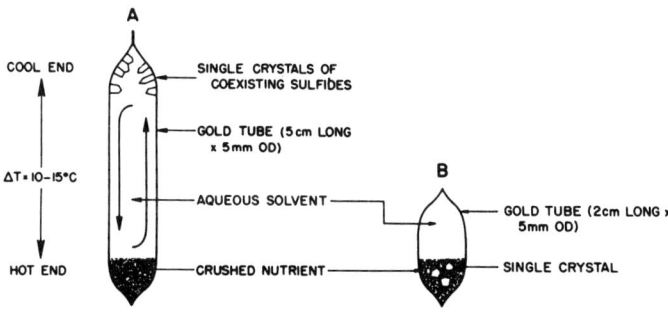

Fig. S-5. Design for hydrothermal recrystallization experiments: (A) with temperature gradient; (B) *in situ*.

Usually better results are obtained with the use of a temperature gradient in which the nutrient sulfides are dissolved at the hotter end of the capsule, transported by convection, and precipitated at the cooler end. However, the gradient must be small (10-15°C) to avoid metastabilities induced by supersaturation (see Scott and Barnes, 1972, p. 1285). *In situ* recrystallization is used at pressures greater than 1 kb because it is difficult to maintain an open capsule for convection at such high pressures.

During a recrystallization experiment across a temperature gradient a steady state is established between the dissolving nutrient and the precipitating crystals. In order for equilibria to be maintained, the experiment must be terminated before all of the nutrient is transported; otherwise, the solution being depleted in one or more components will precipitate a new assemblage during subsequent crysta growths. The nutrient material should preferably be finely crushed metallic sulfides rather than the uncombined native elements. Crystals can be grown from the elements but it was the experience of Chernyshev *et al* (1968) that Fe-sphalerite crystals grown from elemental Zn, Fe and S were compositionally zoned. Those grown from ZnS and FeS are not so zoned although they do contain small, crystallographically-controlled, metastable, Fe-rich patches (Scott and Barnes, 1971).

The choice of aqueous solution as the recrystallizing agent is governed by several factors:

1. It must be stable at the P and T of the experiment. For example, despite the stability of metal ammine complexes, aqueous ammonia is not a useful fluid because NH_3 partitions very strongly to the gas phase leaving the aqueous phase too dilute for effective solubility and transport. This problem can be overcome by using ammonium halides which disproportionate in the manner

$$NH_4I \rightleftarrows NH_3 + H^+ + I^-$$

providing two potentially useful ligands (NH_3 and I^-) for metal transport. However, at temperatures above 530°C NH_3 is irreversibly lost to the system, probably by dissociation to N_2 and H_2, resulting in a very acid HI solution.

2. The sulfide mineral to be synthesized must be stable within the range of f_{O_2}, f_{S_2} and pH provided by the solution. For example, in Figure S-6 magnetite is stable at high pH in place of pyrite or pyrrhotite precluding growth of iron sulfides in highly alkaline solutions.

3. The sulfide minerals must be sufficiently soluble in the solution and have a sufficiently large temperature coefficient of solubility to effect dissolution of nutrient, transportation and precipitation of crystals. Two problems can be encountered. One, although neither common nor insurmountable, is retrograde solubility as found, for example, by Barnes *et al* (1967) for HgS in acid sulfide solutions. The second, which is quite common for ternary compounds, is nonstoichiometric solubility. For example, we have not yet successfully synthesized by hydrothermal recrystallization either arsenopyrite or Ni-Fe sulfides. In both cases, the Fe and S are selectively removed from the nutrient to grow pyrrhotite or pyrite crystals and the remaining component is left behind.

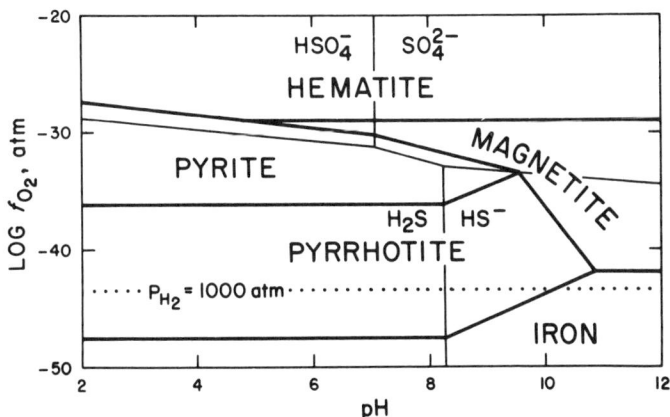

Fig. S-6. Distribution of solid phases of the Fe-S-O system and predominant, sulfur-containing aqueous species as a function of log f_{O_2} and pH at 300°C and $\Sigma S = 0.1$. Calculated from the data of Helgeson (1969), Robie and Waldbaum (1968), and Ellis and Giggenbach (1971).

4. The aqueous solute anion must not be soluble in the sulfide, otherwise it becomes a component in the system, creating one more extensive variable which may affect sulfide phase relations. This fortunately does not appear to be a major problem for most sulfide systems. Scott and Barnes (1972) could detect no (<175 ppm) OH^- in zinc sulfide grown in 6.2m NaOH. Halides also have very low solubility (a few ppm) in sulfides under conditions considered here. However, because the ionic radius of I^- (2.16Å) differs more from S^{2-} (1.84Å) than does Cl^- (1.81Å), iodide solutions are less likely to contaminate sulfides than are chloride solutions.

An important advantage of the hydrothermal recrystallization technique is that the aqueous phase can, under certain conditions, provide stepless control for f_{O_2}, f_{S_2}, and pH over a wide range of values. These variables, which are not independent, along with temperature and to a lesser degree pressure, are the important ones governing stability and solubility of sulfide minerals (Barnes and Czamanske, 1967). Their interrelationships at a fixed temperature, pressure, and total activity of sulfur-containing aqueous species (ΣS) is shown in Fig. S-7. By fixing two of f_{O_2}, f_{S_2}, and pH, the third variable is determined. For example, in their study of sphalerite-wurtzite equilibria, Scott and Barnes (1972) controlled f_{S_2} by buffering f_{O_2} near the $(SO_4)^{2-}/HS^-$ boundary via the reaction $HS^- + 2O_2 \rightleftarrows (SO_4)^{2-} + H^+$ and by varying pH with different concentrations of NaOH solutions. In these experiments the NaOH solution had two roles; it was the growth medium for large zinc sulfide single crystals and it established an f_{S_2} at each temperature.

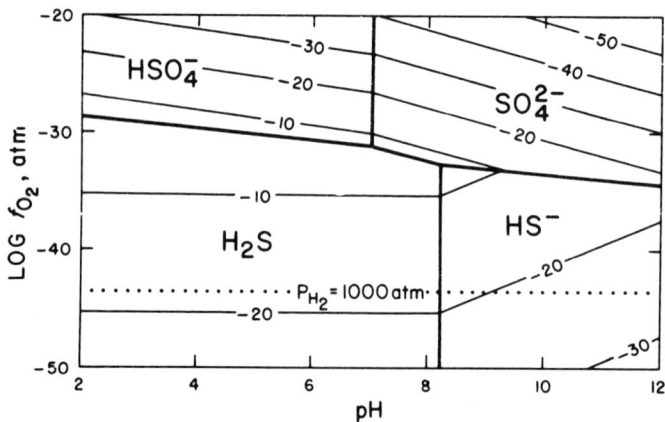

Fig. S-7. Distribution of predominant, sulfur-containing aqueous ions and molecules at 300°C and $\Sigma S = 0.1$. Contours are fugacity of S_2 (g) in atm. Calculated from the data of Helgeson (1969), Robie and Waldbaum (1968), and Ellis and Giggenbach (1971).

The lower limit of f_{O_2} which can be attained in an aqueous system is dictated by the dissociation of water: $H_2O \rightleftarrows H_2 + 1/2 \ O_2$. At 300°C, water dissociation generates 1,000 atm. of hydrogen pressure, an effective upper limit for most experiments, at an f_{O_2} of $10^{-43.5}$ atm (Fig. S-7) which, from Fig. S-6 lies within the stability field of pyrrhotite in a moderately alkaline to acid solution. Inasmuch as P_{H_2} is inversely proportional to f_{O_2} at a given temperature, it is clearly not possible to buffer f_{O_2} at significantly lower values and, therefore, not possible to study an equilibrium such as FeS + Fe in such solutions. However, this equilibrium could be studied at pH's 10.2 - 10.8 provided pH could be buffered within this range.

The pH of the transporting fluid will be controlled by ionic reactions within the concentrated solution but because of large shifts in equilibria with temperature, many of which are not well known at high temperatures, it is often difficult to predict the true pH under run conditions. For example, salt solutions undergo hydrolysis reactions of the type $NaCl + H_2O \rightleftarrows HCl + NaOH$ and the pH of the solution at elevated temperature and pressure is governed by the relative degree of ionization of HCl and NaOH. In some cases, pH can be controlled by solid phase buffers such as $Ag + HCl + 1/4 \ O_2 \rightleftarrows AgCl + 1/2 \ H_2O$ developed by Frantz and Eugster 1973. By buffering f_{O_2} the activity of HCl is fixed, thereby fixing the pH of the solution via the ionization $HCl \rightleftarrows H^+ + Cl^-$. However, this pH buffer has limited use in most sulfide systems because it is operable only where Ag is stable relative to Ag_2S which, at 300°C, is at $f_{S_2} < 10^{-12.4}$ atm.

In most crystal growth experiments sulfur fugacity is a variable dependent on f_{O_2}, pH and T. If a particular value of f_{S_2} is sought which cannot be readily achieved in this manner, solid buffers are used which become thermodynamic components in the system. For example, in their study of the Zn-Fe-S system by hydrothermal recrystallization, Scott and Barnes (1971) used pyrrhotite and pyrrhotite + pyrite to buffer f_{S_2} and produced Fe-rich sphalerite crystals.

Hydrothermal recrystallization experiments across a temperature gradient can be conducted successfully in standard cold seal pressure vessels of the type found in most experimental petrology laboratories, but best results are obtained from equipment especially designed for the task. At the University of Toronto we use pressure vessels and gradient furnaces similar to those designed by H. L. Barnes. As these have been described in detail by Barnes (1971), only a brief discussion is presented here.

The pressure vessel which contains up to seven run capsules is machined from 316 stainless steel and can be used safely to 600°C at 1,000 atm pressure. It is suspended vertically in a snuggly-fitting furnace (Fig. S-8) with separate top and bottom heating elements whose windings are spaced so as to provide a linear temperature gradient. The temperature of one heating element is maintained by a solid-state controller with stepless proportional output and the other by a second triac with an output proportional to that of the main controller. In this

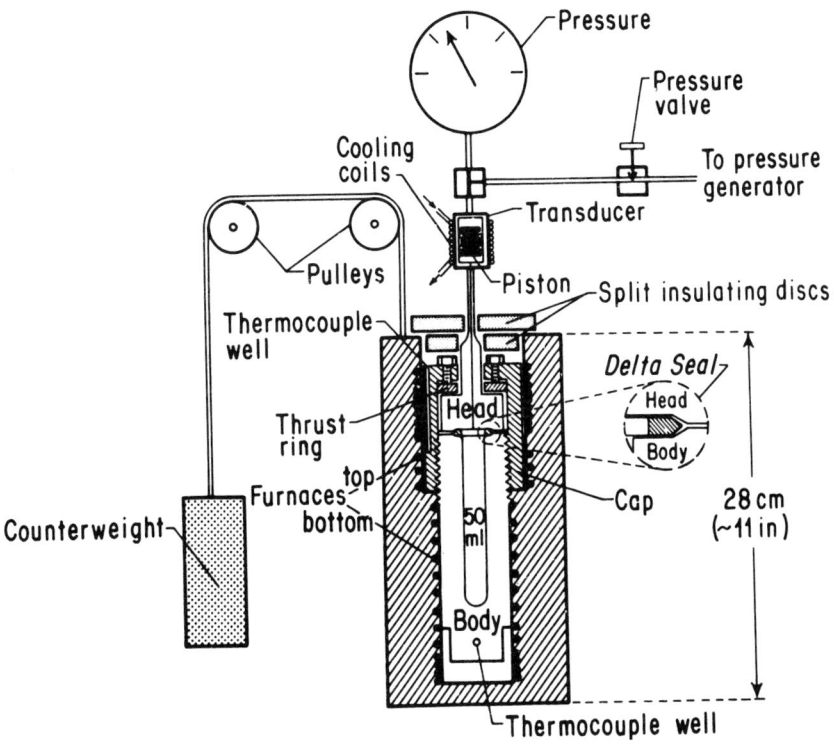

Fig. S-8. Pressure vessel and furnace assembly for hydrothermal recrystallization experiments. Inset shows details of the seal for the pressure vessel. (From Barnes, 1971, *Research Techniques for High Pressure and High Temperature*, Fig. 8, p. 349).

manner, temperatures at the measurement points are controlled within ±1°C. The
furnace is raised and lowered about the pressure vessel on vertical guide rods
and is held in place by a counterweight. The furnace is designed to have a high
heat loss by using only a thin layer of inefficient insulation (e.g. vermiculite)
in order to achieve steep thermal gradients when desired. At the termination of
an experiment, the furnace is lowered and the pressure vessel cooled rapidly to
room temperature in an air blast. Temperatures within the reaction chamber are
usually interpolated between two thermocouples, one inserted in the head of the
pressure vessel and one in the base. Pressure is generated by filling the
reaction chamber with a measured volume of water. At temperatures below the
critical temperature of water (374°C), pressures greater than those on the
boiling curve can be generated by means of a pump or by connecting the vessel
to a reservoir maintained at the desired pressure. Pressure is measured with a
bourdon tube gauge isolated from the vessel by a transducer which transmits
pressure to the liquid water-filled gauge and protects it from corrosive
solutions should a capsule rupture.

The sulfide nutrient and recrystallizing fluid are encapsulated within an
inert container, usually silver, gold or platinum tubes welded closed at both
ends (Fig. S-2, I to M). Pressure within the collapsible tubes is balanced
with that of the pressure vessel by filling each with carefully measured amounts
of solution and water, respectively, as discussed by Barnes (1971, p. 351).
In order for a tube to be a useful container it must not react with either the
recrystallizing fluid or the solid phases. Gold tubes 5mm O.D. x 25 or 50mm
long are most commonly used. Silver, besides being difficult to seal, sulfidizes
readily to Ag_2S under most conditions accessible by hydrothermal recrystallization,
alloys with many metals, and is readily attacked by hydroxide and alkali halide
solutions. Platinum is prohibitively expensive, is more difficult to work with
than gold, and even it reacts with some solids such as arsenopyrite.

The resistance of gold to chemical attack by the recrystallizing solution
appears to be related inversely to f_{O_2} (Hill, 1968). At and below f_{O_2}'s of
the pyrite + pyrrhotite boundary (see Fig. S-6), gold is stable in 5m NH_4I to
400°C, in 5m NH_4Cl to 500°C, in 6m NaCl to at least 500°C, and in 4.5m KCl to
at least 710°C. Gold is not severely attacked by 2-15m NaOH solutions to 700°C
or by 11.5m KOH solutions to at least 500°C at f_{O_2}'s as high as the $(SO_4)^{2-}/HS^-$
boundary (see Fig. S-7). Gold has been found to be unstable in systems containing
the sulfides of lead, mercury, silver and arsenic. In such cases silica tubes
sealed in a high-temperature flame have been used, but they are difficult to seal,
are easily recrystallized by slightly alkaline solutions, and permit only a
limited range of pressures to be attained.

Experiments at pressures and temperatures beyond the capability of the
temperature gradient apparatus are conducted by *in situ* recrystallization in
cold seal vessels capable of holding 10,000 atm at 700°C. Pressure is commonly

transmitted by means of a hydraulic intensifier using argon as the pressure medium. Scott (1973), for example, used such an apparatus to study the sphalerite + pyrrhotite + pyrite equilibrium to 8,000 atm.

ANALYSIS OF RUN PRODUCTS

Microscopy

A microscope is an indispensable tool for examining the products of synthesis experiments. A simple binocular microscope will sometimes suffice for the separation and identification of mm-sized crystals, but most products are sufficiently fine-grained to require the use of polished grain mounts and a petrographic microscope equipped with reflected-light optics. It might be tempting to save time by simply identifying run products by powder x-ray diffraction and, in complex systems, this is a useful adjunct to microscopy. However, whereas x-ray diffractometry requires 10 to 15% of a phase to be present in a mixture for reliable identification, the human eye can detect a fraction of a percent using a standard reflecting-light microscope. This is often the degree of discrimination required for closely bracketed phase boundaries or for decisions on attainment of equilibrium.

Grain mounts are easily made by drilling shallow holes in a one-inch diameter disc of cold-set dental plastic, inserting a fraction of the run products, topping the holes with newly mixed plastic and, after it has hardened, grinding and polishing as for a normal polished section. The result is a high-quality section which can be used both for optical study and microprobe analysis. For the person in a hurry, Kullerud (1971) describes a rapid and simple method for preparing individual grain mounts in Kadon resin.

Textural features such as zoning as well as dislocations and twin boundaries can be enhanced by lightly staining the polished surface of the grains. Most stains are legacies of the old "etch tests" (Short, 1940) which were standard practice until recently for the identification of opaque minerals in polished sections. In fact, HI and chromic acid solutions are still used to distinguish similar phases such as hexagonal and monoclinic pyrrhotite. Response to etching must be interpreted with caution though, because conflicting results have been obtained. For example, Arnold (1967) found that monoclinic pyrrhotite was etched more strongly by chromic acid than was hexagonal pyrrhotite, whereas Schwarz and Harris (1970) observed the opposite when using a 50% HI solution. Naldrett and Kullerud (1967) observed that in some cases hexagonal pyrrhotite assumed the darker stain under chromic acid, but in other cases monoclinic pyrrhotite did.

A preferable way of distinguishing monoclinic from hexagonal pyrrhotite is to apply a thin film of magnetic colloid. The colloid is essentially a dilute suspension of magnetic iron oxide in water which is preferentially attracted to the ferromagnetic monoclinic pyrrhotite imparting a dark brown coloration. The colloid is also useful for revealing magnetic domains in other pyrrhotite phases (Kissin, 1974). A recipe for magnetic colloid, kindly provided by L. A. Taylor, is as follows:

1. Dissolve 2 gm $FeCl_2 \cdot 4H_2O$ and 5.4 gm $FeCl_3 \cdot 6H_2O$ in 300cc distilled water at 70°C.

2. Dissolve 5 gm NaOH in 50cc distilled water.

3. Mix solutions 1 and 2 and stir vigorously. Filter black precipitate and rinse several times with distilled water and finally with 0.01N HCl. Place black precipitate in 500cc 0.05% sodium oleate solution and boil for a short time to mix the soap solution and precipitate. The resulting colloid is stable for many months.

X-ray Diffractometry

Besides identifying run products, x-ray powder diffraction can also be used to measure the composition of some solid solution phases. The latter relies on measureable shifts with composition in the d-spacings of one or more x-ray peaks. The peaks chosen for measurement are those which are particularly sensitive to the small changes in unit-cell volume attending the substitution of elements. Before the proliferation of electron microprobes, this was the usual method for nondestructive and is still preferred for hexagonal pyrrhotite and for finely intergrown or "spongy" run products. The results of x-ray analyses are acceptable so long as the phases are homogeneous and there are not more than two (occasionally three) components in the solid solution. More complicated solid solutions result in too many uncontrolled variables for deciphering the x-ray data.

The usual method for accurately measuring the d-spacing of the desired peak is to compare it against a nearby peak of an internal standard which is thoroughly mixed with the run product in the preparation of the smear mount. The x-ray goniometer is oscillated at 1/8° to 1/4° per minute through two or more cycles between the standard and unknown peaks and the results averaged. More reliable results are obtained if the smear mount is rotated 180° about a vertical axis between cycles. For sharp x-ray peaks, 2θ can be measured to a precision of ±0.01 (e.s.d.). For greatest accuracy, care must be taken to use either the same internal standard and cell edge as the authors of the determinative curve or else a calibrated standard whose cell edge can be related to theirs; otherwise, a systematic error will result. Boorman (1967) has discussed this problem in detail as well as other sources of error in x-ray analysis.

Some of the most intensive analytical work by x-ray diffractometry has been applied to sphalerite, arsenopyrite, and pyrrhotite which will serve as examples.

Sphalerite. Several determinative curves for the iron content of sphalerite have been presented over the years but the most useful is that of Barton and Toulmin (1966) which is reproduced in Fig. S-9. They determined the cell edge of synthetic Fe-sphalerites by comparing the d-spacing of the 531 peak of sphalerite against 531 of CaF_2 as an internal standard. In theory all one needs to do is measure the sphalerite cell edge to obtain the FeS content. However, in practice some difficulties are encountered:

1. Barton and Toulmin (1966) found that the room-temperature cell edges of iron-rich sphalerites synthesized at low sulfur fugacities controlled by the Fe + FeS buffer decreased systematically from those in Fig. S-9 as a function of temperature of synthesis. They were unable to explain this anomalous behavior but suggested that it might be caused by nonstoichiometry in high-temperature, iron-rich sphalerite or by Fe/Zn ordering during quenching. The latter seems to be ruled out by their observation that $\underline{a}$ did not vary with quench time from one second to several hours.

2. Other elements besides Fe may enter sphalerites and will contribute to the cell edge according to this formula (Barton and Skinner, 1967):

$\underline{a}$ = 5.4093 + 0.000564(mole % FeS) + 0.00424(mole % CdS) + 0.00202 (mole % MnS) − 0.000700(mole % CaS) + 0.002592(mole % ZnSe) − 0.003(mole % ZnO).

Clearly, a single x-ray measurement of products containing more than one substituting element will be indeterminant for composition.

3. X-ray diffraction provides a bulk analysis. If the sphalerite is heterogeneous and the resulting broadening of the x-ray peaks is not recognized, a misleading result will be obtained. For example, Chernyshev et al (1968) found reasonable agreement between sphalerite analyses done by x-ray diffraction and by wet chemistry, both bulk methods, on which basis they presented sphalerite compositions coexisting with pyrite and pyrrhotite. Their results were in conflict with those of Boorman (1967) and Scott and Barnes (1971) until subsequent microprobe analyses by Chernyshev (in Boorman et al, 1971) showed that the product sphalerites of his 1968 experiments were heterogeneous by as much as 16.6 mole % FeS.

Arsenopyrite. Clark's (1960a) experimental work on the system Fe-As-S was presented without benefit of actual analyses of the arsenopyrite products. Instead, results for arsenopyrite were quoted in terms of the d-spacing of the 131 reflection. d_{131} was known to be proportional to the As/S ratio in arsenopyrite although the exact functional relationship had not been determined and, consequently, it became common practice in the 1960's to present arsenopyrite analyses as d_{131} values. In 1961, Morimoto and Clark published a calibration curve for arsenopyrite, $d_{131} = 1.6006 + 0.00098(\text{at. % As})$, which has recently been revised by Kretschmar (1973) as at. % As = 845.00 d_{131} − 1345.84 [e.s.d. = 0.45 at. % As].

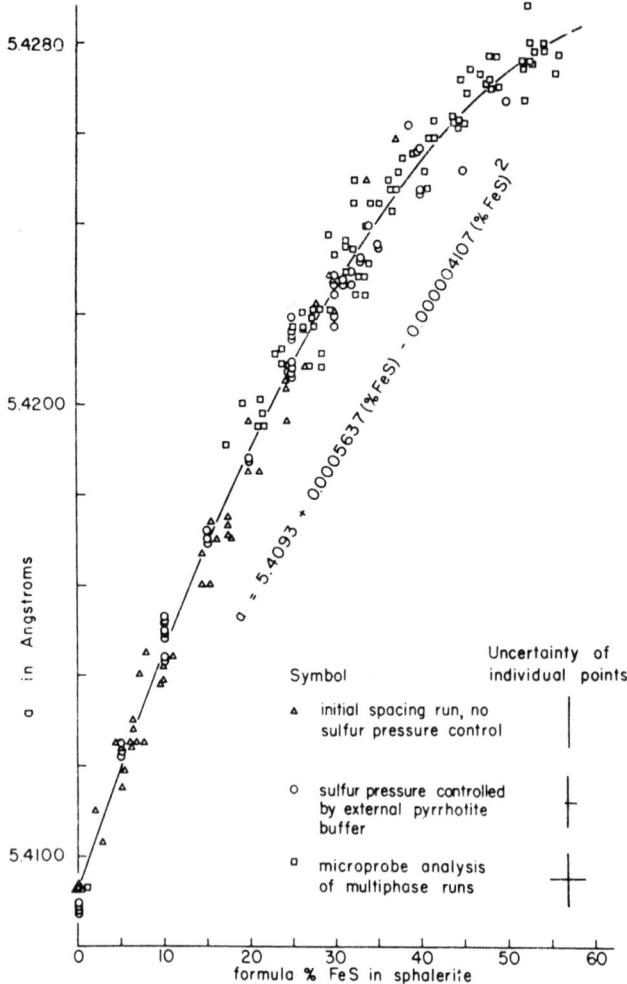

$$a = 5.4093 + 0.0005637(\%\,FeS) - 0.000004107\,(\%\,FeS)^2$$

Symbol		Uncertainty of individual points
▲	initial spacing run, no sulfur pressure control	
○	sulfur pressure controlled by external pyrrhotite buffer	
▫	microprobe analysis of multiphase runs	

Fig. S-9. X-ray determinative curve for the composition of iron-bearing sphalerites as a function of cell edge. (From Barton and Toulmin, 1966, *Econ. Geol*, _61_, 820).

Pyrrhotite. Arnold (1958) was the first to demonstrate that d_{102} of hexagonal pyrrhotite was a sensitive indicator of its iron content so long as the combined concentration of nickel, cobalt, and copper in solid solution was less than 0.6 weight percent (Arnold and Reichen, 1962). Subsequently, determinative curves based on this principle were offered by Arnold and Reichen (1962), Arnold (1962), Toulmin and Barton (1964), Sugaki and Shima (1966), Boorman (1967, but the equation is misprinted) and Yund and Hall (1969). The curve from Toulmin and Barton (1964), shown in Fig. S-10, is commonly accepted and gives reliable

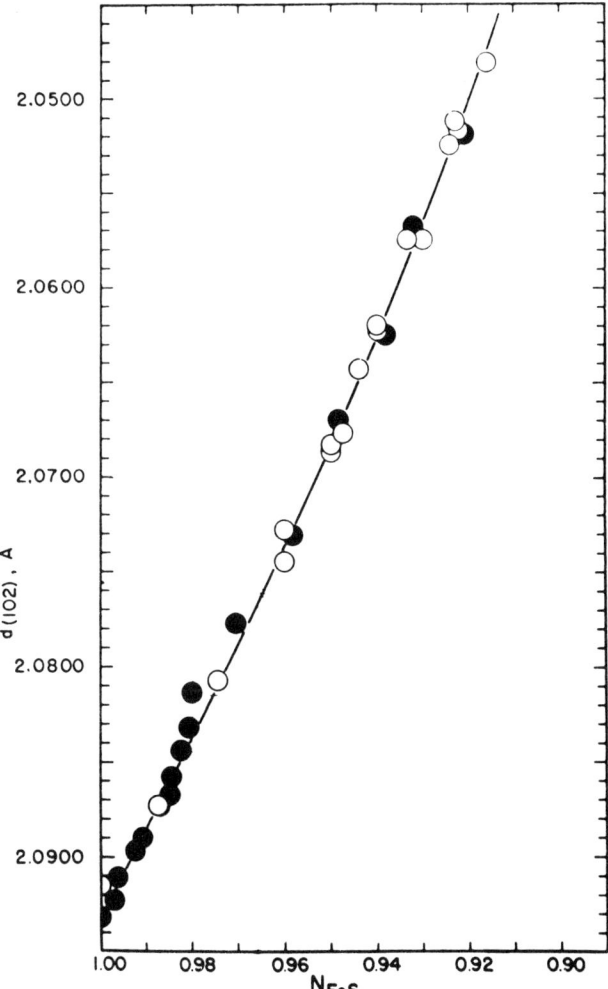

Fig. S-10. X-ray determinative curve for the composition of hexagonal pyrrhotite as a function of the d-spacing of the 102 reflection. (From Toulmin and Barton, 1964, *Geochim. Cosmochim. Acta 28*, 645).

results provided their internal standard is used (halite calibrated against metallic silicon for which $\underline{a}$ was taken as 5.4301Å at 26°C). The importance of the internal standard stems from the fact that the cell edge of Si used by Toulmin and Barton is 0.0005Å smaller than the preferred value (Boorman, 1967). If the larger value of Si is used directly or indirectly for measuring d_{102}, Figure S-10 will give a systematic error of 0.02 at. % Fe. In order to overcome such difficulties, Yund and Hall (1969) have selected previous data plus a few points of their own, normalized them all to $\underline{a}$ = 5.4306Å for silicon and calculated a determinative curve for pyrrhotite whose equation is:

at. % Fe = $45.212 + 72.86(d_{102} - 2.0400) + 311.5(d_{102} - 2.0400)^2$ [e.s.d. = 0.06].

The above methods do not apply to monoclinic pyrrhotite which is recognized by having two peaks at the location of the 102 reflection of hexagonal pyrrhotite (Fig. S-11). However, the monoclinic pyrrhotite can be isochemically converted to hexagonal pyrrhotite by heating it in an evacuated silica tube at 325°C for 3 to 5 minutes and rapidly quenching (Yund and Hall, 1969). It has also been suggested (Arnold, 1966; Graham, 1969) that peak-intensity ratios of intergrown monoclinic and hexagonal pyrrhotite can be used to calculate the amounts of the two phases present, and positions of the overlapping peaks to calculate the composition of the hexagonal pyrrhotite and of the bulk. However, Kissin (1974) found neither method to be very precise as a quantitative technique.

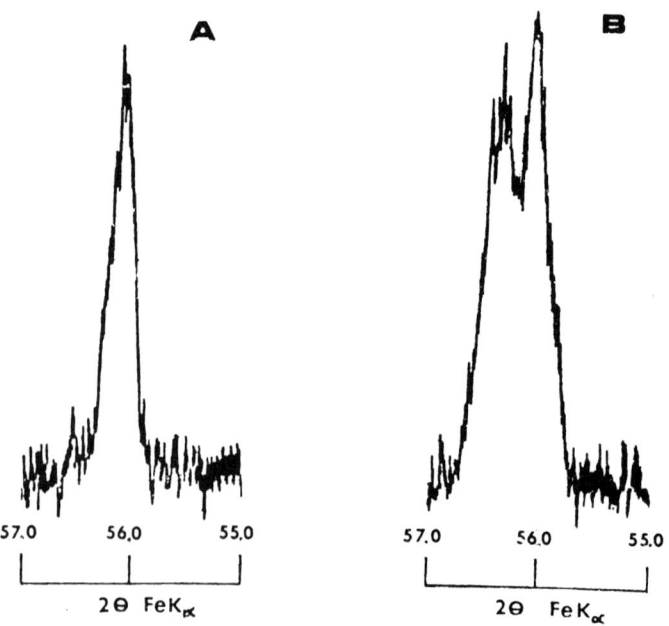

Fig. S-11. Comparison of the 102 x-ray peak of hexagonal pyrrhotite (A) with (B) the 408-40$\overline{8}$ doublet of monoclinic pyrrhotite. (From Kissin, 1974).

Microprobe Analysis

Chemical analysis *in situ* has obvious advantages in experimental studies and can be performed by three techniques of microprobe analysis - the familiar electron probe microanalyzer and two newer developments, the laser microprobe and the ion microprobe. Details of their instrumentation and application are described in several recent books (e.g. Andersen, 1973a).

As in other branches of analytical science, the advent of the electron microprobe has increased our capability of analyzing the products of mineral synthesis experiments to the point that it has now become a routine tool in most laboratories. Its nondestructive analysis of very small areas is ideally suited to checking homogeneity and determining compositions of fine-grained products or intricate intergrowth provided that, as a rule of thumb, the area to be analyzed is 2.5 times the beam diameter (3-5 μm, routinely). Automation permits very rapid analyses of sulfides; nonetheless the microprobe is not a panacea. For example, x-ray diffraction methods, although more laborious, still provide more precise analyses of hexagonal pyrrhotite.

Most of the problems encountered in electron microprobe analyses of sulfides stem from the choice of standards, and some minerals are more sensitive to this than others. To illustrate, I have analyzed a homogeneous, synthetic Fe-bearing sphalerite by two methods. One, following Barton and Toulmin (1966), involved using a wide range of synthetic Fe-sphalerite standards to construct a calibration curve (Fe/Zn x-ray intensity ratio *vs* FeS/ZnS mole ratio) against which the test sample was compared with corrections only for background. The other used metallic zinc and pyrite as standards and the EMPADR VII program of Rucklidge and Gasparrini (1969) for data reduction and matrix corrections. Repeated analyses by the first method spanning several days time gave 26.3 ± 0.4 ($\underline{n}$ = 4) mol % FeS in the sphalerite and by the second method, spanning several years time, gave 26.5 ± 0.2 ($\underline{n}$ = 18) mol % FeS. Thus sphalerite analyses may be performed either way with equal reliability. On the other hand, analyses in which sulfur must be accurately determined, as is the case of nonstoichiometric compounds such as pyrrhotite ($Fe_{1-x}S$), require the use of standards which closely approximate the composition of the unknown. For example, I have encountered errors in sulfur analyses as large as 5 percent compared to x-ray diffraction results when analyzing pyrrhotites against a pyrite standard. Similarly, Czamanske (in preparation) found 38.2 rather than the correct value of 34.9 wt. % S in $CuFeS_2$ when using Cu_5FeS_4 as a standard. Desborough *et al* (1971) have recommended a low accelerating voltage of 6kV in doing sulfur analyses in order to overcome such difficulties, but this voltage is too low to do simultaneous analyses of the other elements present.

The laser and ion probes are relatively new instruments which have not yet attained their full potential of application to sulfide research. In the laser probe a small intense laser beam vaporizes a microvolume of the sample. The laser plume is then analyzed by emmission spectrography, mass spectroscopy, atomic absorption or gas chromatography. Its chief advantages in sulfide research over the electron microprobe are the ability to measure trace as well as minor and major elements and to discriminate isotopes *in situ*. In addition, a laser probe costs considerably less ($20,000 - $50,000) than an electron microprobe ($120,000 - $250,000).

In the ion probe an energetic ion beam is used to sputter away a small surface area of the test sample. Some of the sputtered particles are ionized and can be collected and analyzed by mass spectrometry. This instrument has been very successful in determining the concentration and distribution of isotopes in the surface of a sample (Andersen, 1973b) but its high cost ($\sim$$300,000) precludes routine use in most laboratories.

PRESENTATION OF EXPERIMENTAL RESULTS

Experimental data can be presented diagramatically in a number of ways depending upon the variables measured. Figure S-12 shows several possibilities

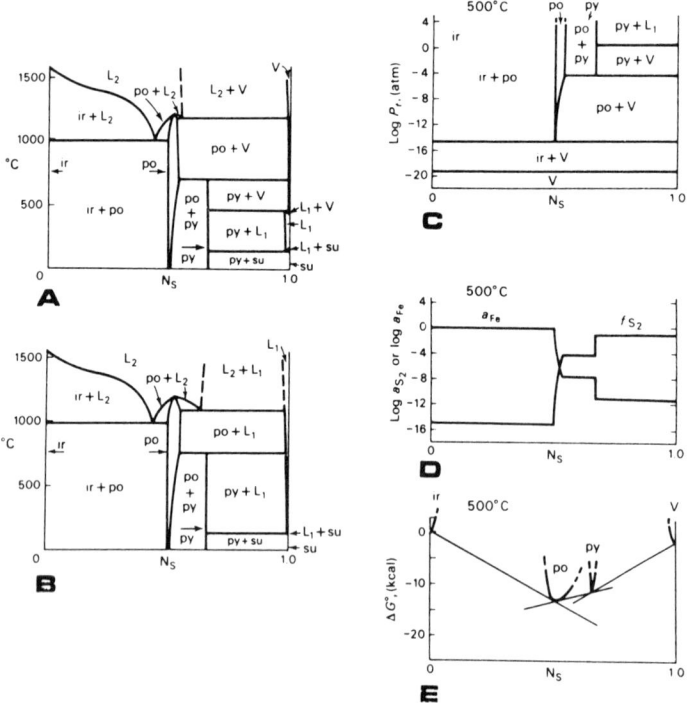

Fig. S-12A. Various ways of representing the phase relations in the binary system Fe-S. N_s = mole fraction of sulfur, ir = iron, po = pyrrhotite, py = pyrite, su = rhombic sulfur, L = liquid. (From Barton and Skinner, 1967, *Geochemistry of Hydrothermal Ore Deposits, p. 282).*

for the Fe-S system. Barton and Skinner (1967) discuss the rationale behind
each type of diagram but the most commonly used are T-X and log a_{S_2} (or log f_{S_2})
- 10^3/T plots. Experiments in evacuated silica tubes have an equilibrium vapor
present, and their results are presented on condensed T-X diagrams. Such diagrams
are polybaric, but because most solids and liquids are relatively incompressible,
the variation in pressure has little effect on phase relations as shown by
Fig. S-12, A(C). This is not always the case, however. In some systems (e.g. Zn-
Fe-S) confining pressure is an important variable, as will be discussed in the
next chapter.

The significance of log a_{S_2} - 10^3/T diagrams arises from the fact that many
sulfide reactions can be expressed as a sulfidation of one solid to another as
discussed below.

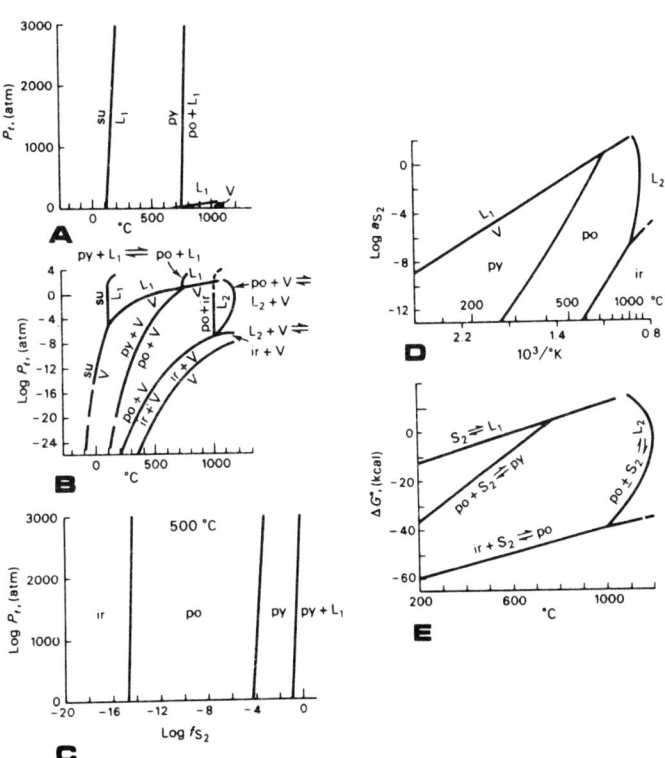

Fig. S-12B. Additional ways of representing phase relations in the system Fe-S.
Symbols as for Fig. S-12A. (From Barton and Skinner, 1967, *Geochemistry of
Hydrothermal Ore Deposits, p. 285*).

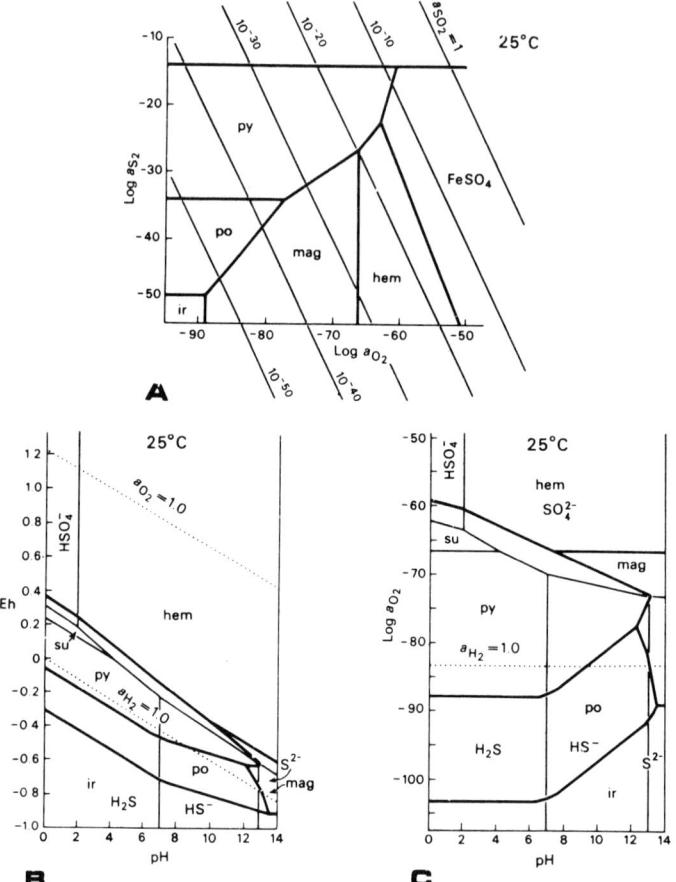

Fig. S-12C. Further ways of representing phase relations in the system Fe-S in the presence of additional components. mag = magnetite, hem = hematite. Other symbols as in Fig. S-12A. (From Barton and Skinner, 1967, *Geochemistry of Hydrothermal Ore Deposits, p. 288*).

SULFUR ACTIVITY: ITS MEASUREMENT AND CONTROL

Many sulfide equilibria can be represented by sulfidation reactions of the type

$$2M_xS_y + S_2 \rightleftarrows 2M_xS_{y+1}$$

where M_xS_y and M_xS_{y+1} are solids, $x = 1,2,3 \ldots$ and $y = 0,1,2 \ldots$. If the solids are in their standard states, their activities will be unity and the equilibrium constant, K, for the reaction is a function only of temperature and activity (or fugacity) of $S_2(g)$:

$$K = a_{S_2}^{-1} = f_{S_2}^{-1} \quad \text{(for a standard state fugacity}$$
$$\text{of 1 atm. for } S_2)$$

From the van't Hoff expression,

$$\frac{d \log K}{d (1/T)} = \frac{\Delta H°}{2.303R} = \frac{d \log a_{S_2}}{d (1/T)}$$

Therefore, plots of $\log a_{S_2}$ against $(1/T, °K)$ for a sulfidation reaction will be linear with a slope of $\Delta H°/2.303R$. If one or both of the solid phases are not in their standard states, the sulfidation reaction will plot as a curved line as a consequence of including their activities as well as a_{S_2} in the equilibrium expression, i.e.

$$\frac{d \log a_{S_2}}{d(1/T)} = \frac{\Delta H°}{2.303R} \cdot \boldsymbol{\pi}_i a_i$$

A large number of equilibria that have been investigated in this manner are summarized by Barton and Skinner (1967). Figure S-13, taken from their paper, illustrates the wide range of sulfur activities and temperatures covered by sulfide reactions. Such diagrams can be used as "sulfidation grids" (Barton, 1970), in a manner analogous to petrologists' petrogenetic grids in which conditions of ore formation are deduced by outlining stability fields of naturally-occurring assemblages. Barton (1970) gives some applications of this concept. See Ch. 6.

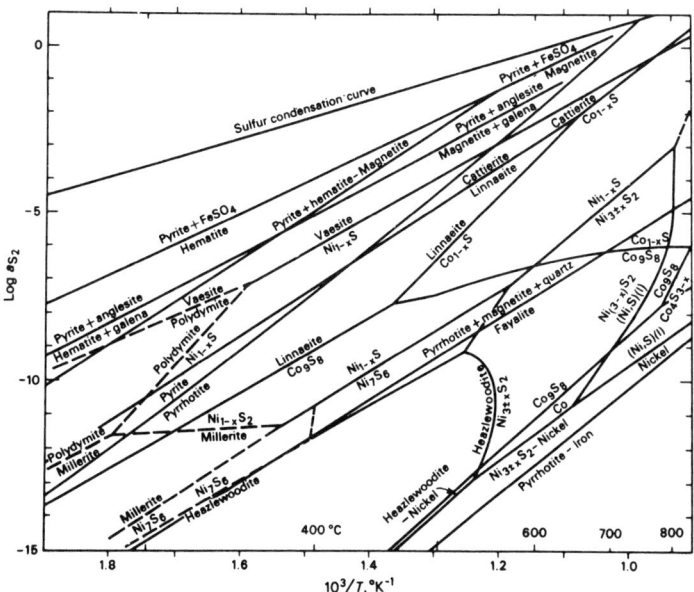

Fig. S-13. Examples of some sulfidation reactions on a plot of S_2 activity vs $10^3/T$ (From Barton and Skinner, 1967, *Geochemistry of Hydrothermal Ore Deposits*, p. 311).

Before turning to actual measurement of sulfur activities in the laboratory, let us first examine the constitution of sulfur vapor. Sulfur vapor is a complex mixture of polymers of the type S_i ($i = 1,2,3,\ldots 8$) whose relative proportions vary with sulfur pressure and temperature. Saturated vapor in equilibrium with liquid sulfur below 600°C is predominantly S_8 (Fig. S-14). The concentration

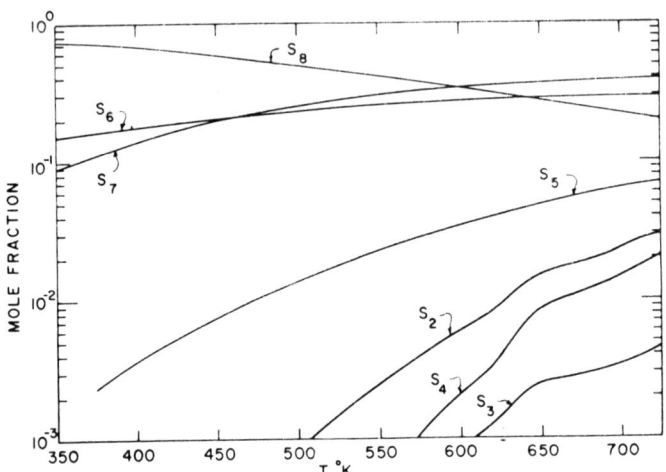

Fig. S-14. Mole fraction of sulfur species in the saturated vapor. (From Berkowitz, 1965)

of less-polymerized species increases with T along the liquid + vapor equilibrium curve and with decreasing P into the region where only vapor is present (Fig. S-15). As a consequence most, but by no means all, sulfidation reactions lie within the region where S_2 is the dominant species and it is for this reason they are written in terms of $S_2(g)$ rather than some other sulfur species.

Until recently the standard data for relating total sulfur pressure to S_2 activity was due to Braune *et al* (1951). However, these authors considered only the even-numbered species S_2, S_4, S_6, and S_8 in their calculations and their results have been superceded by the mass spectrometric data of Berkowitz and Marquart (1963), Berkowitz (1965), and Detry *et al* (1967). The equilibrium expressions recommended by Mills (1974) relating the various species are summarized in Table S-2.* Total vapor pressure of sulfur is given by the expression (Mills, 1974):

$$\log P \text{ (atm)} = -6109.6411T^{-1} + 16.64157 - 0.01705358T + 7.9769\times10^{-6}T^2.$$

----

*Haas and Potter (1974) have questioned the validity of these mass spectrometric data and have derived what they feel to be an internally consistent set of equilibria.

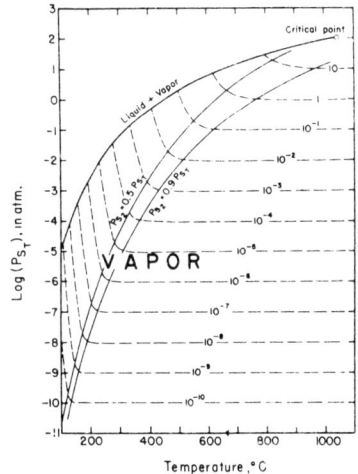

Fig. S-15. P_{S_2} contoured in atmospheres (dashed lines) in the undersaturated region where total sulfur pressure (P_{S_T}) is less than that over liquid or solid sulfur. S_2 is the dominant species below the curve where $P_{S_2} = 0.5\ P_{S_T}$. (From Barnes and Kullerud, 1961, based, in part, on the data of Braune et al, 1951, *Econ. Geol, 56, 668*).

Measurement and control of sulfur activity has tested the ingenuity of a number of researchers. Some of the methods that they have developed are described below and their relative merits are summarized in Table S-2.

S_{L-V} Buffer

This is perhaps the easiest way to control sulfur activity. By mixing excess sulfur either directly with the charge in an evacuated silica tube or separated from the charge by a tube-in-tube arrangement (Fig. S-2, F and G), sulfur activity is fixed by the equilibrium between liquid and vapor at the temperature of the experiment. Activity of S_2 in the saturated vapor can be calculated from the total sulfur pressure and the data in Table S-3.

A variant of the method is to use a long tube with sulfur at a

Table S-2. Comparison of Various Methods for Measuring S_2 Activity as a Function of Temperature (Barton and Toulmin, 1964, *Geochim. Cosmochim. Acta 28, 636.*).

Feature compared	Total pressure	Dew point	e.m.f.	Gas mixtures	Calori-metric	Gas mixture-tarnish	Electrum tarnish	Pyrrhotite indicator
Speed for individual determination	3	3	2	2	1	2	0	2
Speed for many determinations	2	2	1	1	1	1	3	3
Ability to reach low temperatures	1	2	3	1	3+	2+	3	3
S_2 range	1	2	3	2	3	3	2	2
Ability to quench reactants	0	1	0	0	v	1	3	3
Precision	2	2	3	2	3	2	2	1
Ability to avoid thermal segregation	v	0	3	1	3	2	3	3
Tolerance for other components in gas	0	2	3	0	3	0	3	3
Ability to make many determinations on single preparation	3	3	3	2	3	2	0	0
Simplicity of apparatus	1	2	1	1	0	1	3	3

Relative ratings: 3—excellent
2—
1—
0—poor
v—extremely variable.

constant temperature at one end and the charge at a higher temperature at the other end. This configuration permits investigation of a wider range of sulfur activities and temperatures than given by the univariant liquid + vapor curve for sulfur.

Table S-3. Equilibrium Constants for Gaseous
Sulfur Species (Mills, 1974)

$$S_n(g) = (n/2)S_2(g)$$

$$\log K = AT^{-1} + B + C \log T$$

Gaseous Species	A	B	C
S	11,219	- 2.026	-0.308
S_2	---	---	---
S_3	- 2,432	3.807	-0.138
S_4	- 3,607	8.293	-0.586
S_5	-10,880	14.933	-0.747
S_6	-14,790	20.347	-1.166
S_7	-17,930	24.874	-1.47
S_8	-21,690	30.58	-1.968

Dew Point

As described by Dickson et al (1962) and illustrated in Figure S-16, this method utilizes the temperature of the meniscus between liquid sulfur and sulfur vapor to compute total sulfur pressure and, from the equations in Table S-3, activity of S_2. The reaction tube containing the charge at a constant temperature is connected to a silica capillary which lies in a temperature gradient. The equilibrium sulfur vapor over the charge condenses in the capillary to liquid sulfur at a point whose temperature is interpolated between closely-spaced thermocouples.

Gas Mixing

Mixing of gases to control activities of O_2, CO_2, SO_2, etc. is a common practice in petrological studies and can be equally applied to sulfides. For the equilibrium, $2H_2S \rightleftarrows 2H_2 + S_2$, the sulfur activity is given by

$$a_{S_2} = P_{S_2} = K_T(P_{H_2S}/P_{H_2})_T^2$$

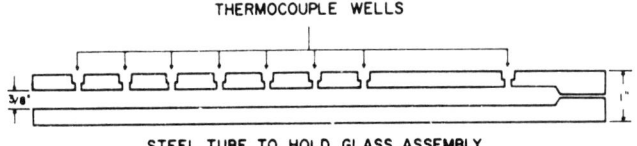

THERMOCOUPLE WELLS

STEEL TUBE TO HOLD GLASS ASSEMBLY

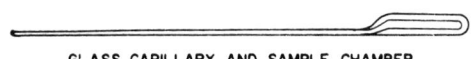

GLASS CAPILLARY AND SAMPLE CHAMBER

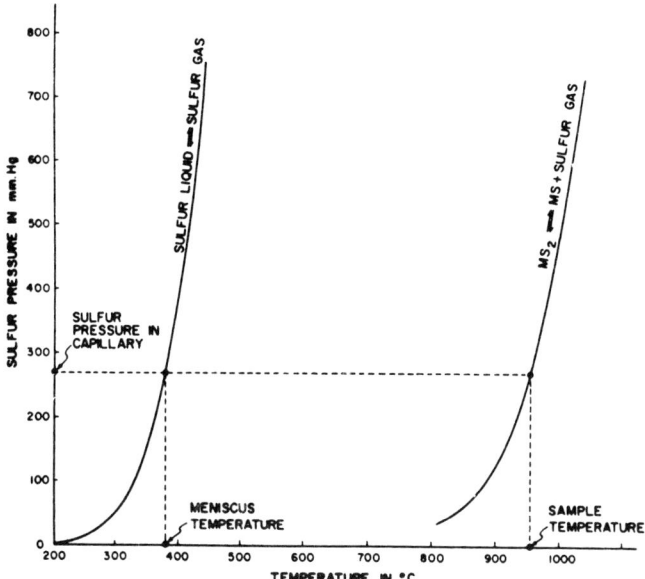

Fig. S-16. Sulfur vapor pressure curve and a curve representing a sulfidation reaction illustrating the dew-point method. Upper diagrams show the stainless steel sheath and glass capillary sample holder. (From Dickson *et al*, 1962, *Econ. Geo* *57*, 1023.)

where (R = room temperature):

$$P_{H_2S(T)} = (1 + P_{S_2(T)}) \ (P_{H_2S(R)}) - 2 \ P_{S_2(T)}$$

and

$$P_{H_2(T)} = (1 - P_{S_2(T)}) \ (P_{H_2(R)}) + 2 \ P_{S_2(T)}$$

By mixing H_2S and H_2 in various proportions in an apparatus shown in Fig. S-17,

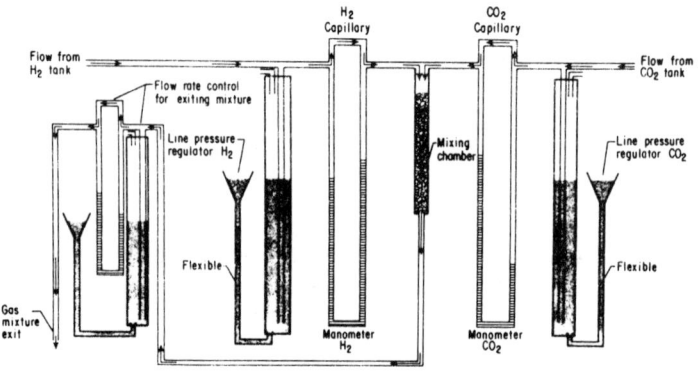

Fig. S-17. Apparatus for regulating the mixing ratio of two gases. For CO_2, read H_2S. (From Nafziger et al, 1971, *Research Techniques for High Pressure and High Temperature,* Fig. 1, p. 29).

a_{S_2} can be regulated in a furnace. A design for a furnace assembly is shown in Fig. S-18, and typical mixing curves are shown in Fig. S-19.

This technique can also be used in reverse to measure the equilibrium S_2 activity over a sulfide assemblage. A small amount of H_2S is sealed into a reaction chamber with the sulfides and after equilibration the gas is removed and analyzed. The equilibrium S_2 activity in the reaction chamber is found by interpolating on Fig. S-18 the H_2 and H_2S mixture analyzed at room temperature.

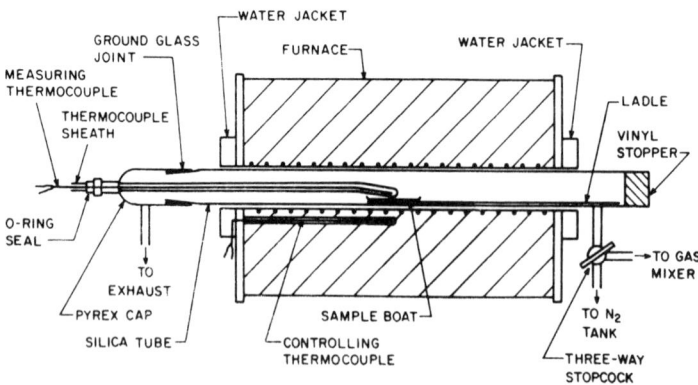

Fig. S-18. Furnace assembly for gas-mixing experiments. Boat containing sample is inserted into and withdrawn from furnace by means of the silica ladle. Nitrogen is used to purge the system and to quench the run. See Scott (1968) for details.

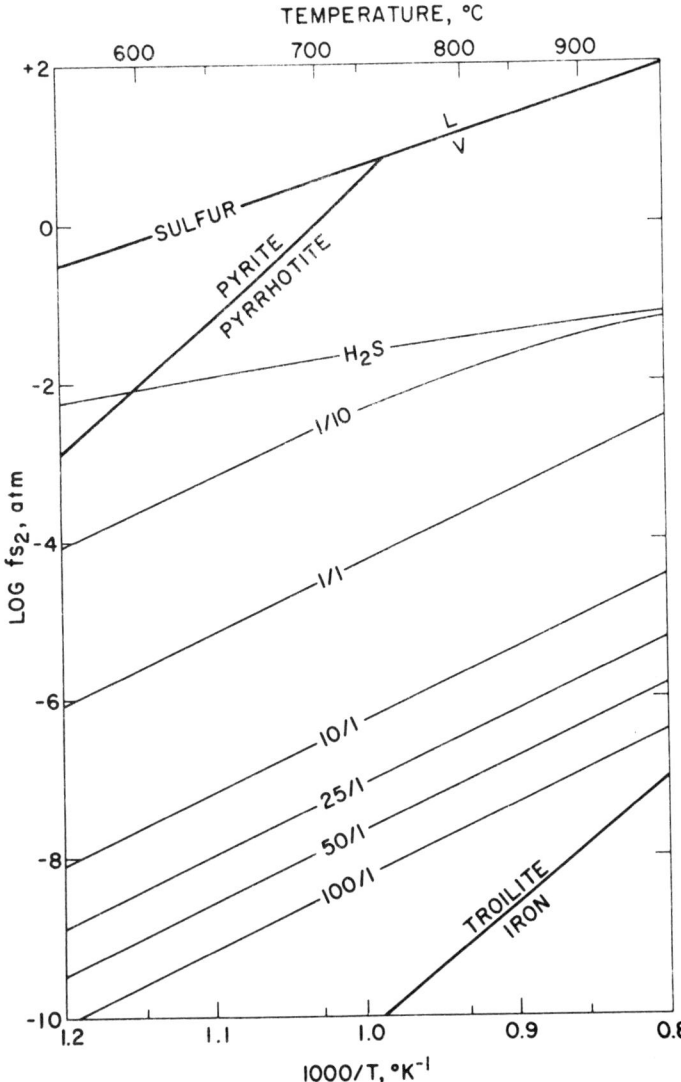

Fig. S-19. Log f_{S_2} vs. temperature calculated for several H_2/H_2S mixing ratios and for pure H_2S. Three sulfide equilibria are shown for reference. (From Scott, 1968).

Electrum Tarnish

This method, developed by Barton and Toulmin (1964), relies on the composition of a Au-Ag solid solution (electrum) in equilibrium with a tarnish (Ag_2S) as an indicator of sulfur activity. For the equilibrium $4Ag + S_2 \rightleftarrows 2Ag_2S$, the activity of S_2 is given by $a_{S_2} = (K_T \cdot a_{Ag}^4)^{-1}$. In pure Ag, a_{Ag} is unity but when alloyed with gold, becomes less than unity resulting in a concomittant rise in a_{S_2} for the formation of a tarnish. Sulfur activities at which various electrum compositions tarnish are shown in Fig. S-20.

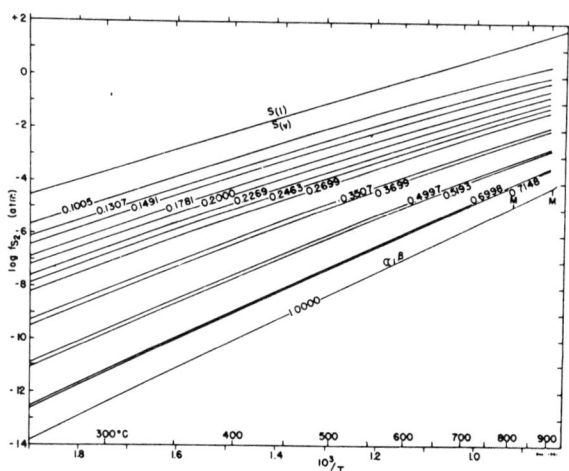

Fig. S-20. Tarnish curves for electrum compositions given in atom fraction Ag. (From Barton and Toulmin, 1964, *Geochim. Cosmochim. Acta* **28**, *635*).

In practice a small piece of electrum is placed in an evacuated silica tube and separated from the charge by silica wool. The system is brought to equilibrium, quenched, and without opening the tube the electrum is examined under a binocular microscope for evidence of a tarnish. If none is found, the system is heated at progressively lower temperatures until the equilibrium vapor over the sulfide charge is also in equilibrium with a tarnish, at which point the temperature can be raised slightly to remove the tarnish and reverse the equilibrium. A major shortcoming of the method lies in the difficulty of deciding whether the electrum is indeed tarnished; this can lead to overshooting of the equilibrium sulfur activity.

Pyrrhotite Indicator

Studies by Toulmin and Barton (1964) and Burgmann *et al* (1968) on the Fe-S system have determined the relationships among sulfur activity, temperature and composition of pyrrhotite. These are discussed in considerable detail in the next chapter. For now, it is sufficient to observe that a small amount of pyrrhotite, added to an evacuated silica tube but physically separated from the sulfide charge, will come to equilibrium with the vapor phase. By analyzing the composition of the pyrrhotite at the conclusion of the experiment, the sulfur activity of the equilibrium vapor can be determined.

Pyrrhotite (or any other sulfide buffer, for that matter) can also be used in a similar manner to <u>control</u> the sulfur activity of an experiment via its equilibrium vapor. The amount of pyrrhotite must be sufficiently large such that the small adjustments made between the test charge and vapor do not change the composition of the pyrrhotite.

<u>Electrochemical Cells</u>

Solid-state electrochemical cells which are reversible with respect to S_2 vapor are in an early stage of development but already show promise of supplanting all other methods for measuring sulfur activities. The most widely used cell at present is

$$Ag\,|\,AgI\,|\,Ag_{2+x}S,\ S_2(g)$$

in which an ionic conductor, AgI, separates pure silver and silver sulfide electrodes. Schneeberg (1973) has recently built and tested such a cell for sulfide research and his design is shown in Fig. S-21. The principle of operation is that of a silver concentration cell in which the chemical potential of Ag is fixed at the left electrode by the pure metal and at the right electrode by $Ag_{2+x}S$. The composition of the silver sulfide is controlled by the activity of S_2 in the ambient vapor. Sulfur fugacity (or activity) is calculated from the expression

$$\ln f_{S_2} = (E-E^\circ)\,\frac{4F}{RT} + \ln f_{S_2}^{\ \circ}$$

where E is the measured emf, E° is the cell emf at the fugacity ($f_{S_2}^\circ$) of the sulfur condensation curve at temperature T, and F is Faraday's constant. The operational limits of the $Ag\,|\,AgI\,|\,Ag_{2+x}S$ cell are shown in Fig. S-22. The low-temperature limit is dictated by the acanthite-argentite transition in Ag_2S and the high limit by the onset on significant electronic conduction in the AgI electrolyte. Another cell described by Sato (1971)

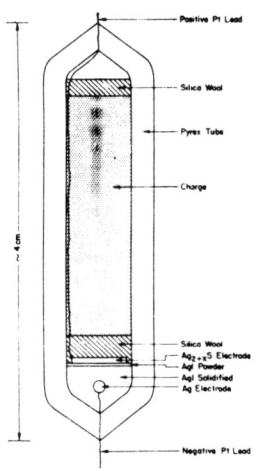

Fig. S-21. Typical $Ag\,|\,AgI\,|\,Ag_{2+x}S$ cell used to measure sulfur fugacity buffered by a mineral assemblage. (From Schneeberg, 1973, *Econ. Geol.* <u>68</u>, 509).

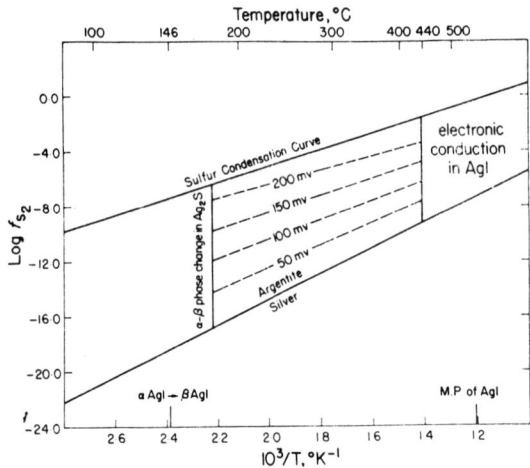

Fig. S-22. Limits of operation of the
$Ag|AgI|Ag_{2+x}S$ cell. Contours show the
emf response. (From Schneeberg, 1973, *Econ. Geol.*
68, 508).

utilizes Ag-saturated
beta-alumina as an electrolyte and can be operated to
considerably higher temperatures. It has been used
to measure sulfur activities
in natural processes but
has not yet been adapted for
routine laboratory use.

Kissin and Scott (in
preparation; see also,
Kissin, 1974) describe a
novel application in which
the $Ag|AgI|Ag_{2+x}S$ cell is
used on a precession x-ray
camera to measure sulfur
activity of a heated crystal
while x-ray photographs are
being taken.

GROWTH OF LARGE SINGLE CRYSTALS

Growth of sulfide single crystals has received considerable attention from
crystallographers seeking suitable material for x-ray study and from solid-state
physicists in their research for compounds with specific electrical and optical
properties. It would be futile to review the vast literature on the subject here.
Instead, I will briefly describe six techniques with which I am familiar without
delving far into their science and art. The proceedings of the international
conferences on crystal growth (Frank *et al*, 1968; Laudise *et al*, 1972) are good
sources for finding details and applications of these and other methods.

Growth from Melts

Sulfides which are stable to their melting points can be grown from their
melts by the Bridgeman technique in which a crucible containing the molten sulfide
and a seed crystal is cooled slowly through a temperature gradient. The method
is not unlike that used in synthesis of single-crystal rods of oxides. The main
difference lies in the use of a pressure vessel as a consequence of the high vapor
pressure of sulfide melts (usually not more than a few hundred atm) and the
necessity of excluding oxygen from the system. It is an easy matter to dope the
crystals by adding small amounts of contaminants to the melt.

Addamiano and Aven (1960) and Kozielski (1967) have grown single crystals
of hexagonal zinc sulfide (mp = 1722°C) of very high purity and optical clarity in
the form of boules several millimeters in diameter and centimeters in length.

Kozielski's experiments were done at 1850°C under an argon pressure of 50 atm. and required about two hours to complete the melting and growth steps.

Sublimation

The simplest technique for growing single crystals of sulfides is by sublimation from the vapor. A long evacuated silica tube is subjected to a temperature gradient with the nutrient, in the form of a polycrystalline sulfide powder or the elements, at the hot end. The single crystal grows at the cool end of the tube. This method is not always successful, however. For example, Ripley (1971) was unable to grow zinc sulfide crystals below 1200°C, presumably because the concentrations of zinc and sulfur in the vapor were too low for efficient transfer of material.

Chemical Vapor Transport

By adding a carrier gas to react with the nutrient in the sublimation experiment, the amount of sulfide transported to the cool end of the tube can be markedly increased resulting, in some cases, in growth of cm-sized single crystals. Typical transport reactions for zinc sulfide are:

$ZnS + I_2 \rightleftharpoons ZnI_2 + 1/2S_2$ and $ZnS + 2HCl \rightleftharpoons ZnCl_2 + H_2S$.

The reactions go to the right at the hot end of the tube and then in the reverse direction, precipitating ZnS, at the cool end. The temperature to which the nutrient must be heated and the gradient vary with the sulfide being grown and the carrier gas but values in excess of 1000°C and 200°C, respectively, are not uncommon.

Growth in Halide Fluxes

I have already discussed the recrystallization of sulfides in halide fluxes. Heating the flux to a much higher temperature than is usual for phase equilibrium studies (but below its boiling point) and then slowly cooling or simply maintaining a temperature gradient can promote growth of single crystals. Metal halides as well as alkali halides have been used as fluxes.

Growth in Gels

Growing sulfide crystals of mm-size near room temperature in a test tube has obvious attractions and is beginning to receive considerable attention. The success of Blank and Brenner (1971; see also Blank et al, 1968) in synthesizing single crystals of galena in gels at 38°C suggests that other sulfides may also be grown this way.

The technique employs an acid sodium metasilicate gel in which is dissolved a soluble salt containing the cation of the crystal to be grown. The simplest method is to hold the gel in a test tube and add a sulfur-containing solution to

the top. Crystal growth takes place at isolated nuclei in the gel by slow reaction
between the cations and sulfur-containing solution. For example, Blank and Brenner's
(1971) recipe for growing galena crystals was 20 ml of sodium metasilicate
containing 20 ml of 2.95N HCl and 0.5 ml of 1M lead acetate and topped with 10 ml
of 0.1M thioacetamide. The galena crystals were 1-1.5 mm on an edge and contained
low densities of etch pits compared to crystals grown at higher temperatures by
other methods.

Hydrothermal Growth

The hydrothermal recrystallization technique described earlier has produced
single crystals of sphalerite, wurtzite, galena, pyrrhotite and pyrite in the
mm-size range as a "byproduct" of phase equilibrium studies. Better results can
be obtained by increasing the size of the container and the time allowed for
growth; by carefully choosing the solution, gradient, and temperature of growth;
and by suspending a seed crystal into the solution. For example, Laudise et al,
(1965) have described the growth of sphalerite crystals more than 2 cm long in
10.8m KOH solutions at a dissolution temperature of 360°C and a crystallization
temperature of 350°C.

Knowledge of solution chemistry is very important in hydrothermal crystal
growth. For example, drawing on the solubility measurements of Barnes et al
(1967) on cinnabar in reduced sulfide solutions enabled Scott and Barnes (1969)
to grow single crystals of cinnabar up to 2mm in diameter and of very high
chemical purity. Rau and Kabenau (1967) had grown larger crystals from acid
chloride solutions but there are two distinct advantages of using alkaline sulfide
solutions: the crystals are free of potentially-contaminating halide anions, and
it is possible to grow cinnabar crystals of different stoichiometries by varying
the sulfur fugacity of the alkaline solution in the manner discussed earlier.

Ch. 5

SULFIDE PHASE EQUILIBRIA

J.R. Craig and S.D. Scott

INTRODUCTION

In recent years experimental studies of phase equilibria have contributed immensely to the growth of sulfide mineralogy, petrology, and geochemistry. Ideally, natural occurrences of minerals provide guidelines for laboratory studies, which in turn elucidate the mode of formation, and perhaps the post-depositional history of mineral assemblages. In reality, there are, of course, complicating factors: (1) natural mineral systems are usually far more complex than synthetic ones; (2) mineral systems have had much longer times to reach equilibrium than can be afforded by experimentalists in their laboratories; and (3) there often exists the plague of metastability either in forming or preserving phases in fields in which they are not stable or in preventing synthesis in realms in which they are stable.

This, of course, raises the question of how one determines the presence or absence of equilibrium. In truth it is often not possible to unequivocably prove equilibrium; however, there are tests which are commonly applied: (1) persistence of an assemblage unchanged through time under a static set of conditions; (2) reversibility of a reaction, *i.e.*, definition of some boundary which when crossed causes change in an assemblage but which when recrossed to the initial conditions, results in reestablishment of the initial assemblage; and (3) synthesis of the same assemblage from different sets of reactions, *i.e.*, $Fe + S_{(liq)} \rightarrow FeS$ or $Fe + FeS_2 \rightarrow 2FeS$. Although these "tests" may be administered in the laboratory, their application to natural assemblages is often impractical if not impossible. Hence the evidence for equilibrium in minerals is often taken to be the absence of evidence for disequilibrium, *i.e.*, zoned crystals. The reader is referred to the chapter on "Experimental Methods in Sulfide Synthesis."

Fortunately for the sulfide mineralogist, synthetic phase equilibria studies frequently have passed the tests for equilibrium and have yielded data sufficiently similar to natural occurrences to provide some assurance of their reliability. In fact, the data have proven useful in understanding mineral genesis and beneficiation, and metal extraction, and they have proven invaluable in the development of phases with important electrical, optical, and physical properties.

The presentation of data pertinent to phase equilibria is generally in the form of a "phase diagram" -- occassionally referred to disrespectfully by the terms "fuzz diagram" or "cartoon" -- which is intended to illustrate the relationships of phases in P-T-X (or some other parameter) space. Although it is possible to construct an almost endless variety of phase diagrams by applying different parameters to the axes, the most commonly employed types are T-X (temperature - composition), P-T (pressure - temperature), a_{S_2} - 1/T (activity of diatomic sulfur gas - inverse temperature), and Eh - pH (redox potential - acidity). We shall not elaborate the

theoretical bases of these or other types of diagrams. Instead, we refer the reader
to the following books:

Alper, A.M. (1970) *Phase Diagrams: Materials Science and Technology*,
   Vol. I. Academic Press.

Ehlers, E.G. (1972) *The Interpretation of Geological Phase Diagrams*.
   W.H. Freeman.

Findlay, A. (1951) *The Phase Rule* (9th ed. by A.N. Campbell and
   N.O. Smith). Dover Publications.

Garrels, R.M. and C.L. Christ (1965) *Solutions, Minerals, and
   Equilibria*. Harper and Row.

Gordon, P. (1968) *Principles of Phase Diagrams in Materials Systems*.
   McGraw-Hill.

Haase, R. and H. Schonert (1969) "Solid and Liquid Equilibria,"
   International Encyclopedia of Physical Chemistry and Chemical Physics,
   Vol. I, Topic 13. Macmillan.

Levin, E.M., C.R. Robbins, and H.F. McMurdie (1964; 2nd ed. 1969)
   Phase Diagrams for Ceramists. American Ceramic Society.

Masing, G. (1944) *Ternary Systems* (translated by B.A. Rogers).
   Dover Publications.

Palatnik, L.S. and A.I. Landau (1964) *Phase Equilibria in Multicomponent Systems* (English trans.). Holt, Rinehart, and Winston.

Prince, A. (1966) *Alloy Phase Equilibria*. Elsevier.

Reisman, A. (1970) *Phase Equilibria*. Academic Press.

Rines, F.N. (1956) *Phase Diagrams in Metallurgy*. McGraw-Hill.

Ricci, J.E. (1951) *The Phase Rule and Heterogeneous Equilibria*.
   D. Van Nostrand and Company.

Zen, E-An (1966) "Construction of Pressure-Temperature Diagrams for
   Multicomponent Systems After the Method of Schreinemakers --
   A Geometric Approach," *U.S.G.S. Bull. 1225*.

In the following portions of the notes, we attempt to present some insight into
sulfide mineralogy by providing: (1) a summary of data sources of information on
important phase equilibria and mineralogic studies in a large number of sulfide
systems, (2) a listing of important sources for thermodynamic data pertinent to
sulfide minerals, (3) rather detailed discussions on a few of the most important
sulfide systems -- Fe-S, Fe-Zn-S, Cu-S, Cu-Fe-S, Ni-S, Fe-Ni-S, (4) a brief
discussion of sulfosalts, and (5) general aspects of the stoichiometry of sulfides.

The sulfide minerals and their synthetic analogs have received considerable attention in recent years not only from mineralogists and geologists, but also from chemists, physicists, metallurgists and materials scientists. As a result, the literature related to sulfide phase equilibria is widely dispersed in publications of these many disciplines. Although a complete listing of published data is beyond the scope of these notes, the authors felt that some sort of listing of major sources of data would be of value. Thus on the following pages we list the important references pertaining to systematic phase equilibria and/or mineralogic studies of a number of the sulfide systems. These are presented alphabetically by metal for binary, ternary, quaternary, and quinary systems. In a general sense the total amount of literature devoted to a particular system is proportional to the number of minerals it contains, their abundance, and their economic value. The references we have arbitrarily included are ones which present up-to-date views, important singular aspects, and/or good bibliographies which will provide the reader with well marked trails into the topic he wishes to pursue.

Although few in number, there are other compilations which frequently serve as valuable sources of data regarding sulfide phase equilibria. These include the following:

(1) Barton, P.B. and B.J. Skinner (1967) Sulfide mineral stabilities in *Geochemistry of Hydrothermal Ore Deposits,* Holt, Rinehart, and Winston, pp. 236-333.

(2) Kullerud, G. (1966) Phase relations in sulfide-type systems. *Handbook of Physical Constants, Geol. Soc. Amer. Mem. 97,* p. 323-343.

(3) Kullerud, G. (1964) Review and evaluation of recent research on geologically significant sulfide-type systems. *Fortschr. Mineralogie, 41,* 221-270.

(4) Hansen, M., and K. Anderko (1958) *Constitution of Binary Alloys* First supplement by R.P. Elliott (1965); Second supplement by F.A. Shunk (1969), McGraw-Hill.

(5) Levin, E.M., C.R. Robbins, and H.F. McMurdie (1964; 2nd ed. 1969) *Phase Diagrams for Ceramists.* American Ceramic Society.

(6) Sorokin, V.I. and N.I. Besmen (1966) Experimental studies of sulfide systems of interest in geology. *Itogi Nauki, Geokhim., Mineral., Petrogr.,* 125-171.

(7) Projected for the future -- Kullerud, G., Physical Chemistry of Sulfide Systems. Chapter FF in *Data of Geochemistry, U.S. Geol. Surv. Prof. Paper 440.*

Each of these papers, with the exception of *Phase Diagrams for Ceramists,* contains annotations as well as actual phase diagrams regarding works in each sulfide system. *Mineralogical Abstracts, Chemical Abstracts, Metal Abstracts* and *Ceramic Abstracts* should be consulted for literature more recent than this volume (summer, 1974).

System	Compound	Mineral name	Max. thermal stability,°C	Remarks	References
Ag–S	Ag_2S	acanthite	177		Kracek (1946)
	Ag_2S	argentite	586–622	inversion temp. comp. dependent	Roy et al (1959)
					Majumdar & Holbe (1959)
	Ag_2S	--	838		Taylor (1969)
	Ag_2S	--	---		Bell & Kullerud (1970)
					Kullerud (1970)
As–S	AsS	realgar	307		Clark (1960)
	AsS	--	265		Barton (1969)
					Kirkinskiy et al
	As_2S_3	orpiment	315		(1967)
	As_2S_3	--	170		Hall (1966)
	As_4S_3	dimorphite	?		Nagoo (1955)
	AsS_2	--	?	amorphous, probably metastable	
Bi–S	Bi_2S_3	bismuthinite	760		Van Hook (1960)
	BiS_2	--	---	high P form	Cubicciotti (1962)
					Silverman (1964)
					Craig et al (1971)
					Cubicciotti (1963)
Ca–S	CaS	oldhamite			Larimer (1968)
					Skinner & Luce (1971)
Cd–S	CdS	hawleyite		metastable cubic form	Corrl (1964)
					Miller et al (1966)
	CdS	greenockite	1475		O'sugi et al (1966)
					Trail & Boyle (1955)
	CdS	--		high P cubic form	Yu & Gielisse (1971)
					Boer & Nalesnik (1969)
Co–S	Co_9S_8	cobalt pentlandite	832		Rosenqvist (1954)
					Curlook & Pidgeon
	Co_4S_3	--	780–930		(1953)
					Lafitte (1959)
	$Co_{1-x}S$	jaipurite?	464–1182	464°C questionable	Kuznetsov et al (1965)
	Co_3S_4	linnaeite	660		
	CoS_2	cattierite	950		
Cr–S	CrS	--	330	inverts to inter. form	Bunch & Fuchs (1969)
					Jellinek (1957)
	CrS	--	570	inverts to high T form	Hansen & Anderko
					(1958)
	CrS	--	~1650		
	Cr_2S_8	--			
	Cr_5S_6	--			

	Cr_3S_4	brezinaite			
	$Cr_{2.1}S_3$	--	?	inverts to high T	
	$Cr_{2.1}S_3$	--	∿1100		
Cu-S	Cu_2S	chalcocite	103		Roseboom (1966)
	Cu_2S	--	∿435		Morimoto & Koto (1970)
	$Cu_2S-Cu_9S_5$	--	1129		Kullerud (1965)
	Cu_2S	--	>500		Skinner (1970)
	$Cu_{1.97}S$	djurleite	93		Barton (1973)
	Cu_9S_5	digenite	83		Taylor & Kullerud (1972)
	Cu_7S_5	anilite	70		Luquet et al (1972)
	$Cu_{1+x}S$	blue-remaining covellite	157		
	CuS	covellite	507		
	CuS_2	--	>550		
Fe-S	FeS	troilite	140		Arnold (1962)
	FeS_{1-x}	mackinawite	?		Barton & Toulmin (1964)
	$Fe_{1-x}S$	hexagonal pyrrhotite	1190	stable only above ∿100 C	Kullerud & Yoder (1959)
	$Fe_{1-x}S$	MC pyrrhotite	308		Kissin (1974)
	$Fe_{1-x}S$	NA pyrrhotite	266		Nakazawa & Morimoto (1971)
	$Fe_{1-x}S$	NC pyrrhotite	213		
	Fe_9S_{10}	5C pyrrhotite	∿100		
	$Fe_{10}S_{11}$	11C pyrrhotite	∿100		
	$Fe_{11}S_{12}$	6C pyrrhotite	∿100		
	$Fe_{7\pm x}S_8$	monoclinic pyrrhotite	254		
	$Fe_{7+x}S_8$	anomalous pyrrhotite	?		
	Fe_2S_3	γ-iron sulfide	?		
	Fe_9S_{11}	smythite	∿ 75		
	Fe_3S_4	greigite	---		
	FeS_2	pyrite	743		
	FeS_2	marcasite	---		
Hg-S	$Hg_{1-x}S$	cinnabar	316-345	inversion comp. dependent	Dickson & Tunell (1959)
	$Hg_{1-x}S$	metacinnabar	572		Potter & Barnes (1971)
	HgS	--	?	high P form	Kullerud (1965)
					Mariano & Warekois (1963)
					Scott & Barnes (1969)
					Barnes (1973)
					Potter (1973)

SEE TABLE CS-6 AND TEXT

SEE TABLE CS-1 AND TEXT

Mg-S	MgS	niningerite		contains Fe, Mn, Co	Skinner & Luce (1971)
					Keil & Snetsinger (1967)
Mn-S	MnS_2	haverite	423		Blitz & Wiechmann (1936)
	MnS	alabandite	1610		Rooymans (1963)
	MnS	--		High P form	Skinner & Luce (1971)
Mo-S	MoS_2	molybdenite	∿1350	Hexagonal	Bell & Herfert (1957)
	MoS_2	molybdenite	∿1350	Rhombohedral	Morimoto & Kullerud (1962)
					Clark (1970)
	Mo_2S_3	--	>610	stable at high T	Zelikman & Belyaerskaya (1956)
	MoS_2	jordisite	---	amorphous	Graeser (1964)
					Clark *et al* (1971)
Ni-S	Ni_3S_2	heazlewoodite	556	SEE TABLE CS-8 AND TEXT	Kullerud & Yund (1962)
	$Ni_{3\pm x}S_2$	--	806		Lafitte (1959a)
	α-Ni_7S_6	godlevskite	400		Lafitte (1959b)
	Ni_7S_6	--	573		Arnold & Malik (1974)
	NiS	millerite	379		Ariya *et al* (1971)
	α-$Ni_{1-x}S$	--	992		Rosenqvist (1954)
	Ni_3S_4	polydymite	356		
	NiS_2	vaesite	1007		
Os-S	OsS_2	erlichmanite	---		Snetsinger (1971)
					Sutarno & Reid (1967)
					Ying-chen & Yu-jen (1973)
Pb-S	PbS	galena	1115		Kullerud (1969)
					Bloem & Kroger (1956)
					Stubbles & Birchenall (1959)
Pt-S	PtS	cooperite			Richardson & Jeffes (1952)
	PtS_2	--			Hansen & Anderko (1958)
					Cabri (1972)
					Gronvold *et al* (1960)
Ru-S	RuS_2	laurite			Ying-chen & Yu-jen (1973)
					Leonard *et al* (1969)
					Sutarno & Reid (1967)
Sb-S	Sb_2S_3	stibnite	556		Pettit (1964)
	Sb_2S_{3-x}	meta-stibnite		amorphous, prob. meta-stable	Barton (1971)
					Clark (1970)
Sn-S	β-SnS	herzenbergite	600	inverts to α	Moh (1969)
	α-SnS	--	880		Albers & Schol (1961)
					Karakhanova *et al* (1966)

	δ-Sn_2S_3	ottemannite	661-675	inversion to comp. dependent	Rau (1965) Moh & Berndt (1964) Mootz & Puhl (1967)
	δ-Sn_2S_3	--	710-715	inversion to. β comp. dependent	
	β-Sn_2S_3	--	744-753	inversion to α comp. dependent	
	α-Sn_2S_3	--	760		
	β-SnS_2	berndtite	680-691	inversion to α comp. dependent	
	α-SnS_2	--	865		
Ti-S	Ti_2S	--		$Pnnm$ $\underline{a}$ = 11.35 $\underline{b}$ = 14.06, $\underline{c}$ = 3.32	Franzen $et\ al$ (1967) Viaene $et\ al$ (1971) Viaene & Kullerud (1971) Wiegers & Jellinek (1970)
	TiS	--	> 800		
	Ti_8S_9	--			
	Ti_4S_5	--			
	Ti_3S_4	--			
	TiS_2	--			
	TiS_3	--	610		
V-S	V_3S_4	--			Chevreton & Sapet (1965) Baumann (1964) Shunk (1969)
	VS_4	patronite			
	VS	--			
	V_3S	--			
	$\sim V_2S_3$				
Zn-S	ZnS	sphalerite		See text	Scott & Barnes (1972) Nickel (1965) Skinner & Barton (1960) Mardix $et\ al$ (1967)
	ZnS	wurtzite		See text	
Ag-As-S	Ag_7AsS_6	As-billing-sleyite	575		Toulmin (1963) Hall (1966) Hall (1968) Roland (1968) Roland (1970) Wehmeier $et\ al$ (1968)
	Ag_5AsS_4	As-stephanite	361		
	Ag_3AsS_3	proustite	495		
	Ag_3AsS_3	xanthocanite	192	inverts to proustite	
	δ-$AgAsS_2$	smithite	415	inverts to $\alpha AgAsS_2$	
	$\delta AgAsS_2$	trechmannite	255	inverts to smithite	
	$\alpha AgAsS_2$	--	415-421		
Ag-Au-S	Ag_3AuS_2	-	181	inverts to high T form	Graf (1968) Barton & Toulmin (1964)
Ag-Bi-S	$\beta AgBiS_2$	matildite	195	inverts to α	Schenck $et\ al$ (1939) Van Hook (1960) Craig (1967) Karup-Moller (1972)
	$\alpha AgBiS_2$	--	195-801		
	$AgBi_3S_5$	pavonite	732		

Ag-Cu-S					
	$Cu_{0.45}Ag_{1.55}S$	jalpaite	117	inverts to bcc phase which has s.s. with Ag_2S	Djurle (1958) Skinner (1966) Graf (1968)
	$Cu_{0.8}Ag_{1.2}S$	mckinstiyite	94	inverts to fcc phase with complete s.s. from Cu_2S-Ag_2S	Skinner *et al* (1966) Suhr (1965) Krestovnikov *et al* (1968) Valverde (1968)
	$Ag_{1*}Cu_{1+x}S$	stromeyerite	93	inverts to hcp phase which has complete s.s. with Cu_2S	Werner (1965)
Ag-Fe-S	$AgFe_2S_3$	argentopyrite	152	low T polymorph	Czamanske (1969)
	$AgFe_2S_3$	sternbergite	152	high T polymorph	Czamanske & Larson (1969) Taylor (1970)
	$Ag_3Fe_7S_{11}$	argyropyrite			Taylor (1970)
	$Ag_2Fe_5S_8$	frieseite			
Ag-Pb-S					Vogel (1953) Van Hook (1960) Craig (1967)
Ag-Sb-S	Ag_7SbS_6	Sb-billings-leyite	475		Barstad (1959) Somanchi (1963)
	Ag_5SbS_4	stephanite	197		Toulmin (1963) Chang (1963)
	Ag_3SbS_3	pyrargyrite	485		Cambi & Elli (1965)
	Ag_3SbS_3	pyrostilpnite	192		Hall (1966) Hall (1968)
	$AgSbS_2$	miargyrite	380		Keighin & Honea (1969)
	$\beta AgSbS_2$	--	510		Wehmeier *et al* (1968)
As-Co-S	CoAsS	cobaltite			Bayliss (1969) Gammon (1966) Klemm (1965)
As-Cu-S	Cu_3AsS_4	luzonite	275-320	inverts to enargite	Maske & Skinner (1971) Gaines (1957)
	Cu_3AsS_4	enargite	671		
	$Cu_{24}As_{12}S_{31}$	--	578		
	$Cu_6As_4S_9$	sinnerite	489		
	$Cu_{12}AsS_{13}$	tennantite	665	s.s. of Cu & As	
	CuAsS	lautite	574		
As-Fe-S	FeAsS	arsenopyrite	702	comp. is sensitive to P	Clark (1960a) Morimoto & Clark (1961) Barton (1969) Kretschmar (1973)
As-Ni-S	NiAsS	gersdorffite	>700	variable As/S ratio	Yund (1962) Klemm (1965) Bayliss (1968)

System	Formula	Mineral	Temp	Notes	References
As-Pb-S	$Pb_9As_4S_{15}$	gratonite	250	inverts to jordanite	Rosch & Hellner (1959)
	$Pb_9As_4S_{15}$	jordanite	549		Le Bihan (1963)
					Roland (1968a)
	$Pb_2As_2S_5$	dufrenoysite	485?		Kutoglu (1969)
	$Pb_{19}As_{26}S_{58}$	rathite II	474		Burkart-Baumann et al (1972)
	$Pb_3As_4S_9$	baumhauerite	458		Chang & Bever (1973)
	$PbAs_2S_4$	sartorite	305		
As-Sb-S	$AsSbS_3$	getchellite	345		Weissberg (1965)
	$(As,Sb)_{11}S_{18}$	wakabayashilite	?		Kato (1970)
	$AsSb_2S_2$	--	538		Moore & Dixon (1973)
					Craig et al (1974)
					Dickson et al (1974)
					Radtke et al (1973)
As-Tl-S	Tl_3AsS_3				Canneri & Fernandes (1925)
	$Tl_4As_2S_5$				Graeser (1967)
	$Tl_6As_4S_9$				Radtke et al (1974a)
	$TlAsS_2$	lorandite	~ 300		Radtke et al (1974b)
Bi-Cu-S	Cu_9BiS_6	X	$\sim 375-650$		Vogel (1956)
	Cu_3BiS_3	wittichenite	527		Buhlmann (1965)
					Sugaki & Shima (1972)
	$Cu_6Bi_4S_9$	klaprothite	?	doubtful validity	Godovikov & Ptitsyn (1968)
					Buhlmann (1971)
	$Cu_{24}Bi_{26}S_{51}$	emplectite	~ 360	inverts to cuprobismutite	Sugaki & Shima (1970)
					Sugaki et al (1972)
	$Cu_{24}Bi_{26}S_{51}$	cuprobismutite	474		Godsvikov et al (1972)
	$Cu_3Bi_5S_9$	D	442-620		
	$CuBi_3S_5$	E	649		
	$Cu_3Bi_3S_7$	Y	~ 498		
Bi-Fe-S	$FeBi_4S_7$		608-719		Urazov et al (1960)
					Craig et al (1971)
					Sugaki et al (1972)
					Ontoev (1964)
Bi-Mo-S					Stemprok (1967)
Bi-Ni-S	$Ni_3Bi_2S_2$	parkerite	>400		Schenck & von der Forst (1939)
					Fleet (1973)
					DuPreez (1945)
					Peacock & McAndrew (1950)
					Brower et al (1974)

System	Compound	Mineral Name	Max. thermal stability, °C	Remarks	References
Bi-Pb-S	$6Pb_{1-x}Bi_{2x/3}S \cdot Bi_2S_3$	heyrov-skyite	829	Contains minor Ag & Cu	Schenck *et al* (1939)
	$3Pb_{1-x}Bi_{2x/3}S \cdot Bi_2S_3$	lillianite	816		Van Hook (1960) Craig (1967) Salanci (1965)
	$PbBi_2S_4$	galenobismutite	750		Otto & Strunz (1968)
	$Pb_2Bi_2S_5$	cosalite	<450		Salanci & Moh (1969)
	$Pb_{1-x}Bi_{2x/3}S \cdot 2Bi_2S_3$	--	680-730		Klominsky (1971)
	$Pb_5Bi_4S_{11}$	bursaite	?		Chang & Bever (1973)
	$PbBi_4S_7$	bonchevite	?		
	$PbBi_6S_{10}$	ustarasite			
Bi-Sb-S	(complete s.s. Bi_2S_3-Sb_2S_3 at >200°C; natural specimens indicate a gap from $Bi_{1.16}Sb_{0.84}S_3$ to Sb_2S_3)				Hayase (1955) Springer (1969) Springer & LaFlamme (1971) Sugaki & Shima (1965)
Bi-Se-S	$Bi_4(S,Se)_3$	ikunolite			Kato (1959) Markham (1962) Godovikov & Il'yasheva (1971)
Bi-Te-S	$\sim Bi_2Te_{1.5+x}S_{1.5-x}$	δ-tetradymite			Beglaryan & Abrikasov (1959)
	$\sim Bi_2Te_{2+x}S_{1-x}$	β-tetradymite			Godovikov *et al* (1970)
	$Bi_8Te_7S_5$	--			
	$Bi_{18}(TeS_3)_3$	joseite-C			
	$Bi_9(Te_2S_2)$	--			
	Bi_3TeS	--			
	$Bi_{15}(TeS_4)$	--			
Ca-Fe-S					Skinner & Luce (1971)
Ca-Mg-S					Skinner & Luce (1971)
Ca-Mn-S					Skinner & Luce (1971)
Cd-Pb-S	(extensive s.s. of CdS in PbS)				Bethke & Barton (1961,1971)

Cd-Zn-S	(nearly complete s.s. between CdS and ZnS)				Hurlbut (1957) Bethke & Barton (1971)
Co-Cu-S	$CuCo_2S_4$	carrollite	>600	complete s.s. with Co_3S_4	Craig & Higgins (1973) Williamson & Grimes (1974) Clark (1974)
Co-Fe-S	(complete s.s. between CoS_2 and FeS_2 and $Co_{1-x}S$ and $Fe_{1-x}S$ at high T)				Curlook & Pidgeon (1953) Klemm (1962) Vogel & Hillner (1953) Straumanis et al (1964) Riley (1965) Bouchard (1968) Springer et al (1964) Bartholome et al (1971) Riley (1968)
Co-Ni-S	(complete s.s. between Co_3S_4 and Ni_3S_4 and CoS_2 and NiS_2)				Klemm (1962) Klemm (1965) Craig & Higgins (1974) Delafosse & Can Hoang Van (1962) Bouchard (1968)
Co-Sb-S	CoSbS CoSbS	paracostibite costibite	876 <100°C?		Lange & Schlegel (1951) Cabri et al (1970a,b)
Cr-Fe-S	Cr_2FeS_4 Cr_2FeS_4 (extensive s.s. between $Fe_{1-x}S$ and $Cr_{1-x}S$)	daubreelite --	>740	high P form stable >14 kbar	El Goresy & Kullerud (1969) Buzek & Prabhala (1965) Vogel (1968) Bell et al (1970)
Cu-Fe-S	$(Cu,Fe)_9S_5$ Cu_5FeS_4 Cu_5FeS_4 Cu_5FeS_4 $Cu_5FeS_{4.05}$ $Cu_{5.5}FeS_{6.5}$ Cu_3FeS_8	digenite bornite -- -- x-bornite idaite fukuchilite	83 228 -- -- 125 501 ∿200	SEE TABLE CS-7 AND TEXT	Merwin & Lombard (1937) Yund & Kullerud (1966) Kullerud et al (1969) Mukaiyama & Izawa (1970) Cabri & Harris (1971) Cabri & Hall (1972)

	$CuFeS_2$	chalcopyrite	557		Cabri (1973)
	$CuFe_2S_3$	cubanite	200–210		Barton (1973)
	wide range	iss	960		
	$Cu_9Fe_8S_{16}$	talnakhite	186		
	$Cu_9Fe_8S_{16}$	iss-1	186–230		
	$Cu_9Fe_8S_{16}$	iss-11	230–520	SEE TABLE CS-6 AND TEXT	
	$Cu_9Fe_9S_{16}$	mooihoekite	167		
	$Cu_9Fe_9S_{16}$	A	167–236		
	$Cu_4Fe_5S_8$	haycockite	?		
	wide range	pc	20–200		
Cu-Ga-S	$CuGaS_2$	gallite			Strunz *et al* (1958)
Cu-Mo-S					Grover & Moh (1969)
Cu-Ni-S	$CuNi_2S_6$	villamaninite	503		Kullerud *et al* (1969)
					Bouchard (1968)
Cu-Pb-S	$Cu_{14}Pb_2S_{9-x}$		486–528		Schuller & Wohlmann (1955)
					Craig & Kullerud (1968)
					Clark & Sillitoe (1970)
Cu-Sb-S	Cu_3SbS_4	famatinite	627		Avilov *et al* (1971)
	$CuSbS_2$	chalcostibite	553		Skinner *et al* (1972)
	$Cu_{12+x}Sb_{4+y}S_{13}$	tetrahedrite	543		Tatsuka & Morimoto (1973)
	Cu_3SbS_3	skinnerite	607		Karup-Moller & Makovicky (1974)
Cu-Sn-S	Cu_2SnS_3				Wang (1974)
					Roy-Choudhury (1974)
Cu-V-S	Cu_3VS_4	sulvanite			Dolanski (1974)
Cu-W-S					Moh (1973)
Cu-Zn-S	$\sim Cu_3ZnS_4$	--	?		Craig & Kullerud (1973)
					Clark (1970)
Fe-Ge-S	Fe_2GeS_4	---	>800		Viaene (1972)
					Viaene (1968)
Fe-In-S	$FeIn_2S_4$	indite	>1000		Genkin & Murav'eva (1963)
					Hahn & Klinger (1950)

Fe-Mg-S	(Mg,Fe)S	niningerite	>1200	Skinner & Luce (1971)
				Keil & Snetsinger (1967)
				Kurash *et al* (1973)
Fe-Mn-S	(extensive s.s. of FeS in MnS)			Shibata (1926)
				Sugaki & Kitakaze (1972)
				Skinner & Luce (1971)
Fe-Mo-S	$FeMo_3S_4$	--	?	Kullerud (1967)
				Grover & Moh (1966)
				Lawson (1972)
				Grover *et al* (1973)
Fe-Ni-S	$(Fe,Ni)_9S_8$	pentlandite	610	Kullerud *et al* (1969)
	$(Fe,Ni)_{1-x}S$	--	1192-992	Naldrett *et al* (1967)
	$FeNi_2S_4$	violarite	461	Shewman & Clark (1970)
	$(Fe,Ni)S_2$	bravoite	137	Craig (1971)
				Misra & Fleet (1973)
				Craig (1973)
				Clark & Kullerud (1963)
				Vaughan & Craig (1974)
				Scott *et al* (1974)
Fe-Pb-S				Brett & Kullerud (1967)
Fe-Sb-S	$FeSb_2S_4$	berthierite	563	Barton (1971)
	FeSbS	gudmundite	280	Clark (1966)
Fe-Sn-S				Stemprok (1971)
Fe-Ti-S	$FeTi_2S_4$	--	>1000	Plovnick *et al* (1968)
	$FeTi_2S_4$	--	540	Hahn *et al* (1965)
	$Fe_{1+x}Ti_2S_4$	heideite	?	Keil & Brett (1974)
	(possibly related to low T form of $FeTi_2S_4$)			Viaene *et al* (1971)
				Viaene & Kullerud (1971)
Fe-W-S				Stemprok (1971)
				Vogel & Weizenkorn (1961)
				Grover & Moh (1966)
Fe-Zn-S	(extensive s.s. of FeS in ZnS)			Barton & Toulmin (1966)
				Boorman (1967)
	(see text for discussion)			Boorman *et al* (1971)
				Scott & Barnes (1971)
				Scott & Kissin (1973)
				Scott (1973)
				Chernyshev & Anfilogov (1968)

System	Formula	Mineral	Temp	References
Hg–Sb–S	$HgSb_4S_8$	livingstonite	451	Craig (1970) Learned (1966) Tunell (1964) Learned *et al* (1974)
Hg–Se–S	(considerable s.s. between HgS and HgSe)			Boctor & Kullerud (1973)
Mg–Mn–S				Skinner & Luce (1971)
Mn–Zn–S				Kroger (1939) Juza *et al* (1956) Bethke & Barton (1971)
Mo–W–S				Holl & Weber-Diefenbach (1973) Moh (1972)
Ni–Pb–S	$Ni_3Pb_2S_2$	shandite	?	Peacock & McAndrew (1950) Brower *et al* (1974)
Ni–Sb–S	$NiSbS$	ullmanite	752	Schenck & von der Forst (1939) Williams & Kullerud (1970) Bayliss (1969)
Ni–Sn–S	$Ni_3Sn_2S_2$	Sn-shandite	?	Brower *et al* (1974)
Pb–Sb–S	$Pb_3Sb_2S_6$	--	625–642	Kitikaze (1968)
	$Pb_5Sb_4S_{11}$	boulangerite	638	Jambor (1967) Salanci & Moh (1970)
	$Pb_5Sb_8S_{14}$	--	576–603	Wang (1973)
	$Pb_6Sb_{10}S_{21}$	robinsonite	582	Garvin (1973) Craig *et al* (1973)
	$PbSb_2S_4$	zinckenite	545	Chang & Bever (1973) Sugaki *et al* (1969)
Pb–Se–S	(complete s.s. between PbS and PbSe)			Kovlenker *et al* (1971) Bethke & Barton (1961) Coleman (1959) Heier (1953) Wright *et al* (1965)
Pb–Sn–S	$PbSnS_2$	teallite	complete s.s. with SnS	Chang & Brice (1971) Stemprok & Moh (1969) Yamaoka & Okai (1970)
	$PbSnS_3$		730 complete s.s. with Sn_2S_3	Sachdev & Chang (in prep)
	$PbSn_4S_5$	"montesite"	part of teallite-herzenbergite s.s.	Kuznetsov & Ch'ih-Fa (1964) Nekrasov *et al* (1974)

Pb-Te-S				Darrow *et al* (1966)
Pb-Zn-S				Maurel (1973)
Pd-Pt-S				Ying-chem & Yu-jen (1973)
Sb-Sn-S	$Sn_3Sb_2S_6$	--	565	Vogel & Gilde (1949)
	$SnSb_2S_4$	--	493	Sachdev & Chang (in prep)
	$\sim Sn_{.85}Sb_{.30}S_{1.45}$	--	627	Parravano & DeCesaris (1912)
Se-Zn-S	(complete s.s. between ZnS and ZnSe)			Wright *et al* (1965) Bethke & Barton (1961)
Ag-As-Cu-S	$(Ag_{14.7-x}Cu_{1.3+x})$-As$_2$S$_{11}$	pearcite	$\sim$500	Frondel (1963) Hall (1967)
	$(Ag_{15.4-x}Cu_{0.6+x})$-As$_2$S$_{11}$	arsenpoly-basite	350	
	$(Ag_{15.4-x}Cu_{0.6+x})$-As$_2$S$_{11}$	X	$\sim$450	
Ag-As-Sb-S	(complete s.s. of proustite Ag$_3$AsS$_3$ - pyrorgyrite (Ag$_3$SbS$_3$)			Toulmin (1963)
Ag-Bi-Pb-S	$AgBi_3Pb_2S_7$	schirmerite	?	Wernick (1960) Van Hook (1960) Craig (1967) Hoda & Chang (1972) Nedachi *et al* (1973) Karup-Møller (1972) Karup-Møller (1973)
	$AgPbBi_3S_6$	gustavite	?	
Ag-Bi-Sb-S	(complete s.s. between AgBiS$_2$ and AbSbS$_2$)			Chen & Chang (1971)
Ag-Cu-Sb-S	$(Ag_{15.4-x}Cu_{0.6+x})$Sb$_2$S$_{11}$	polybasite	$\sim$400	Frondel (1963) Hall (1967) Cambi *et al* (1966) Riley (1974)
	$(Ag_{14.3-x}Cu_{1.7+x})$Sb$_2$S$_{11}$	antimonpear-cite	$\sim$480	
	(complete s.s. between Cu$_{12}$Sb$_4$S$_{13}$ and Ag$_{12}$Sb$_4$S$_{13}$)			
Ag-Fe-Ni-S	$(Fe,Ni)_8AgS_8$	argentian-pentlandite		Vuorelainen *et al* (1972) Scott & Gasparrini (1973) Hall & Steward (1973)

Ag-Pb- Sb-S	$Ag_3PbSb_3S_7$	--		Godovikov & Nenasheva (1969a)
	$Ag_2Pb_2Sb_6S_{12}$	andorite	?	Godovikov & Nenasheva
	$Ag_2Pb_3Sb_8S_{13}$	ramdohrite	?	(1969b)
	$Ag_3Pb_2Sb_3S_8$	diaphorite	?	Wernick (1960) Hoda & Chang (1972)
	$AgPbSbS_3$	--	?	Godovikov et al
	$Ag_2Pb_5Sb_6S_{15}$	owyheeite	?	(1972)
	$Ag_6Pb_2Sb_6S_{14}$	brongniardite	?	
	$Ag_2Pb_2Sb_2S_6$	freislebenite	?	
	(complete s.s. of PbS and $AgSbS_2$)			
As-Bi- Pb-S	$27PbS \cdot 7(As_{0.7}Sb_{.3})_2S_3$ A		>400°	Walia & Chang (1973)
	$2Pbs \cdot (As_{0.6}B_{0.4})_2S_3$ B		>400	
As-Co- Fe-S	$Co_{0.75}Fe_{0.25}AsS$	alloclasite		Klemm (1962) Klemm (1965)
	(extensive s.s. between CoAsS and FeAsS)			Gammon (1966) Kingston (1970)
As-Cu- Fe-S				Gustavson (1963) McKinstry (1963)
As-Cu- Pb-S	$PbCuAsS_3$	seligmannite	?	Wernick & Geller (1958)
As-Cu- Sb-S	(complete s.s. between tetrahedrite and tennantite)			Feiss (1974) Skinner (1960)
	(extensive s.s. between enargite and famatinite)			Barton & Skinner (1967) Sakharova (1966)
As-Hg- Tl-S				Radtke et al (1974a)
As-Pb- Sb-S	$3PbS \cdot (As,Sb)_2S_3$	madocite	>400	Walia & Chang (1973) Jambor (1967,68)
	(variable Pb/(As,Sb) and As/Sb composition)			Roland (1968)
	$2PbS \cdot (Sb,As)_2S_3$	veenite	>400	Burkart-Baumann
	(complete s.s. with dufrenoysite, $Pb_2As_2S_5$)			et al (1966)
	$PbS \cdot (As_{.5}Sb_{.5})_2S_3$	guettardite	>400	all formulas questionable
	$16PbS \cdot 9(Sb,As)_2S_3$	playfairite	?	
	$12PbS \cdot (Sb,As)_2S_3$	sterryite	?	
	$17PbS \cdot 11(Sb,As)_2S_3$	sorbyite	?	
	$PbS \cdot (Sb,As)_2S_3$	twinnite	?	
	$27PbS \cdot 7(As_{.45}Sb_{.55})_2S_3$	geochronite	?	(complete s.s. with jordanite)

Bi-Cu- Fe-S	$Cu_{8.12}Bi_{11.54}Fe_{0.29}S_{22}$	hodrushite	?	Sugaki & Shimall (1970) Onteov (1964)
	$Cu_{8.4}Bi_{10.8}Fe_{1.2}S_{22}$	--	525	Sugaki et al (1972) Kodera et al (1970)
Bi-Cu- Pb-S	$BiCuPbS_3$ (complete s.s. with Bi_2S_3)	aikinite	540	Springer (1971)
Bi-Cu- Sb-S	(extensive s.s. between Cu_3BiS_3 - Cu_3SbS_3 and $CuSbS_2$ - $CuBiS_2$) 			Chen & Chang (1971)
Ca-Fe- Mg-S				Skinner & Luce (1971)
Ca-Fe- Mn-S				Skinner & Luce (1971)
Ca-Mg- Mn-S				Skinner & Luce (1971)
Cd-Mn- Zn-S				Kroger (1939)
Cd-Pb- Zn-S				Bethke & Barton (1971)
Co-Fe- Ni-S	(extensive s.s. among mono and di-sulfides) (complete s.s. between $(Fe,Ni)_9S_8$ and Co_9S_8)			Knop & Ibrahim (1961) Springer & Schachner- korn (1964) Demirsoy (1969) Nickel (1970) Klemm (1965)
Co-Ni- Fe-S	(Co,Ni)SbS (considerable s.s. toward CoSbS and NiSbS)	willyamite	>550	name retained where Co>Ni)	Cabri et al (1970) Bayliss (1969)
Cu-Fe- Ge-S	Cu_2FeGeS_4	briartite	∿640	inverts to α-form	Francotte et al (1965) Bente (1974)
	$\alpha\text{-}Cu_2FeGeS_4$	--	∿990		
Cu-Fe- Ni-S				Craig & Kullerud (1969) Kushima & Asano (1953
Cu-Fe- Pb-S	$Cu_{10}(Fe,Pb)S_6$	betektinite	?	Schuller & Wohlmann (1955) Craig & Kullerud (1967a,b)

Cu-Fe- Sn-S	β-Cu$_2$FeSnS$_4$	stannite	680	Inverts to α-form	Springer (1972) Bernhardt (1972) Bente (1974)
	α-Cu$_2$FeSnS$_4$	--	878		
	Cu$_7$Fe$_2$SnS$_{10}$	mawsonite			
	(considerable s.s. of Cu$_2$FeSnS$_4$ in CuFeS$_2$ above 462°C)				
Cu-Fe- Zn-S	(limited s.s. between CuFeS$_2$ and ZnS)				Toulmin (1960) Fuji (1970) Buerger (1934) Jankovic (1953) Wiggins (1974)
Cu-Pb- Sb-S	CuPbSbS$_3$	bournonite	>530		Harada et al (1970)
	CuPb$_{13}$Sb$_7$S$_{24}$	meneghinite	>615		Fredricksson & Anderson (1964)
Cu-Sn- Zn-S	Cu$_2$SnZnS$_4$	kesterite	1002		Springer (1972) Roy-Choudhury (1974)
Fe-Mg- Mn-S					Skinner & Luce (1971)
Fe-Ni- Zn-S					Scott et al (1974) Scott et al (1972) Czamanske & Goff (1973)
Fe-Pb- Zn-S					Avetisyan & Gnaty- shenko (1956)
Ir-Os- Ru-S	(complete s.s. among RuS$_2$-OsS$_2$-IrS$_2$)				Ying-chen & Yu-jen (1973)
Mn-Pb- Zn-S					Bethke & Barton (1971)
Pb-Sb- Sn-S	Pb$_5$Sn$_3$Sb$_2$S$_{14}$	franckeite	>500		Sachdev & Chang (in prep)
	Pb$_3$Sn$_4$Sb$_2$S$_{14}$(?)	cylindrite	>500	Fe necessary?	
	Pb$_{.6}$Sn$_{.23}$Sb$_{.34}$S$_{1.57}$	III	617		
Pb-Se- Zn-S					Bethke & Barton (1971)
As-Co- Fe-Ni-S	(extensive s.s. among FeAsS-CoAsS-NiAsS)				Nickel (1970) Klemm (1965)
Ca-Fe- Mg-Mn-S	(extensive s.s. series)				Skinner & Luce (1971)
Cu-Fe- Ge-Sn-S	(complete s.s. between Cu$_2$FeSnS$_4$ and Cu$_2$FeGeS$_4$)				Bente (1974)

Cu-Fe-   (complete s.s. of Cu_2FeSnS_4 and Cu_2ZnSnS_4 >680°C)   Springer (1972)
Sn-Zn-S   Lee (1972)
          (complete s.s. of $\alpha-Cu_2FeSnS_4$ and ZnS above ~860°C)   Bernhardt *et al*
                                                                          (1972)
                                                                       Petruk (1973)
                                                                       Harris & Owens (1972)

Ag-As-Cu- (considerable s.s. of Cu and Ag, Fe and Zn, and   Riley (1974)
Fe-Sb-S   As and Sb in the tetrahedrite-freibergite series)

MAJOR SOURCES OF THERMOCHEMICAL DATA ON SULFIDES

A thorough understanding of the behavior of the sulfide minerals requires knowledge of their thermochemical parameters. Presently the state of our knowledge of these parameters varies from highly refined for many binary sulfides at high temperatures to nonexistent for numerous ternary and quaternary sulfides. A complete listing of all of the data available is beyond the scope of the present work; we are limited to the brief thermochemical considerations included in our discussions of some of the specific sulfide systems. The following is a short list of some other major sources of thermochemical data on sulfides:

Adami, L.H., and E.G. King (1964) Heats and free energies on formation of sulfides of manganese, iron, zinc, and cadmium. *U.S. Bureau of Mines, Report of Investigations, 6495.*

Astakhov, K.V. (1970) *Thermodynamic and Thermochemical Constants.* Science Publishing House, Moscow.

Barton, P.B.,Jr., and B.J. Skinner (1967) Sulfide mineral stabilities. In *Geochemistry of Hydrothermal Ore Deposits,* H.L. Barnes, ed. Holt, Rinehart, and Winston, pp. 236-333.

Bulletin of Thermodynamics and Thermochemistry

Freeman, R.D. (1962) Thermodynamic properties of binary sulfides. *Research Foundation Rept. 60,* Oklahoma State University, Stillwater.

Garrels, R.M. and C.L. Christ (1965) *Solutions, Minerals, and Equilibria.* Harper and Row, New York.

Karapet'yants, M.K. and M.L. Karapet'yants (1970) *Thermodynamic Constants of Inorganic and Organic Compounds.* Humphrey Science Publishing, Ann Arbor.

Kelley, K.K., (1960) Contributions to the data on theoretical metallurgy XIII. High-temperature heat-content, heat-capacity, and entropy data for the elements and inorganic compounds. *U.S. Bureau of Mines Bull. 584.*

Kubaschewski, O., E.L. Evans, and C.B. Alcock (1967) *Metallurgical Thermochemistry,* 4th ed. Pergamon, New York.

Mills, K.C. (1974) *Thermodynamic Data for Inorganic Sulphides, Selenides, and Tellurides.* Butterworth, London.

Reed, T.B. (1971) *Free Energy of Formation of Binary Compounds: An Atlas of Charts for High-temperature Chemical Calculations.* Mass. Inst. Tech., Cambridge Press.

Richardson, F.D. and J.H.E. Jeffes (1952) The thermodynamics of substances of interest in iron and steel making, III -- Sulphides. *Jour. Iron Steel Inst.* (London), *171,* 165-175.

Robie, R.A., and D.R. Waldbaum (1968) Thermodynamic properties of minerals and related substances at 298.15°K (25°C) and one atmosphere (1.013 bars) pressure and at higher temperatures: *U.S. Geol. Survey Bull. 1259.*

Timmermans, J. (1959) *Physico-Chemical Constants of Binary Systems.* Interscience, New York.

Wagman, D.D., *et al* (1968, 1969, 1971) Selected values of chemical thermodynamic properties: *NBS Tech.Notes 270-3, 270-4, 270-5, 270-6.*

Besides containing two of the most common sulfide minerals, pyrite and pyrrhotite, the Fe-S system is a cornerstone to the understanding of phase relations and thermochemistry of many other important systems including Zn-Fe-S, Cu-Fe-S, Fe-Ni-S and Fe-As-S. The many complexities of the system, some of which are yet unresolved, make it a good place to begin our systematic review of sulfide phase relations.

The basic phase diagram above 400°C is shown in Fig. CS-1. Details at high temperature (>700°C) are in Fig. CS-2 and at low temperatures (<350°C) in Fig. CS-3. Note that the compositions in Fig. CS-1 and CS-2 are in weight percent whereas Fig. CS-3 is in atomic percent Fe. We will focus our attention between the compositional limits FeS to FeS_2 by first describing the stoichiometry and stability of the phases in the condensed system followed by discussions of thermodynamics and the effect of pressure on phase relations. The properties of solid phases encountered in the system are summarized in Table CS-1.

Throughout the discussion, superlattice dimensions <u>a</u> and <u>c</u> of pyrrhotites and related phases are presented in terms of <u>A</u> and <u>C</u>, the dimensions of the simple hexagonal subcell with a NiAs (1<u>C</u>) structure.

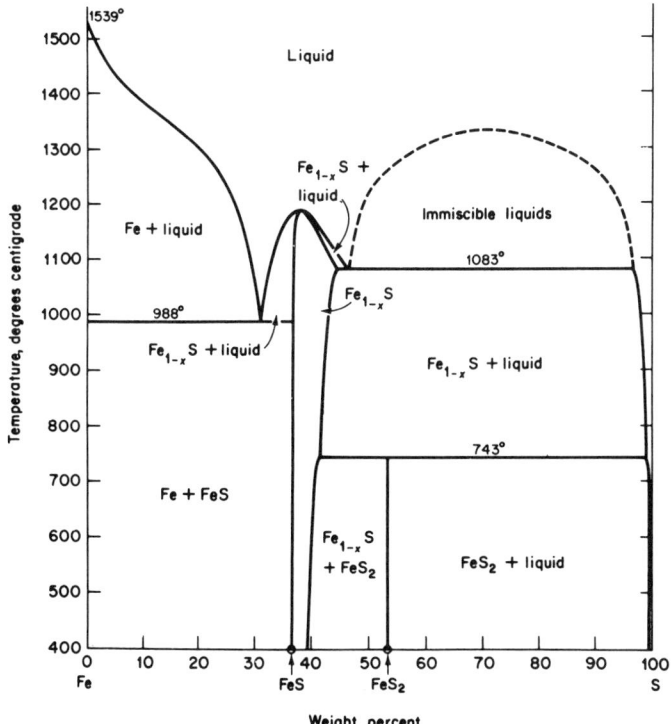

Fig. CS-1. Relations among condensed phases in the system Fe-S above 400°C. (From Ehlers, 1972, after Kullerud, 1967, *The Interpretation of Geological Phase Diagrams,* Fig. 217, p. 232).

Table CS-1. Minerals and Phases in the Central Portion of the Fe-S System.*

Mineral Name	Composition	Thermal stability,°C Maximum	Minimum	Structure Type (Cell edges in Å)	Remarks
Troilite	FeS	140 (1)*		Hexagonal 2$\underline{C}$ (2) $\underline{a}$ = 3$\underline{A}$ = 5.965 (1) $\underline{c}$ = 2$\underline{C}$ =11.750 (1)	Inverts to hexagonal 1$\underline{C}$ pyrrhotite
Mackinawite	FeS$_{1-x}$ 0.07>x >0.04 (3)	?		Tetragonal P4/nmm $\underline{a}$ = 3.675, $\underline{c}$ = 5.030	Often contains Ni and Co
Hexagonal pyrrhotite	Fe$_{1-x}$S 44.9-50.0 at. % Fe (5)	m.p. = 1190 (6)	100 (7)	Hexagonal 1$\underline{C}$ (7) $\underline{a}$ = 3.45, $\underline{c}$ = 5.75	NiAs structure
M$\underline{C}$-type pyrrhotite	Fe$_{1-x}$S 47.4-44.7 at. % Fe (8)	308 (8)	262 (8)	Hexagonal? (7) $\underline{a}$ = 2$\underline{A}$, $\underline{c}$ = M$\underline{C}$	3.0<M<4.0 (7)
N$\underline{A}$-type pyrrhotite	Fe$_{1-x}$S 47.2-47.8 at. % Fe (8)	∿266 (8)	209 (8)	Hexagonal? (7) $\underline{a}$ = N$\underline{A}$, $\underline{c}$ = 3$\underline{C}$	40<N<90 (7)
N$\underline{C}$-type pyrrhotite	Fe$_{1-x}$S 47.2-48.1 at. % Fe (8)	∿213 (8)	∿100 (7)	Hexagonal? (7) $\underline{a}$ = 2$\underline{A}$, $\underline{c}$ = N$\underline{C}$	3.0<N<6.0 (7)
5$\underline{C}$ pyrrhotite	Fe$_9$S$_{10}$ (7)	∿100 (7)		Hexagonal (9) $\underline{a}$ = 2$\underline{A}$ = 6.88, $\underline{c}$ = 5$\underline{C}$ = 28.70	
11$\underline{C}$ pyrrhotite	Fe$_{10}$S$_{11}$ (7)	∿100 (7)		Orthorhombic (9) $Cmca$ or $C2ca$ $\underline{a}$ = 2$\underline{A}$ = 6.892, $\underline{b}$ = 1$\overline{1}$.952, $\underline{c}$ = 11$\underline{C}$ = 63.184	
6$\underline{C}$ pyrrhotite	Fe$_{11}$S$_{12}$ (7)	∿100 (7)		Orthorhombic? (8) Hexagonal cell: (10) $\underline{a}$ = 2$\underline{A}$ = 6.892, $\underline{c}$ = 6$\underline{C}$ = 34.476	
Metastable pyrrhotite	Fe$_{1-x}$S 0.06>x >0.03 (7)	Meta- stable		Hexagonal? (8) See text $\underline{a}$ = 2$\underline{A}$, 3$\underline{C}$<$\underline{c}$<7$\underline{C}$	
4$\underline{C}$ monoclinic pyrrhotite	Fe$_{7+x}$S$_8$ 46.4-47.3 at. % Fe (8)	254 (8)		Monoclinic $C2/c$ or Cc $\underline{a}$ = 11.90, $\underline{b}$ = 6.87, $\underline{c}$ = 22.88, β = 90°30' Hexagonal supercell $\underline{a}$ = 2$\underline{A}$ = 6.88, $\underline{c}$ = 4$\underline{C}$ = 22.90 (11)	

Mineral Name	Composition	Thermal stability, °C Maximum	Minimum	Structure type (Cell edge in Å)	Remarks
Anomalous pyrrhotite	$Fe_{7+x}S_8$ (12) 46.4 at. % Fe	?		Triclinic? (12)	
Gamma iron sulfide	Fe_2S_3 (13)	?		Spinel *Fd3m* (13) a = 9.87	
Symthite	Fe_9S_{11} (14)	∿75 (15)		Pseudo-rhombohedral Hexagonal or (14) monoclinic? a = 3.47, c = 34.4	Contains 0.4 to 7.5 wt. % Ni in nature (14
Greigite	Fe_3S_4	Meta-stable (?)		Spinel *Fd3m* a = 9.876	
Pyrite	FeS_2	743 (21)		Cubic *Pa3* (17) a = 5.417	Breaks down t(1$\underline{C}$ pyrrhotite + S (liq)
Marcasite	FeS_2 slightly S-deficient? (18)	Meta- (19) stable		Orthorhombic *Pnnm* a = 4.445, $\underline{b}$ = 5.425 $\underline{c}$ = 3.388 (20)	

*Numbers in brackets are references:

(1) Yund & Hall (1968)

(2) Hagg & Sucksdorff (1933)

(3) Takeno & Clark (1967)

(4) Evans *et al* (1964)

(5) Arnold (1971)

(6) Jensen (1942)

(7) Nakazawa & Morimoto (1971)

(8) Kissin (1974)

(9) Morimoto *et al* (1971)

(10) Fleet & MacRae (1969)

(11) Mukherjee (1969)

(12) Clark (1966)

(13) Yamaguchi & Wada (1973)

(14) Taylor & Williams (1972)

(15) Taylor (1970a, b)

(16) Skinner *et al* (1964)

(17) Bragg (1914)

(18) Buerger (1934)

(19) Rising (1973)

(20) Buerger (1931)

(21) Kullerud & Yoder (1959)

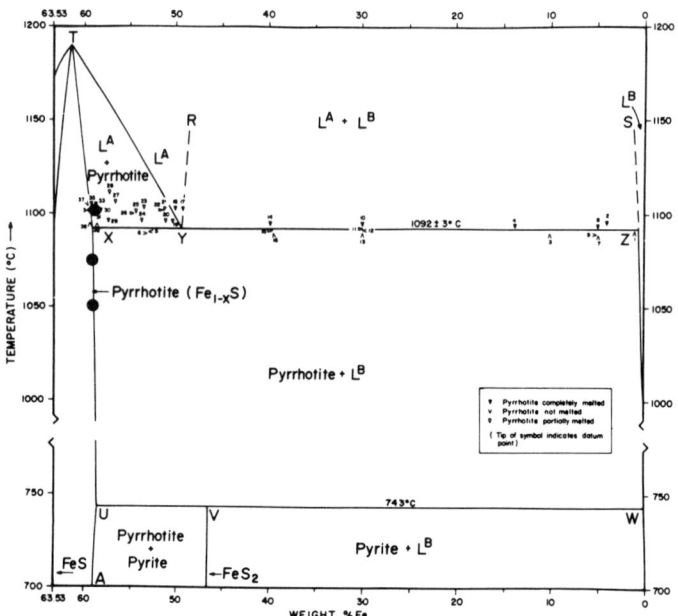

Fig. CS-2. Relations among condensed phases in the sulfur-rich portion of the Fe-S system above 700°C. The closed circles are runs which used coexisting sphalerite to determine the phase boundaries as described in the section on Zn-Fe-S. (Modified from Arnold, 1971).

Phases

Troilite (FeS). Strictly speaking, the name troilite applies only to the polymorph of stoichiometric FeS which is stable below 140°C (Gronvold and Haraldsen, 1952; Yund and Hall, 1968). Above 140°C, FeS has the NiAs ($1\underline{C}$) structure of high-temperature hexagonal pyrrhotite whose composition range extends for a considerable distance across the system.

Although a common constituent of meteorites, troilite is found only occasionally in terrestrial environments, usually with low-temperature hexagonal pyrrhotite. The solvus separating troilite from other pyrrhotites has been determined most recently by Yund and Hall (1968).

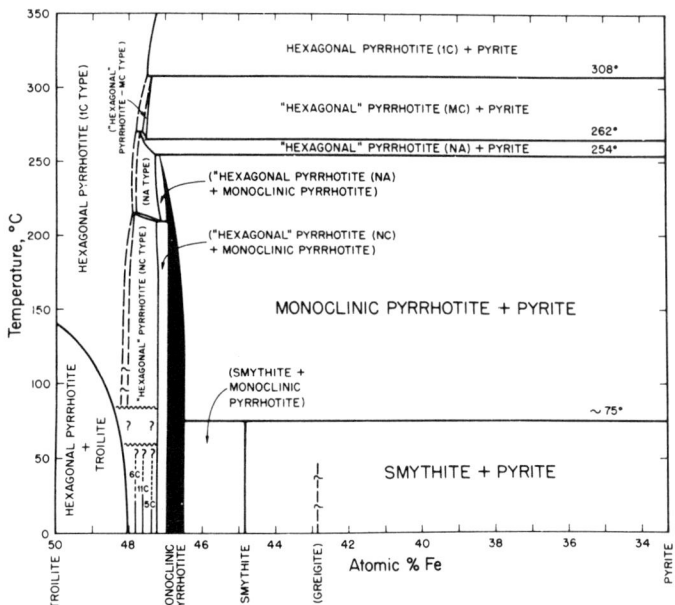

Fig. CS-3. Relations among condensed phases in the central portion of the Fe-S system below 350°C. (Kissin, 1974, modified from Scott and Kissin, 1973).

Mackinawite (FeS_{1-x}). Evans et al (1964) proposed the name mackinawite for a relatively-rare tetragonal mineral of near-FeS composition. It is found either with troilite or with low-temperature pyrrhotites (Kullerud, 1967) in a wide variety of geological environments including: black iron sulfide muds, contact metasomatic and hydrothermal deposits, bedded cupriferous iron sulfide deposits within metamorphic rocks, in certain ultrabasic rocks and in several lunar rocks (Zoka et al, 1973). Mackinawite invariably contains some Co and Ni although these elements are not essential as demonstrated by Berner (1964), who precipitated the phase from aqueous iron sulfide solutions between 20° and 95°C. The metal to sulfur ratio in mackinawite is slightly greater than unity (1.04 to 1.07; Takeno and Clark, 1967) and the formula is usually written $Fe_{1+x}S$. However, Taylor and Finger (1971) have shown that there is a deficiency of sulfur in the structure rather than an excess of metal so the formula is properly FeS_{1-x}.

Very little is known about the thermal stability of mackinawite and, for this reason, it is not shown as a stable phase in Fig. CS-1 and CS-3. Zoka et al (1973) found that natural mackinawite from a variety of localities broke down non-isochemically between 120° and 153°C to pyrrhotite of a more S-rich composition. They concluded that their experiments did not represent equilibrium, however, because the expected breakdown assemblage of metallic Fe, Co, Ni + troilite was not found.

"Hexagonal" Pyrrhotite ($Fe_{1-x}S$). There has probably been more written on pyrrhotite than on any other sulfide mineral but, in spite of this voluminous research effort, many enigmas remain, particularly at low temperatures where slow reaction kinetics tend to obscure the phase relations. However, recent studies at low temperatures by Nakazawa and Morimoto (1971) and by Kissin (1974; cf. also Kissin and Scott, 1972; Scott and Kissin, 1973), in which single crystal x-ray diffraction was used as an analytical tool, have reduced what before was chaos to mere confusion.

Between its maximum melting temperature of 1190°C (point T on Fig. CS-2) and 308°C, the full width of the pyrrhotite phase field is occupied by a single solid solution, $Fe_{1-x}S$, in which iron and vacancies are randomly distributed in the cation sites of the NiAs (1C) structure. This phase also extends to lower temperatures but with a more restricted composition range (Fig. CS-3). Disordered 1C pyrrhotite cannot be quenched (Corlett, 1968) and crystals cooled from its phase field acquire one or more of the many superstructures shown on Fig. CS-3. Earlier diagrams which now pervade the literature placed a phase boundary which was independent of composition across the 1C field near 320°C. This beta-transition, as it is known, manifests itself as a discontinuity in magnetic susceptibility attending the antiferromagnetic to paramagnetic Néel transition (van den Berg, 1972). This transition became further entrenched as a phase boundary with the observations of Desborough and Carpenter (1965) that pyrrhotite quenched from above the beta-transition possessed a 2A,7C superlattice whereas those annealed below the beta-transition possessed a 2A,5C superlattice. However, from the work of Corlett (1968), Nakazawa and Morimoto (1971), and Kissin (1974) we now know that the Néel transition does not accompany a phase change. The superstructures observed by Desborough and Carpenter (1965) correlate with some lower temperature phases and undoubtably formed during quenching.

An indication that the simple solid-solution field of hexagonal pyrrhotite would give way to a far more complicated picture at low temperatures was evident from the observations of Carpenter and Desborough (1964), Mukaiyama and Izawa (1966), and Morimoto et al (1971) that x-ray diffraction patterns of all natural pyrrhotites contain superlattice reflection at room temperature. Furthermore, Arnold (1967) had shown that although the bulk composition of naturally-occurring pyrrhotite mixtures spanned nearly the entire range of hexagonal pyrrhotite solid solutions (Fig. CS-4a), the compositional ranges of the constituent pyrrhotite phases were very narrow (Fig. CS-4b), suggesting the interruption of the high-temperature solid solution by at least two solvi.

The overall picture which has emerged from the single crystal x-ray studies of Nakazawa and Morimoto (1971) and Kissin (1974) is one of increased ordering of vacancies spreading across the pyrrhotite phase field with decreasing temperature

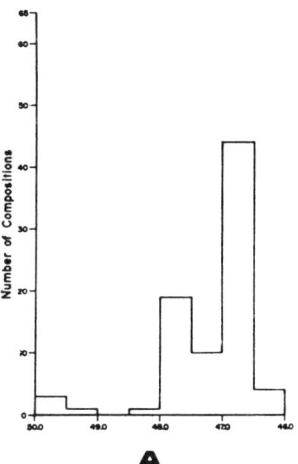
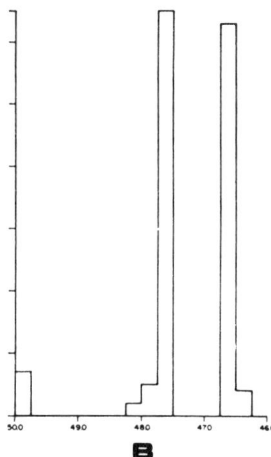

Fig. CS-4. Frequency distribution of natural terrestrial pyrrhotite compositions (From Arnold, 1967). (a) Bulk compositions of pyrrhotites; (b) composition of individual pyrrhotite phases in the bulk. (Composition is in atomic percent Fe.)

(Fig. CS-3) and giving rise to the various superstructures. Monoclinic pyrrhotite (4$\underline{C}$) also fits this scheme but is discussed separately below. The 5$\underline{C}$, 11$\underline{C}$ and 6$\underline{C}$ types found at room temperature are stoichiometric phases with compositions $Fe_{n-1}S_n$ (n = 10, 11 and 12, respectively) and are separated by two-phase fields (Morimoto et al, 1971). Between the upper stability limit of the three stoichiometric types (which is unknown but probably below 100°C) and 308°C there are three other phases which also possess subcells of the NiAs structure but have supercells of uncertain symmetry. These phases are distinguished by their nonintegral repeats of x-ray reflections along hexagonal $\underline{a}$* or $\underline{c}$* axes of the subcell and have been termed, N$\underline{A}$, N$\underline{C}$, and M$\underline{C}$ types of Nakazawa and Morimoto (1971). Their reciprocal $\underline{a}$*$\underline{c}$* planes are shown in Fig. CS-5. In the N$\underline{C}$ type the periodicity of the superlattice reflections parallel to $\underline{c}$* (Δl) can vary continuously from 1/6 to 1/3 of the repeat of the subcell reflections giving rise to apparent cell dimensions of $\underline{a}$ = 2A and $\underline{c}$ = N$\underline{C}$. N varies from 3.0 to 6.0 as a function of temperature and composition and is generally nonintegral. The 2$\underline{A}$,5$\underline{C}$ superstructure of Carpenter and Desborough (1964) is probably equivalent to the N$\underline{C}$ type (Kissin, 1974). The diffraction pattern of the M$\underline{C}$ type is similar to that of N$\underline{C}$. Its apparent cell dimensions are $\underline{a}$ = 2A and $\underline{c}$ = M$\underline{C}$, where M varies continuous from 3.0 to 4.0. The N$\underline{A}$ type is characterized by pairs of supercell reflections of alternating intensity along the a* axis. The spacing between the pairs (Δh) changes continuous with increasing temperature from 1/20 to 1/45 of the repeat of the subcell reflections. The apparent cell dimensions are $\underline{a}$ = N$\underline{A}$ and $\underline{c}$ = 3$\underline{C}$, where N is 1/2 Δh and varies continuously from 40 to 90.

A final type, "metastable pyrrhotite," is encountered when iron-rich pyrrhotites are quenched into the two phase field of troilite + pyrrhotite. If exsolution does not occur, a metastable, strained structure results which is similar to the M$\underline{C}$ structure.

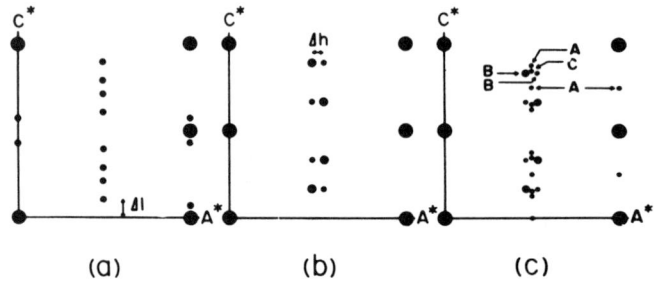

(a)   (b)   (c)

Fig. CS-5. Diagramatic sketches of precession x-ray photographs of the $\underline{a}^*\underline{c}^*$ planes. (a) the N$\underline{C}$ type; (b) the N$\underline{A}$ type; (c) the 4$\underline{C}$ (A), N$\underline{A}$ (B) and M$\underline{C}$ (C) types. The large solid circles represent subcell reflections for the 1$\underline{C}$ type. (From Nakazawa and Morimoto, 1971, *Mater. Res. Bull* $\underline{6}$, *p. 642).*

Monoclinic Pyrrhotite ("Fe$_7$S$_8$"). Ferromagnetic pyrrhotite with a monoclinic superlattice and composition centered about Fe$_7$S$_8$ has been known as a mineral for decades (Bystrom, 1945) and has been synthesized repeatedly in evacuated silica tubes. However, the earlier experimental studies by Clark (1966), Arnold (1969), Yund and Hall (1969) and Taylor (1970a), to name just a few, gave results that were conflicting and were plagued by demonstrable, extensive metastability. Such metastability in evacuated silica tube experiments is not surprising in view of Yund and Hall's (1970) observation that pyrrhotite oversaturated with respect to pyrite by as much as 0.18 at. % Fe did not nucleate pyrite after annealing for more than a year at 325°C. More recently Kissin (1974; see also Kissin and Scott, 1972; Scott and Kissin, 1973) and Rising (1973) have used the method of hydrothermal recrystallization to overcome these kinetic difficulties and establish phase relations for monoclinic pyrrhotite which are internally consistent and devoid of obvious metastabilities (Fig. CS-3). Kissin has synthesized single crystals of monoclinic pyrrhotite in the temperature range of 115-254°C and obtained a close control on the compositional limits of this mineral. As shown in Fig. CS-3, monoclinic pyrrhotite has variable stoichiometry. It is only nominally Fe$_7$S$_8$ and is separated from the N$\underline{A}$- and N$\underline{C}$-type pyrrhotites by narrow solvi. Kissin's upper stability limit of 254°C for monoclinic pyrrhotite going to N$\underline{A}$ pyrrhotite + pyrite has been reversed. It is in good agreement with the indirect determination at 251±3°C by Rising (1973) but is considerably lower than previous estimates in the range 292°-325°C from the aforementioned evacuated silica tube experiments.

"Anomalous" Pyrrhotite (Fe$_7$S$_8$). Clark (1966) has described a pyrrhotite containing 46.4 atomic % Fe in which the intensity of the (40$\bar{8}$) x-ray reflection is greater than (408), the reverse of the usual situation with normal monoclinic pyrrhotite (see Fig. S-11B). This "anomalous" pyrrhotite, as it is called, appears to be widespread in low-temperature, sedimentary environments. Unlike normal monoclinic pyrrhotite which is ferromagnetic, it is anti-ferrimagnetic as is hexagonal pyrrhotite. Its stability field has not been ascertained, although Taylor (1971) has shown that one way in which it can form is by oxidation of hexagonal pyrrhotite.

Gamma Iron Sulfide (Fe$_2$S$_3$). This phase has not been encountered in nature but has been precipitated from an aqueous sulfide solution at 60°C by Yamaguchi and Wada (1973). Electron diffraction patterns indicate that it has a spinel structure similar to greigite, the main difference in their patterns being the intensities of some of the reflections. Like greigite it is magnetic, and Yamaguchi and Wada suggest that γ-Fe$_2$S$_3$ bears the same relation to greigite as γ-Fe$_2$O$_3$ (maghemite) does to magnetite.

Nothing is known about the thermal stability of γ-Fe$_2$S$_3$. Yamaguchi and Wada claim, on the basis of published diffraction intensities, that natural greigite is a solid solution of Fe$_2$S$_3$ and Fe$_3$S$_4$.

Smythite (Fe$_9$S$_{11}$). Smythite was originally described by Erd et al (1957) as having a rhombohedral structure and a Fe$_3$S$_4$ composition which led Kullerud (1967) to believe that it was a polymorph of greigite. However, after an extensive re-examination of naturally occurring smythites, Taylor and Williams (1972) have redefined the mineral as having a pseudorhombohedral structure related to that of monoclinic pyrrhotite and a composition (Fe,Ni)$_9$S$_{11}$ (∼(Fe,Ni)$_{3.25}$S$_4$). This composition was also found by Nickel (1972) and Bennett et al (1972). Thus, smythite is not a polymorph of greigite but may be another ordered pyrrhotite of the Fe$_{n-1}$S$_n$ clan.

Not much is known about the stability of smythite. It occurs as exsolution lamallae in monoclinic pyrrhotite (Bennett et al, 1972) and in geodes which have fluid inclusion filling temperatures of 25 to 40°C, suggesting that it is a phase stable only at low temperatures. Kissin (1974) did not encounter smythite in his hydrothermal recrystallization experiments above 115°C. Rickard (1968) precipitated smythite from aqueous sulfide solutions but Taylor (1970b) was unable to duplicate the experiment. Taylor (1970a,b) found that smythite began to break down at 210°C in nonequilibrium experiments but concluded that it must be stable below 75°C because he was unable to synthesize it at higher temperatures. The fact that all natural smythites contain 0.4 to 7.5 weight % Ni suggested to Taylor and Williams (1972) that it might not be a phase in the pure Fe-S system.

Greigite (Fe_3S_4). The thiospinel, greigite is another problematical phase. It too is found in low temperature environments; for example, in ancient (Skinner et al, 1964) and modern (Dell, 1972) lake sediments, and thus far has resisted all attempts to determine its thermal stability. Skinner et al (1964) found that greigite broke down to pyrrhotite + sulfur (not pyrrhotite + pyrite as expected) between 238° and 282°C in experiments of short duration. However, they cautioned that because of slow reaction rates these temperature limits might not represent the true stability. Uda (1967) synthesized greigite at 180° and 190°C in disequilibrium assemblages as evidenced by coexisting mackinawite and sulfur instead of the intervening stable iron sulfides. According to Berner (1971), greigite is metastable relative to FeS and FeS_2 at 25°C, a viewpoint accepted by Scott and Kissin (1973) at higher temperatures, because Kissin (1974) did not encounter greigite in hydrothermal recrystallization experiments between 115° and 350°C.

Pyrite (FeS_2). Pyrite is stable to 743°C where it undergoes a peritectic breakdown to hexagonal 1C pyrrhotite + sulfur (Kullerud and Yoder, 1959).

The relationship between pyrite and its polymorph marcasite remains a puzzle despite its intensive investigation. Buerger (1934) concluded, on the basis of selected superior chemical analyses, that the minerals had slightly different compositions; marcasite was slightly deficient in sulfur (i.e. FeS_{2-x}) relative to pyrite which was nearly stoichiometric FeS_2. Pabst (1959) and Kullerud and Yoder (1959) have refuted Buerger's (1934) conclusion, claiming that the precision of sulfur analyses does not warrant it. Nevertheless, electrical measurements by Rakcheev and Chernyshev (1968) have demonstrated measureable, albeit small, nonstoichiometry in pyrite. Furthermore, Kullerud (1967) found that marcasite could be inverted to pyrite as low as 150°C in the presence of excess sulfur, but not below 400°C in the absence of sulfur; this suggests that sulfur in excess of the Fe:S ratio in marcasite is a necessary constituent of pyrite. Kullerud (1967) also describes experiments which showed that coexisting marcasite + pyrite could be synthesized up to 432°C (423°C at 2kb) in the presence of water but not in its absence. From this, he concluded H-S bonds might stabilize marcasite although an alternative explanation of his experiments is that the activity of sulfur was buffered within a range where marcasite is stable. More recently, Rising (1973) has investigated all aspects of the problem and concluded that because the inversion rate of marcasite to pyrite above 157°C is directly proportional to temperature and inversely proportional to grain size, marcasite is metastable relative to pyrite and pyrrhotite in this temperature range.

Below 157°C his data are not conclusive but do suggest that marcasite is metastable here as well. In support of this conclusion, Kissin (1974) did not encounter marcasite in his experiments at 115°C and above.

Pyrite-Pyrrhotite Solvus

The composition of hexagonal pyrrhotite on the solvus separating pyrrhotite from pyrite has been determined above 300°C by Arnold (1962) and Toulmin and Barton (1964) (Fig. CS-6). Below 300°C (Fig. CS-3) the appearance of pyrrhotite superstructures interrupts the smooth solvus and causes it to reverse its slope at 262°C.

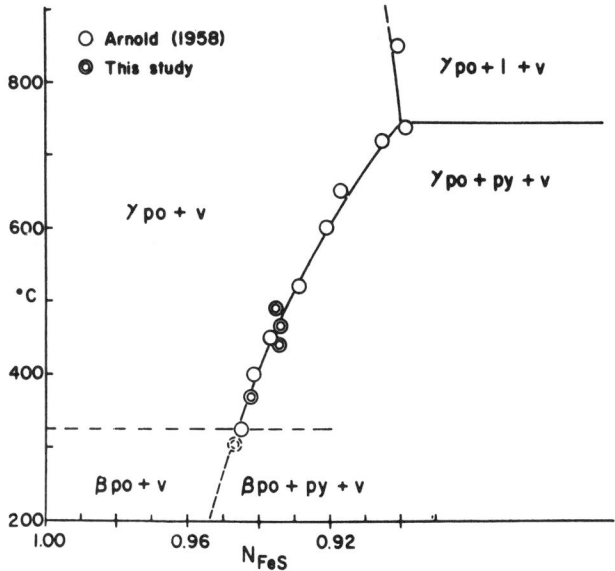

Fig. CS-6. Composition of hexagonal pyrrhotite (N_{FeS}) in equilibrium with pyrite above 300°C. N_{FeS} is the mole fraction of FeS in the system FeS-S_2 and is twice the atomic fraction of Fe in the system Fe-S. Dashed line is the beta-transition. γpo is 1C hexagonal pyrrhotite. βpo is a superstructure but its phase field is misplaced (see text). (From Toulmin and Barton, 1964, *Geochim. Cosmochim. Acta, 28, 642.*)

The variation in pyrrhotite compositions on the solvus above 300°C coupled with the common occurrence of pyrite + pyrrhotite in nature suggested to Arnold (1962) that the assemblage pyrite + hexagonal pyrrhotite would be a useful geothermometer provided the equilibrium composition of the pyrrhotite was preserved during subsequent cooling. However, we now know that natural pyrrhotite does not readily retain its high-temperature composition. In an isochemical system, pyrrhotites from the solvus at high temperatures will continue to exsolve pyrite to near 300°C, depending on its impurity content and degree of super-saturation with respect to pyrite (Yund and Hall, 1970). With further cooling below 254°C, monoclinic pyrrhotite is nucleated and the invariant (at fixed pressure) assemblage persists metastably to room temperature. In an open system, even when temperature is constant, hexagonal pyrrhotite will readily react with surrounding iron and sulfur-bearing solutions and change its composition whereas the pyrite might not adjust to new conditions. Also, simple oxidation will cause pyrrhotite to become more sulfur rich. For example, Genkin (1971) has demonstrated this process at Norilsk where hexagonal pyrrhotite has been oxidized to monoclinic pyrrhotite + magnetite along grain boundaries and cracks. And if these problems are not enough to contend with, Fig. CS-3 shows that between 340° and 254°C each pyrrhotite composition occurs at two different temperatures because of the reversal in the solvus at 262°C. In brief, applications of the pyrrhotite geothermometer should be regarded with suspicion, if not scepticism.

Thermochemistry

The main value of the Fe-S system is the understanding given to us by Toulmin and Barton (1964) and by Burgmann *et al* (1968) of its thermochemistry, particularly the relationships among pyrrhotite composition, temperature, pressure, and activities of FeS and S_2. Armed with this information, experimental mineralogists have been able to make important advances in other systems whose mineral assemblages are more applicable to interpreting natural processes than is Fe-S. As reliable data exist only above 300°C we will confine our remarks to these temperatures and avoid as much as possible the complexities introduced by the multitude of phases at lower temperatures.

FeS activity is given by the expression

$$a_{FeS} = \gamma C_{FeS}$$

where γ is the activity coefficient and C_{FeS} is the concentration of FeS in the phase under consideration. There are many ways of stating C_{FeS} (e.g. weight % Fe, atomic fraction Fe, atomic % Fe, gram formula % FeS, mole fraction FeS, mole % FeS, etc.) and each will have a different value of γ for a given a_{FeS}. We will follow Toulmin and Barton's (1964) convention and give compositions as mole fraction FeS (N_{FeS}) in the system FeS-S_2, i.e.

$$N_{FeS} = \frac{\text{moles FeS}}{\text{moles FeS} + \text{moles } S_2}$$

Mole % FeS is 100 N_{FeS} and is twice the atomic % Fe in the system Fe-S.

The chosen standard states (i.e. arbitrary conditions at which $a_i = 1.0$) are:

    (a) stoichiometric Fe, FeS and FeS$_2$ at the temperature and
           pressure of interest, and

(b) ideal S_2 gas at 1 atm. and the temperature of interest. In this convention $a_{S_2} = f_{S_2}$ ($f^{\circ}_{S_2} = 1$ atm).

Pyrite + Hexagonal Pyrrhotite Solvus. The reaction representing the sulfidation of pyrrhotite to pyrite may be written as $2FeS + S_2 \rightleftarrows FeS_2$ from which the sulfur fugacity (or activity) is given by

$$f_{S_2} = K_T^{-1} (a_{FeS})^{-2} \qquad \text{(cs-1)}$$

where $a_{FeS_2} = 1$ because pyrite is in its standard state. This equilibrium is best represented on a plot of log f_{S_2} vs $1000/T,K$ as described in the previous chapter.

Three recent determinations of the pyrite-pyrrhotite solvus are shown in Figure CS-7. Toulmin and Barton (1964) used the electrum tarnish method; Giletti _et al_ (1968) determined f_{S_2} from the concentration of ^{35}S tracer in the equilibrium vapor; and Scott and Barnes (1971) computed their curve from the composition of sphalerite coexisting with pyrite + pyrrhotite as described below in the section on Zn-Fe-S. A fourth determination by Schneeberg (1973) using an electrochemical cell is coincident with that of Scott and Barnes (1971). Although the agreement is quite good among the curves in Fig. CS-7, their small differences are significant so there are reasons for choosing one. Scott and Barnes' (1971) curve is preferred because it is consistent with Toulmin and Barton's _data_ (but not the curve they choose to draw), it is internally consistent with data on the Zn-Fe-S system, and it has been reproduced by the electrochemical technique which is a superior method for measuring sulfur fugacities.

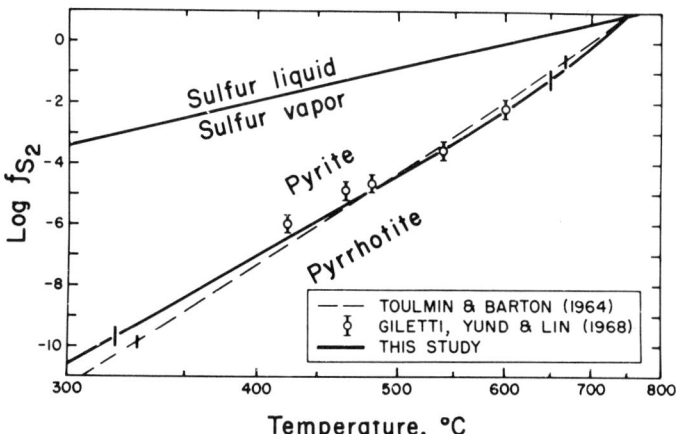

Fig. CS-7. Three determinations of sulfur fugacities on the pyrite + pyrrhotite solvus. (From Scott and Barnes, 1971, _Econ. Geol. 66, 660_).

The pyrite-pyrrhotite curve is a straight line below 500°C indicating, from equation (cs-1), that a_{FeS} is constant, at a value of 0.48 as will be shown later. Above 500°C, the pronounced curvature in Fig. CS-7 indicates that a_{FeS} is not constant but is decreasing with increasing temperature.

Pyrrhotite Field. Using the electrum tarnish method, Toulmin and Barton (1964) determined the variation in the composition of hexagonal pyrrhotite in the condensed system as a function of sulfur fugacity and temperature. Their results in Fig. CS-8 show a wide range in f_{S_2} across the pyrrhotite field between the iron + troilite and pyrite + pyrrhotite boundaries. Because it is a solid solution and its equilibrium f_{S_2}'s lie within the range of a large number of sulfidation reactions, pyrrhotite is an excellent "sliding-scale" buffer whose principle of operation was described in the previous chapter.

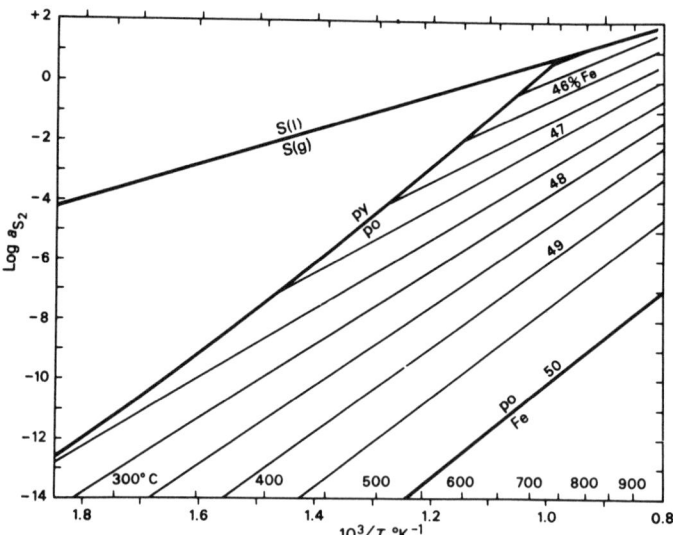

Fig. CS-8. Composition of pyrrhotite in atomic % Fe as a function of S_2 activity and temperature. Uncertainty in pyrrhotite isopleth is approximately ±0.35 in log a_{S_2}. (From Barton and Skinner, 1967, *Geochemistry of Hydrothermal Ore Deposits,* Fig. 7.6, p. 293).

It is a relatively easy matter to calculate FeS activities from the measurements of f_{S_2} for pyrrhotite by means of the Gibbs-Duhem relation

$$\sum_i n_i \, d \, \mu_i = 0 \ (T, P \ \text{constant}) \tag{cs-2}$$

In this expression n_i is the number of moles of component i and μ_i is its chemical potential,

$$\mu_i = \mu_i^o + 2.303 \, RT \log a_i$$

For one gfw of any phase in the system FeS-S$_2$, equation (cs-2) becomes

$$N_{FeS} \, d \log a_{FeS} + (1-N_{FeS})d \log a_{S_2} = 0 \ (T,P \text{ constant}) \hspace{2cm} (cs-3)$$

where

$$d \mu_i = d \log a_i \ (\mu_i^o \text{ is constant}) \text{ and } N_{S_2} = 1-N_{FeS}.$$

The integrated form of equation (cs-3) is evaluated for selected values of N_{FeS} and a_{S_2} from Fig. CS-8 giving FeS activities in Fig. CS-9. There are several features of Fig. CS-9 worth noting:

1. The contours of a_{FeS} are linear whereas isopleths of N_{FeS} in Fig. CS-8 are curved. Therefore, a_{FeS} varies with temperature for a given pyrrhotite composition.

2. a_{FeS} shows a wide variation with pyrrhotite composition and is much less than 1.0 at the pyrite + pyrrhotite solvus.

3. At a given temperature the variation in a_{FeS} with f_{S_2}, as indicated by the spacing of the curves, is greater at high f_{S_2} than at low f_{S_2}, following the trend of the isopleths in Fig. CS-8.

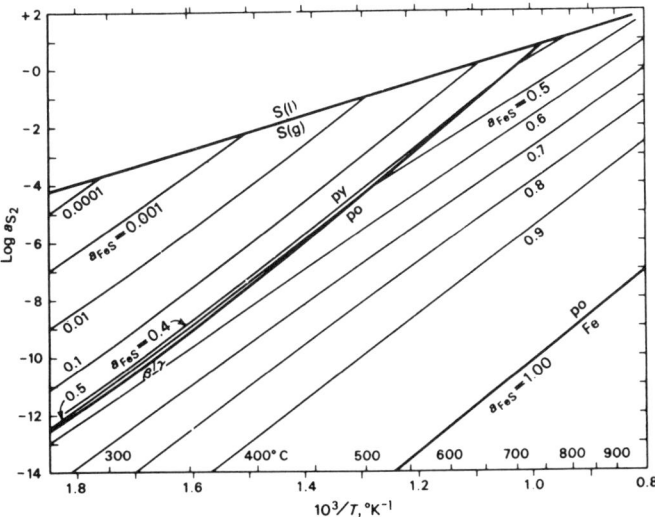

Fig. CS-9. FeS activities calculated from equation (cs-3). (From Barton and Skinner, 1967, *Geochemistry of Hydrothermal Ore Deposits*, Fig. 7.7, p. 294).

CS-35

4. Above 500°C, a_{FeS} changes rapidly with temperature along the pyrite + pyrrhotite solvus giving rise, as discussed earlier, to the curvature of the solvus. Although Fig. CS-9 shows a slight rise in a_{FeS} with declining temperature below 500°C along the solvus, the preferred solvus curve of Scott and Barnes (1971) is coincident with the $a_{FeS} = 0.48$ contour.

5. The activity coefficient, γ ($= a_{FeS}/N_{FeS}$), is a function of temperature and composition. It varies between 0.44 and 1.0 indicating that the FeS-S$_2$ solid solution is very nonideal.

Pyrite Field. Equation (cs-1) shows that a_{FeS} varies rapidly as the square root of S$_2$ activity in the pyrite stability field where $a_{FeS_2} = 1$ (Fig. CS-9).

Free Energy and Heat of Formation of Pyrite. The free energy of formation of pyrite in the reaction Fe (α) + S$_2$ (g, 1 atm) $\vec{\leftarrow}$ FeS$_2$ (py) can be calculated using the expression from Toulmin and Barton (1964, p. 662)

$$\Delta G_f^\circ = RT \ln \left[(f_{S_2}' f_{S_2}'')^{1/2} a_{FeS} \right] \qquad \text{(cs-4)}$$

where ' refers to the Fe + FeS boundary and " to the pyrite + pyrrhotite solvus. Equation (cs-4) has been evaluated using a_{FeS} from Fig. CS-9 above 500°C, a_{FeS} = 0.48 between 300° and 500°C, f_{S_2}' from Fig. CS-8 and f_{S_2}'' from the curve of Scott and Barnes (1971) in Fig. CS-7. The data are represented by a linear equation.

$$\Delta G_f^\circ = 0.04613T - 70.766 \pm 0.80 \text{ kcal/mole.}$$

At 298°K and 1 atm, the standard free energy of formation of pyrite from α-Fe and S$_2$ (gas) is calculated to be -57.013 kcal/mole or -37.893 kcal/mole from α-Fe and rhombic sulfur. The standard heat of formation at 298°K relative to α-Fe and rhombic sulfur obtained by adding $298 \Delta S_{298}^\circ$ (Robie and Waldbaum, 1968) to ΔG_{f298}° is -40.596 kcal/mole. The best value accepted by Mills (1974) from heat capacity measurements is -41.0 ± 2 kcal/mole.

Effect of Pressure on Phase Relations

Because most ore-forming processes take place at pressures greater than 1 atm, we are interested in what effect a high confining pressure will have on phase relationships and their underlying thermodynamic variables. Of particular concern are a_{FeS} in pyrrhotite and the location of the pyrite + pyrrhotite solvus.

FeS activity in pyrrhotite. The pressure coefficient of activity at a fixed temperature is given by

$$\left[\frac{\partial \ln a}{\partial P} \right]_T = \frac{\Delta V}{RT}.$$

For pyrrhotite this becomes

$$\left[\frac{\partial \ln a_{FeS}}{\partial P} \right]_{T, N_{FeS}^{po}} = \frac{\bar{V}_{FeS}^{po} - V^{tr}}{RT} \qquad \text{(cs-5)}$$

where $\bar{V}^{po}_{FeS}$ is the partial molar volume of FeS in pyrrhotite, and V^{tr} is the volume of one mole (molar volume) of troilite which is the standard state for FeS. A partial molar volume is obtained by drawing a tangent to the curve of molar volume *vs* composition at the composition of interest and extrapolating that tangent to 100% of each component. The way in which this is done is explained more fully in the section on the Fe-Zn-S system elsewhere in this chapter. In the case of pyrrhotites, the data have a lot of scatter, but a plot of molar volume of pyrrhotite *vs* composition can be approximated by a straight line (Toulmin and Barton, 1964; Kissin, 1974). A tangent to this line is, therefore, coincident with the line, and the partial molar volume for any pyrrhotite composition is the same as the molar volume of troilite. In other words, the volume term in equation (cs-5) is zero and pressure has no effect on a_{FeS} in pyrrhotite of a given composition.

Pyrite + Pyrrhotite Solvus. The best way to approach the pressure effect on the solvus is to follow Toulmin and Barton (1964, p. 666) and calculate its effect on a_{FeS} under the restriction that pyrrhotite is in equilibrium with pyrite. The treatment given here is not complete but may help to amplify some of the expressions derived by Toulmin and Barton.

All pertinent variables are related at constant temperature by the expression $a_{FeS} = f(N^{po}_{FeS}, P)_T$. Differentiating ($a = a_{FeS}$; $N = N^{po}_{FeS}$):

$$(d \ln a)_T = \left[\frac{\partial \ln a}{\partial P}\right]_{T,N} dP + \left[\frac{\partial \ln a}{\partial N}\right]_{P,T} dN.$$

Dividing by dP:

$$\left[\frac{d \ln a}{dP}\right]_T = \left[\frac{\partial \ln a}{\partial P}\right]_{T,N} + \left[\frac{\partial \ln a}{\partial N}\right]_{P,T} \left[\frac{dN}{dP}\right]_T \qquad (cs-6)$$

which is equation (23) of Toulmin and Barton. As already shown by equation (cs-5) the first term on the right side of equation (cs-6) is zero. The second term is not zero, however:

1. Fig. CS-8 and CS-9 show that $\left[\frac{\partial \ln a}{\partial N}\right]_{P,T}$ is positive.

2. For a given pyrrhotite composition,

$$\left[\frac{\partial \ln a_{S_2}}{\partial P}\right]_{T,N} = \frac{\bar{V}_{S_2}}{RT} = \text{positive}.$$

In other words, pressure raises the pyrrhotite isopleths in Fig. CS-8 to higher sulfur activities. The pyrite-pyrrhotite solvus also moves to higher a_{S_2} with pressure (Fig. CS-10) but not as much as the pyrrhotite isopleths. The net result is a more Fe-rich pyrrhotite on the high-pressure solvus at each temperature.

As a consequence of both the above terms being positive, a_{FeS} along the pyrite-pyrrhotite solvus rises with pressure and can be calculated from equation (24) of Toulmin and Barton:

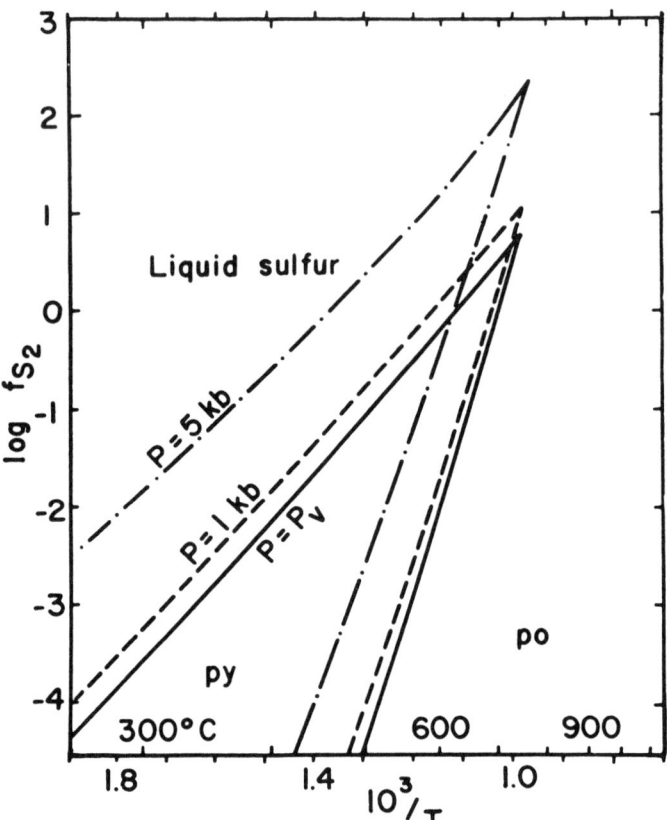

Fig. CS-10. Calculated pressure effect on the sulfur-condensation, and pyrite-pyrrhotite curves. Not corrected for small effects of thermal expansion and compression of phases. P_V = vapor pressure of the system. (From Toulmin and Barton, 1964, Geochim. Cosmochim. Acta 28, 668).

$$\left[\frac{d\ln a_{FeS}^{po}}{dP}\right]_T = \left[\frac{N_{S_2}^{po} (2\bar{v}_{FeS}^{po} + \bar{v}_{S_2}^{po} - 3v^{py})}{(RT)[3N_{FeS}^{po} - 2]}\right]_T \qquad (cs-7)$$

It is convenient to rearrange equation (cs-7) in order to eliminate the logarithm by taking advantage of the differential expression

$$\frac{d\ln u}{dx} = \frac{1}{u}\frac{du}{dx}$$

Equation (cs-7) becomes (Scott, 1973, equation 2):

$$\left[\frac{da_{FeS}}{dP}\right]_T = a_{FeS}\left[\frac{N_{S_2}^{po} (\bar{v}_{FeS}^{po} + \bar{v}_{S_2}^{po} - 3v^{py})}{(RT)[3N_{FeS}^{po} - 2]}\right]_T \qquad (cs-8)$$

The volume data needed to evaluate equation (cs-8) are in Table CS-2. These must be corrected for thermal expansion (Table CS-3) and compression (Table CS-4).

Table CS-2. Volumes at 25°C and 1 Bar. (One mole is 2 gram-atoms)
From Toulmin and Barton, 1964; Scott, 1973

Mineral	Molar Volume cc	Partial Molar volume, cc	
	V	$\bar{V}_{S_2}$	$\bar{V}_{FeS}$
Pyrite (py)	15.962	--	--
Troilite (tr)	18.20	--	--
Pyrrhotite (po)	---	16.420	18.20

Table CS-3. Thermal Expansions. From Scott, 1973.

Mineral	$\frac{V-Vo}{Vo} \times 100$ Temperature, °C			
	400	500	600	700
Pyrite	1.29	1.69	2.09	2.49
Troilite	4.30	4.94	5.58	6.22
Sphalerite	0.898	1.16	1.44	1.72

Table CS-4. Compressions. (P in kilobars). From Scott, 1973

Mineral	$\frac{V-V_o}{V_o} \times 100$ Compression, %
Pyrite	-0.069P
Troilite	-0.12P
Sphalerite	-0.13P

The easiest and a sufficiently precise way to solve equation (cs-8) is to integrate it in increments of 1 kb in the form

$$a''_{FeS} - a'_{FeS} = a'_{FeS} \left[\frac{(\bar{V}^{po}_{FeS} + \bar{V}^{po}_{S_2} - 3V^{py})}{RT (3N^{po}_{FeS} - 2)} \right]'_T (P'' - P')$$

where ' refers to the lower pressure condition and " to the higher pressure.
The pressure effects calculated by evaluating equation (cs-8) are quite small;

$$\frac{d\ a_{FeS}}{dP}$$ on the pyrite-pyrrhotite solvus ranges from +0.0041 kb^{-1} at 400°C, 1 bar to +0.0045 at 700°C, 1 bar and +0.0042 at 700°C, 5 kb. These values are about one-half as large as those given by Toulmin and Barton (1964, p. 667) because they did not correct the volume terms in equation (cs-8) for thermal expansion and compression and they made an error in transposing from base e to base 10 logarithms. The corresponding values of pyrrhotite compositions along the pyrite-pyrrhotite solvus calculated from equation (cs-8) are in Fig. CS-11. This figure shows, as predicted above, that pyrrhotite becomes more iron-rich on the solvus at high pressures but the shift is too small to be measured experimentally, at least at 2 kb (Arnold, 1962).

In summary, increasing pressure on the pyrite-pyrrhotite solvus causes a small increase in both N_{FeS}^{po} and a_{FeS}.

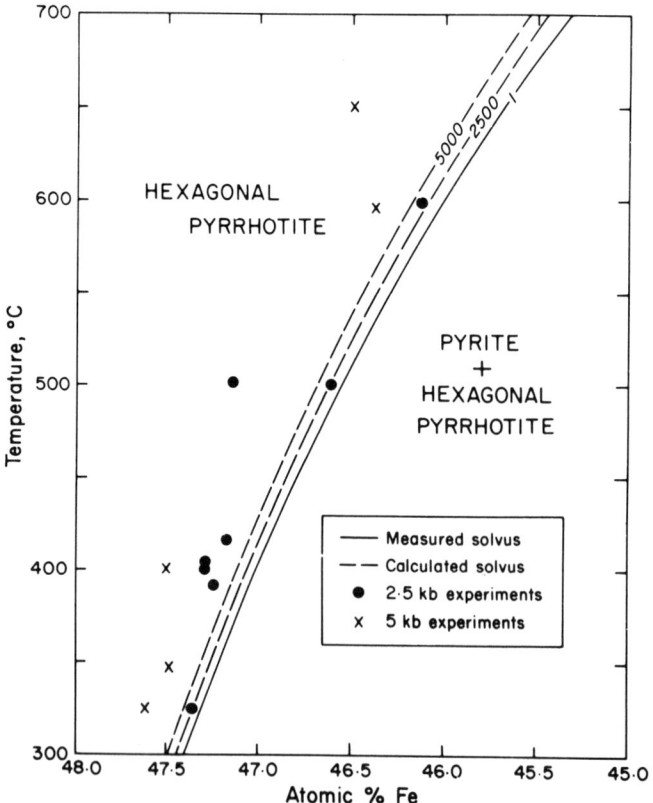

Fig. CS-11. T-X diagram of the pyrite + hexagonal pyrrhotite solvus in equilibrium with vapor (Arnold, 1962; Toulmin and Barton, 1964) and calculated at 2500 and 5000 bars from equation (cs-8). Experimental points were unsuccessful attempts to measure the pressure effect. (From Scott, 1973, *Econ. Geol.* *68*, 470).

In addition to those encountered in the Fe-S system, the only new phases in the Fe-Zn-S system are the wurtzite polytypes and sphalerite. The relation of wurtzite to sphalerite is discussed below in the section on "Stoichiometry of Sulfides;" only sphalerite will be considered here.

Sphalerite has a cubic close packed structure $(F\bar{4}3m)$ in which every other tetrahedral site is occupied by zinc. The remaining tetrahedral sites and the octahedral sites are empty. Iron (also cadmium and manganese) enters the structure by replacing zinc and results in a large increase in the cell volume as shown in Fig. S-9 of the previous chapter. Mössbauer spectra of sphalerites containing very small amounts of iron do not show quadrupole splitting, thus indicating no distortion of the lattice. However, when the sphalerite contains more than one or two atomic percent Fe, iron-iron interaction produces a large quadrupole splitting (Scott, 1971; Gerard *et al*, 1971). Distortion of the lattice is also evident from the nonlinearity of the cell edge with composition at very high Fe contents (Fig. S-9). Manning (1967) claims on the basis of infrared absorption spectra that as much as 10% of the iron in sphalerite is octahedral Fe^{3+}. However, Mössbauer studies (Scott, 1971) indicate that all of the iron is tetrahedral Fe^{2+}.

Because sphalerite is one of the more refractory sulfides and displays a wide range of Fe contents as a function of conditions of formation, it promises to be a useful geochemical tool for deciphering environments of sulfide deposition and deformation. As early as 1953, Kullerud proposed that the Fe content of sphalerite coexisting with pyrrhotite could be used as a geothermometer. An underlying assumption, based on information available at the time, was that FeS activity of pyrrhotite did not depart appreciably from unity. We now know (Toulmin and Barton, 1964) that this assumption is incorrect, and the original concept of a sphalerite geothermometer is invalidated. Nevertheless, their study was the key to complete understanding of phase relations in the Fe-Zn-S system which since have proven valuable to the ore mineralogist. In this section we will examine these pertinent phase relations and their applications above 300°C and then discuss phase relations at lower temperatures.

Schematic isothermal sections illustrating the ZnS-FeS-S portion of the system are in Figure CS-12. These show the extensive solid solution of FeS in ZnS and its dependence on the Fe sulfide(s) coexisting with sphalerite. Because the solid solution is binary, or nearly so, it can be conveniently represented by a T-X projection from sulfur onto the FeS-ZnS join. Such projections are shown in Figures CS-13 and CS-14 for the FeS-rich and ZnS-rich portions of the system, respectively. In these isobaric projections the two-phase fields of Figure CS-12 appear as areas and the three-phase triangles as univariant lines tracing out the equilibrium sphalerite compositions. The breakdown of pyrite to pyrrhotite + sulfur is represented in Figure CS-14 by the point at 743°C from which emanate

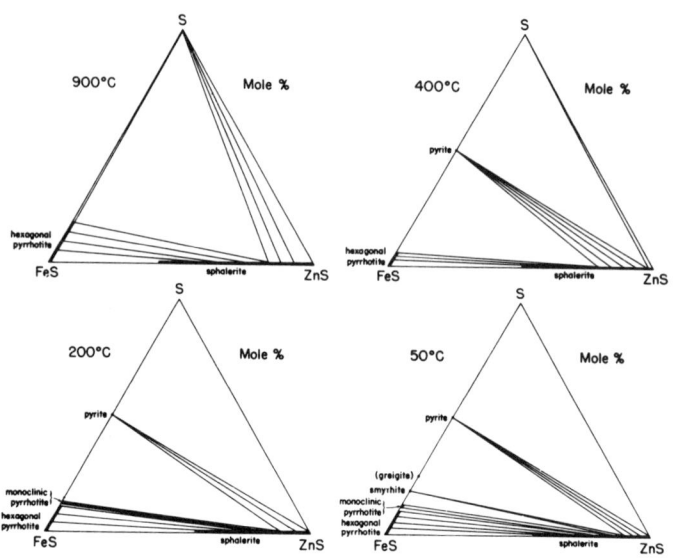

Fig. CS-12. Schematic isothermal sections of the condensed system FeS-ZnS-S.

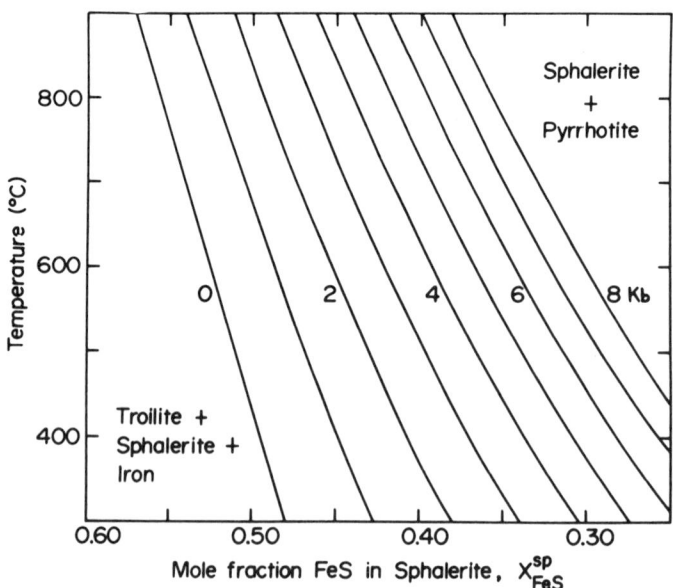

Fig. CS-13. T-X projection of the sphalerite + troilite + iron solvus at the vapor pressure of the system (0 kb) from Barton and Toulmin (1966) and calculated at higher pressures. (From Schwarz, Scott, and Kissin, in preparation).

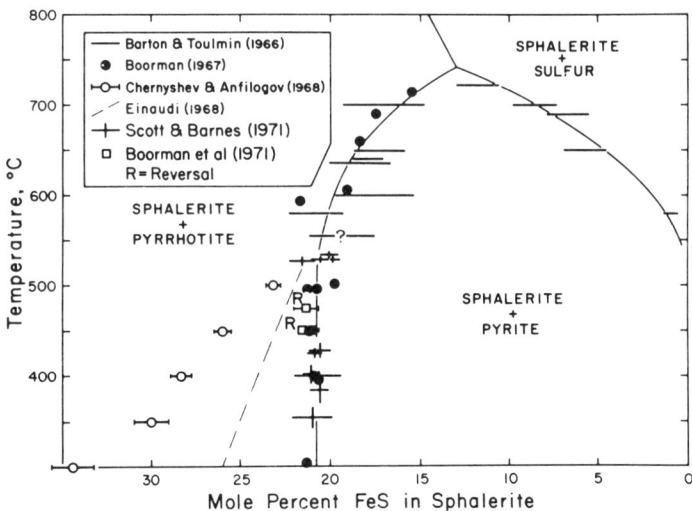

Fig. CS-14. T-X projection of a portion of the system FeS-ZnS-S at 1 bar. (From Scott and Barnes, 1971, *Econ. Geol.* *66*, 659).

the three univariant curves: sphalerite + pyrrhotite + sulfur extending to higher temperatures, and sphalerite + pyrrhotite + pyrite and sphalerite + pyrite + sulfur extending to lower temperatures. Another convenient way of presenting the equilibria, taking cognisance of the variation of FeS activity in the system, is a plot of log f_{S_2} vs temperature (Fig. CS-15). In this representation, the two-phase fields of Figure CS-12 appear as areas in which sphalerite composition can be contoured and the three-phase triangles appear as sloping univariant lines.

FeS Activity and Sphalerite Composition

One of Barton and Toulmin's (1966) more important discoveries in their landmark study of the Fe-Zn-S system is shown in Figure CS-16. Over a wide range of temperatures the iron content of sphalerite is a function only of a_{FeS} except, as noted below, for very Fe-rich compositions. The relationship is not upset by other components in the system such as MnS, CdS, CoS, ZnSe, ZnO and probably NiS. This is demonstrated by the fact that, as described in the previous chapter, the cell volume of sphalerite is a linear function of its FeS, MnS, CdS, ZnSe and ZnO content which requires that the molar volume and hence activity coefficient of FeS in sphalerite is not affected by these other components.

In view of Figure CS-16 and the refractory nature of sphalerite, the FeS content of sphalerite is a reliable indication of a_{FeS} provided ZnS is not soluble in the phases under study. Consider these exemplary applications:

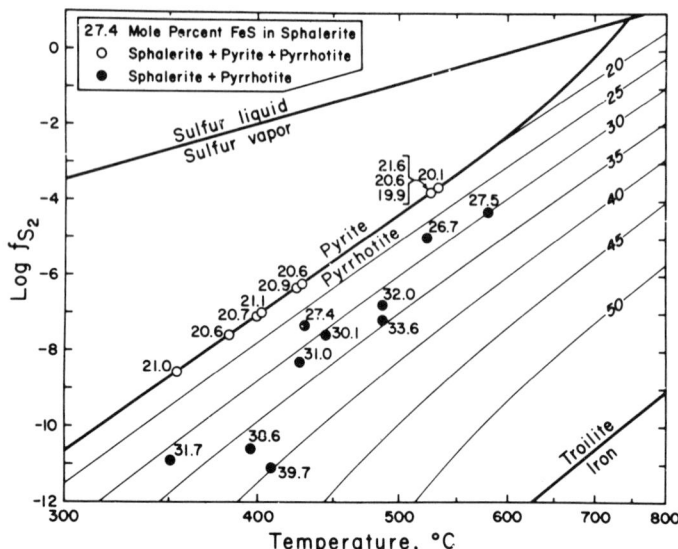

Fig. CS-15. Composition of sphalerite in equilibrium with pyrrhotite (light contours, in mole % FeS) and with pyrite + pyrrhotite (open circles). The contours are from equation (cs-12). (From Scott and Barnes, 1971, *Econ. Geol.* *66*, *658*).

1. In an investigation of the high-temperature portion of the Fe-S system shown in Figure CS-2, Arnold (1971) experienced considerable difficulty in quenching pyrrhotite compositions. One remedy to the problem was to equilibrate sphalerite with the Fe sulfides, analyze its composition, obtain a_{FeS} from Figure CS-16, and determine the corresponding pyrrhotite composition from the data of Figures CS-8 and CS-9 relating N_{FeS}^{po}, a_{FeS}, and temperature. The results of three such experiments, shown by the closed circles in Figure CS-2, agreed with the appearance-of-phase experiments within 0.1 wt. % Fe.

2. Within the Fe-Ni-S system, the $(Fe,Ni)_{1-x}S$ (mss) solid solution extends continuously at high temperatures from the FeS-NiS join to more sulfur-rich compositions. Scott *et al* (1972, 1974) have measured FeS activities for mss at 930°C by equilibrating it with sphalerite which, at this temperature, is not soluble in mss. The sphalerite products contained progressively decreasing amounts of FeS, indicating (from Fig. CS-16), decreasing a_{FeS} in mss passing from stoichiometric FeS (a_{FeS} = 1) to NiS (a_{FeS} = 0) and from the FeS-NiS join to more sulfur-rich $(Fe,Ni)_{1-x}S$ compositions. From this information, Scott *et al* were able to define a regular solution model for mss.

Sphalerite + Troilite + Iron

Sphalerite compositions buffered by troilite and iron have been measured in the condensed system by Barton and Toulmin (1966). A least-squares line fit to their

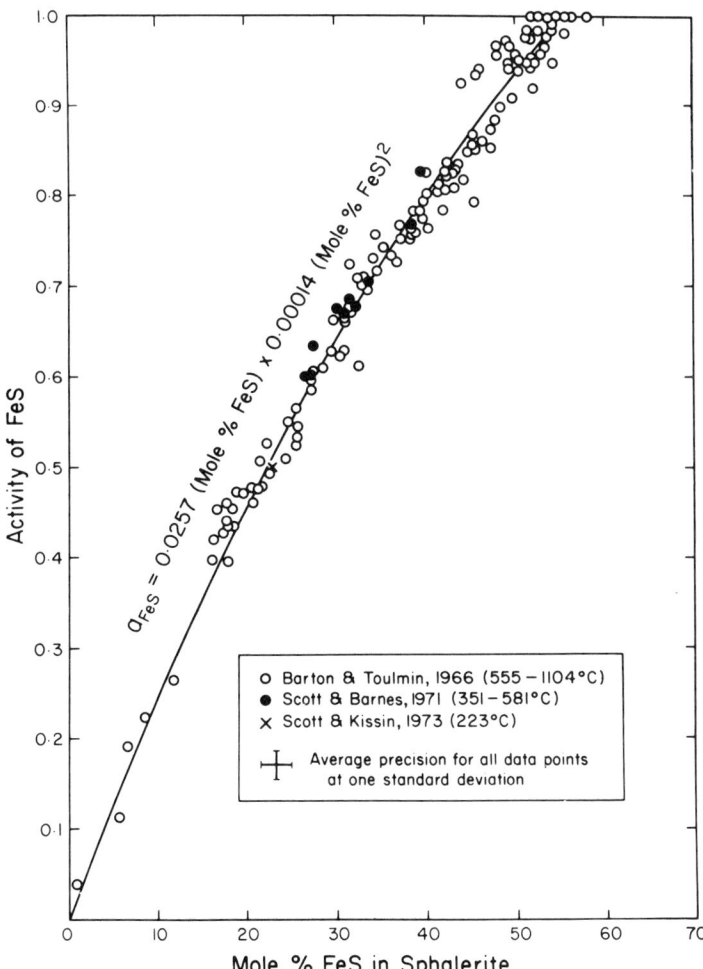

Fig. CS-16. Relation between activity of FeS (relative to troilite) and the composition of sphalerite.

data and extrapolated below 580°C is shown in Figure CS-13. The buffer assemblage fixes the FeS activity at unity for all temperatures. Therefore, the negative slope of the 0 kb line in Figure CS-13 requires that the activity coefficient of FeS ($\gamma = a_{FeS}/N_{FeS}^{sp}$) in sphalerite decreases with increasing temperature, at least for these very Fe-rich sphalerite compositions, and produces most of the scatter in the data of Figure CS-16 near $a_{FeS} = 1$.

The effect of pressure on the sphalerite + troilite solvus is large. It has been calculated by Barton and Toulmin (1966), and their calculation has been refined by Schwarcz, Scott, and Kissin (in preparation) taking into account corrections for thermal expansions and compressibility of the phases. The mole fraction of FeS in sphalerite (N_{FeS}^{sp}) is related to the activity of FeS (a_{FeS}) by the expression $N_{FeS}^{sp} = a_{FeS}/\gamma_{FeS}^{sp}$ where γ_{FeS}^{sp} is the activity coefficient. Differentiating,

$$\frac{dN_{FeS}^{sp}}{dP} = \left[\gamma_{FeS}^{sp} \left[\frac{da_{FeS}}{dP} \right] - (a_{FeS}) \frac{d\gamma_{FeS}^{sp}}{dP} \right] \Big/ (\gamma_{FeS}^{sp})^2 .$$

Rearranging, and noting that $d\gamma/\gamma = d\ln\gamma$

$$\frac{dN_{FeS}^{sp}}{dP} = \frac{1}{\gamma_{FeS}^{sp}} \left[\frac{da_{FeS}}{dP} - (a_{FeS}) \frac{d\ln\gamma_{FeS}^{sp}}{dP} \right] \qquad (cs-9)$$

which is equation (1) of Barton and Toulmin (1966). Furthermore, Figure S-9 of the previous chapter shows that the cell edge (or cell volume) of sphalerite can be represented by a straight line over large compositional intervals so the partial molar volume of FeS in sphalerite ($\bar{V}_{FeS}^{sp}$) is constant over these intervals and the following relation is valid:

$$\frac{d\ln\gamma_{FeS}^{sp}}{dP} = \left[\frac{\partial\ln\gamma_{FeS}^{sp}}{\partial P} \right]_{T,N_{FeS}^{sp}} = \frac{\bar{V}_{FeS}^{sp} - V^{tr}}{RT} \qquad (cs-10)$$

where V^{tr} is the molar volume of troilite. For troilite: $a_{FeS} = 1$, $\gamma_{FeS}^{sp} = 1/N_{FeS}$, and, as shown earlier, $da_{FeS}/dP = 0$. Therefore, substituting these relations and equation (cs-10) into equation (cs-9) we obtain

$$\frac{dN_{FeS}^{sp}}{dP} = -\left[\frac{(\bar{V}_{FeS}^{sp} - V^{tr}) \ N_{FeS}^{sp}}{RT} \right] \qquad (cs-11)$$

which is equation (3) of Barton and Toulmin. The isobars in Figure CS-13 were calculated in increments of 1 kb using the volume data in Tables CS-2, 3, 4, and 5 and the integral form of equation (cs-11):

$$N'' - N' = - \frac{(V - V)''' \ N'}{RT} (P'' - P')$$

where ' refers to conditions at low pressure, " to conditions at high pressure, and "' to conditions at $(P'' + P')/2$. Neuhaus and Cemic (1970) have measured experimentally a much larger pressure effect on the troilite + sphalerite solvus than is shown in Figure CS-13. The reason for the disagreement is uncertain, but it is suspected that they did not properly buffer the troilite in their experiments.

Applications of Figure CS-13 are limited to meteorites. Some meteorites contain small sphalerite blebs within nearly pure FeS (troilite) nodules surrounded by kamacite (Fe,Ni). Schwarcz et al (in preparation) have used the composition of these sphalerites together with Figure CS-13 as a paleomanometer to estimate the size of the parent asteroids which spawned the meteorites (see Schwarcz and Scott, 1974).

Sphalerite + Pyrrhotite

Isopleths of sphalerite composition coexisting with hexagonal pyrrhotite are shown in Figure CS-15 as a function of log f_{S_2} and temperature. They were located using the data obtained by Barton and Toulmin (1966) from evacuated silica tube experiments above 580°C and by Scott and Barnes (1971) from hydrothermal recrystallization experiments at lower temperatures (closed circles). In both studies, f_{S_2} was determined from the composition of the coexisting pyrrhotite (Fig. CS-8). A least-squares regression (trend) surface fit to the experimental data has the equation (Scott and Barnes, 1971):

Mole % FeS = 72.26695 - 15900.5/T + 0.01448 log f_{S_2} - 0.38918

x $(10^8/T^2)$ - (7205.5/T) log f_{S_2} - 0.34486 (log f_{S_2})^2 [e.s.d. = 1.7]   (cs-12)

The effect of pressure on the location of the isopleths in Figure CS-15 is complicated and, except for the pyrite + pyrrhotite and troilite + iron boundaries, has not yet been rigorously evaluated. Qualitatively, the pyrrhotite compositional isopleths in Figure CS-8 and a_{FeS} isopleths in Figure CS-9 shift towards higher f_{S_2} with high confining pressure as discussed previously, which would tend to make sphalerite more Fe-rich at a given log f_{S_2} and T. However, this is more than offset by the rise in activity coefficients of FeS in sphalerite which makes sphalerite less Fe-rich at a given a_{FeS}. For example, at 500°C γ_{FeS}^{sp} increases from 2.3 at 1 bar to 3.6 at 5kb (personal communication from P. Toulmin, III, 1973). Consequently, a sphalerite containing 30 mole % FeS and coexisting with a particular pyrrhotite composition at 500°C and 1 bar would contain only 19.2 mole % FeS coexisting with the same pyrrhotite at 500°C and 5kb. Therefore, without additional information on f_{S_2} and pressure during deposition, the isopleths in Figure CS-15 are of little use in geothermometry. The effect of pressure can be calculated, but the coexisting pyrrhotite usually will have suffered post-depositional changes which will make it an unreliable indicator of f_{S_2}.

Sphalerite + Pyrrhotite + Pyrite

Addition of pyrite to the preceding assemblage buffers a_{FeS} and overcomes the problems of having to preserve pyrrhotite compositions from high temperature in order to use sphalerite as a geochemical tool. For this reason, and because the three-phase assemblage is common in nature, there has been a concentrated effort to determine its T-X relationships over a wide temperature range, initially with a view to salvaging Kullerud's (1953) sphalerite geothermometer. The results of six different investigations, five of which extended below 550°C, are shown in Figure CS-14. Independent studies by Boorman (1967) and Boorman et al (1971) using a salt flux technique and by Scott and Barnes (1971) using hydrothermal recrystallization are in excellent agreement that the solvus remains vertical below 550°C at 20.7+0.6 mole % FeS. Two sets of results disagree with this conclusion but can be discounted. Einaudi's (1968) calculation of the solvus, shown by the dashed line in Figure CS-14 was correct in concept but is rejected because he used thermodynamic data that were slightly in error. The results of Chernyshev and Anfilogov (1968) are also rejected because Chernyshev (in Boorman *et al*, 1971) later found their sphalerite products to be exceedingly heterogeneous. There is no longer any doubt that the bulk composition of sphalerite in the three-phase assemblage remains independent of temperature throughout a major portion of the geologically-interesting temperature range and thus is <u>not</u> a geothermometer.

Iron-Rich Patches. In the course of analyzing by electron microprobe the sphalerite crystals grown in their hydrothermal recrystallization experiments, Scott and Barnes (1971) found small, scattered, Fe-rich patches which were richer by as much as 8.7 mole % FeS than the matrix of the crystals. Several patches could often be found in random sections through crystals as discrete islands 30 to 150 μm across and, surprisingly, all have the same composition. Typically, at the margin of a patch there is a stepwise increase in FeS content and a concomitant decrease in ZnS compared to the homogeneous matrix of the crystal (Fig. CS-17). The measured differences in composition, Δ, between Fe-rich patches and the matrix of sphalerites coexisting with pyrite and pyrrhotite define a simple curve which is dependent on temperature (Fig. CS-18). Values of Δ decrease with rising temperature until, at 525°C and above, no patches were found.

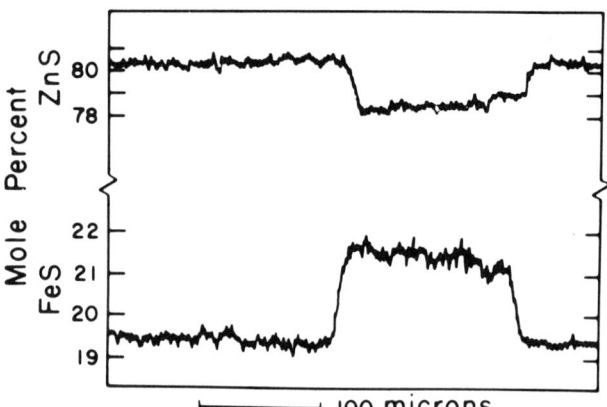

Fig. CS-17. Electron microprobe tracing showing Fe and Zn contents across an Fe-rich patch in a synthetic sphalerite crystal. (From Scott and Barnes, 1971, *Econ. Geol. 66, 665).*

The consistent compositional differences between the patches and matrix within a single crystal and the sharp boundaries between patches and matrix suggest that the patches can be treated thermodynamically as separate phases which grew in metastable equilibrium with the matrix of the crystal. Equilibrium between the growing surface of the crystal and the aqueous solution is very rapid whereas

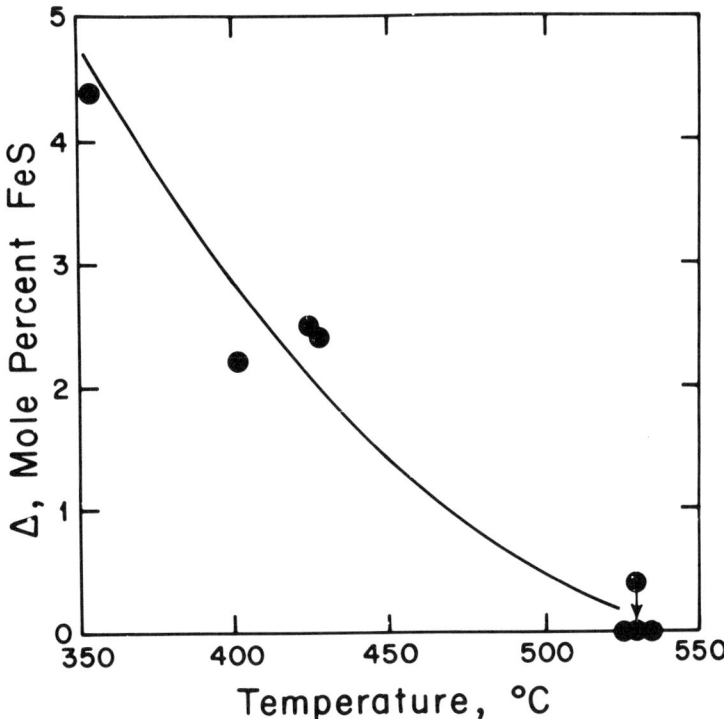

Fig. CS-18 . A polybaric curve relating compositional differences (Δ) between iron-rich patches and matrix in sphalerite crystals coexisting with pyrite and pyrrhotite. (From Scott and Barnes, 1971, *Econ. Geol.* 66, 665).

solid-state diffusion of Fe between patch and matrix is not. Consequently, the surfaces of the matrix and metastable patch are both very probably in equilibrium with the hydrothermal solution but bulk solid equilibrium is not maintained. The fact that it is the patches which are metastable and not the matrix was indicated, but not proven, by Scott and Barnes' calculations of free energy changes in a patch-forming reaction. Scott and Barnes (1971) have suggested two possible mechanisms which may be responsible for the formation of the iron-rich patches:

(1) The patches represent sectoral zones in which FeS is preferentially included on certain faces of the crystal during growth. The zoning may be a consequence of surface equilibrium related to differences in surface bonding on each growing face or to a kinetic process (Hollister and Bence, 1967). Sectoral zoning giving rise to crystallographic control of major element chemistry is not uncommon in silicates (e.g., Hollister and Bence, 1967; Hollister, 1970). Frondel *et al* (1942) similarly explained the heterogeneous distribution of some trace elements in calcite and galena by surface equilibria.

(2) The patches represent a metastable, high-temperature polymorph or polytype of sphalerite. This hypothesis is based on the assumption that sphalerite
coexisting with pyrite and pyrrhotite undergoes a phase change near 525°C
(Fig. CS-19). The matrix composition corresponds to the stable solvus and the
patches represent compositions lying on the metastable extension of the higher-

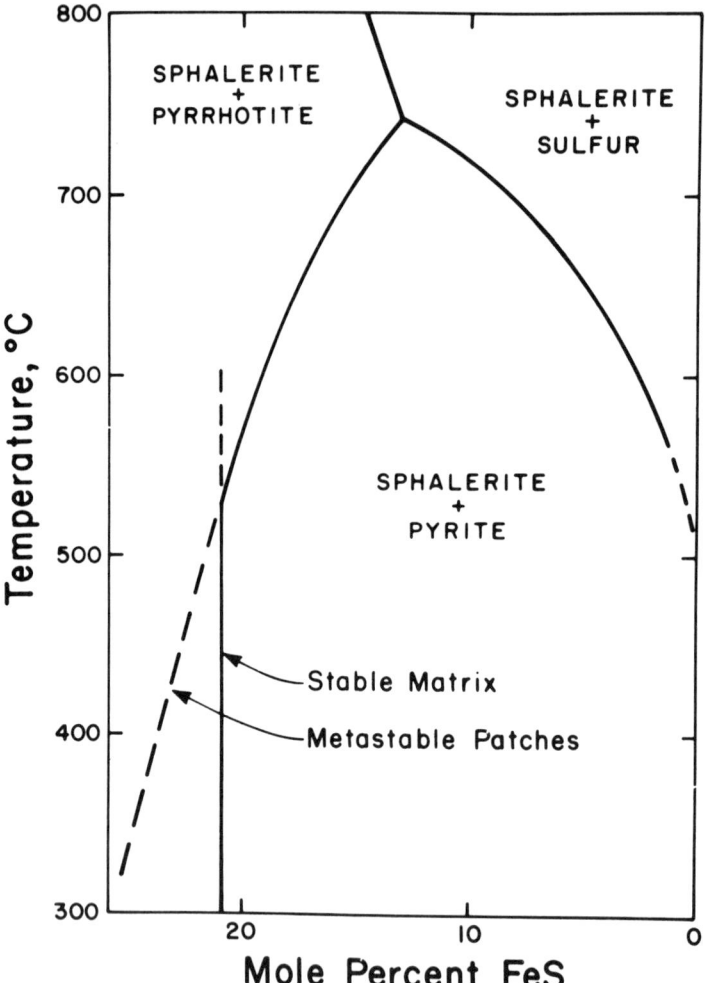

Fig. CS-19. T-X projection showing the effect of a hypothetical phase change in
sphalerite coexisting with pyrite and pyrrhotite. The phase element for two
coexisting sphalerites is omitted but must extend to the left from the intersection
of the univariant curves at 525°C. (Scott and Barnes, 1971, *Econ. Geol.* 66, 667).

temperature phase boundary. A first-order transition is shown in Figure CS-19. If it is second order involving the ordering of Fe and Zn, the transition must occur over a range of temperatures and the patches may be a disordered phase in metastable equilibrium with the ordered matrix.

As yet there is no indication of which, if either, of these explanations is correct, although the second hypothesis readily explains the dependence of the composition of patches on temperature and their disappearance near 525°C. Scott and Barnes (1971) have described microscopic heterogeneities with the appearance of patches in coarsely-crystalline, natural sphalerite crystals from hydrothermal environments, which raises the possibility that the temperature-dependence of the patch compositions relative to their matrix may be a useful geothermometer. Development of this potential geothermometer will require much more experimental study, however. At present, Figure CS-18 applies only to sphalerite coexisting with pyrite + pyrrhotite although patches have been synthesized in sphalerite coexisting with pyrrhotite only. A further restriction is that patches are not likely to be found in metamorphosed ores because they would tend to be annealed to the composition of the matrix during prolonged heating.

Effect of Pressure on the Sphalerite + Pyrite + Pyrrhotite Solvus: The Sphalerite Geobarometer

The partial molar volume of FeS in sphalerite is large compared to that of ZnS as evidenced by the substantial increase in the cell edge of sphalerite with iron content (Fig. S-9). As a consequence, we should expect that in a system in which a_{FeS} is buffered by pyrite + pyrrhotite, sphalerite should become less FeS-rich with increasing pressure. Using equation (cs-9) derived by Barton and Toulmin (1966), Scott and Barnes (1971) calculated this change in sphalerite composition with pressure and showed how it could be used as a geobarometer. However, their calculations were known to be inaccurate because the effects of thermal expansion and compression were not included. More recently, Scott (1973) has measured the composition of sphalerite on the pyrite + pyrrhotite solvus by hydrothermal recrystallization experiments over a wide range of temperatures and pressures. His data, in Figure CS-20, show that sphalerite isobars do indeed shift to lower FeS contents, but the change in composition is not as large nor of the same slope as calculated by Scott and Barnes (1971).

The most significant feature of Fig. CS-20, so far as geobarometry is concerned, is the vertical slope of the 2.5 and 5kb isobars below approximately 600°C. They remain vertical until they encounter the slope reversal of the pyrite + pyrrhotite solvus in Figure CS-3 or the stability field of monoclinic pyrrhotite, neither of which are known at high pressure but are probably well below 300°C at 5 kb. Therefore, this geobarometer is independent of temperature within much of the geologically-important temperature and pressure ranges. The 7.5kb isobar as drawn shows

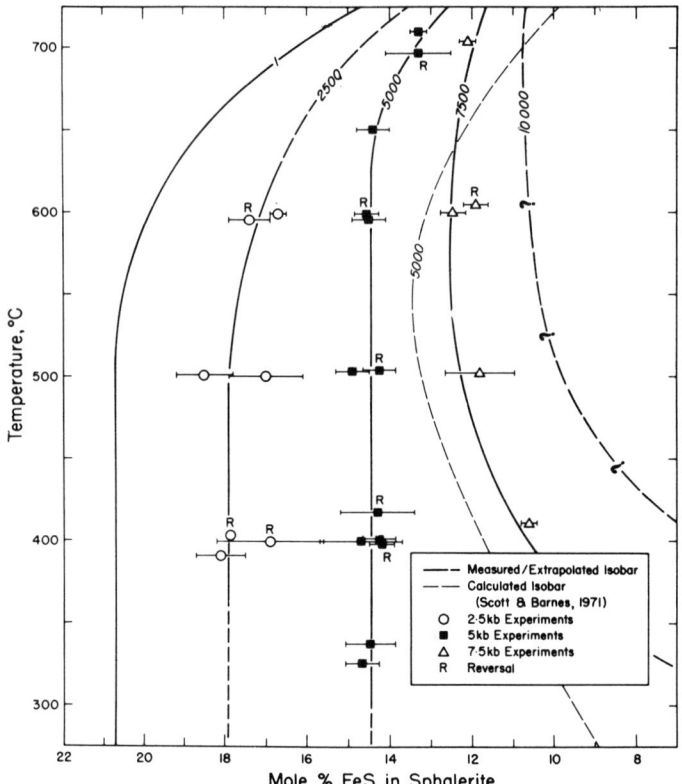

Fig. CS-20. T-X projection of the sphalerite + pyrite + hexagonal pyrrhotite solvus isobars. Pressures are in bars. The 1 bar solvus is from Fig. CS-14. (From Scott, 1973, *Econ. Geol. 68*, 469).

considerable curvature, suggesting that there may be a large viaration with temperature in either a_{FeS} or γ_{FeS}^{sp} at this high pressure. On the other hand, the isobar was drawn to conform with the single data point at 411°C. There is no objective reason for discarding this result but subsequent experiments on the 7.5kb isobar might possibly show it to have a slope that is more in line with the isobars at lower pressure.

New calculations by Scott (1973, with the help of P. Toulmin, III) which included the effects of thermal expansion and compression in the volume terms are in better agreement with the experimental results at 5kb than are those of Scott and Barnes (1971) as shown in Figure CS-21. The calculations were performed by integrating equation (cs-9) in five steps between 0 and 5kb in a manner similar to that described earlier for the sphalerite + troilite calculation. Values of

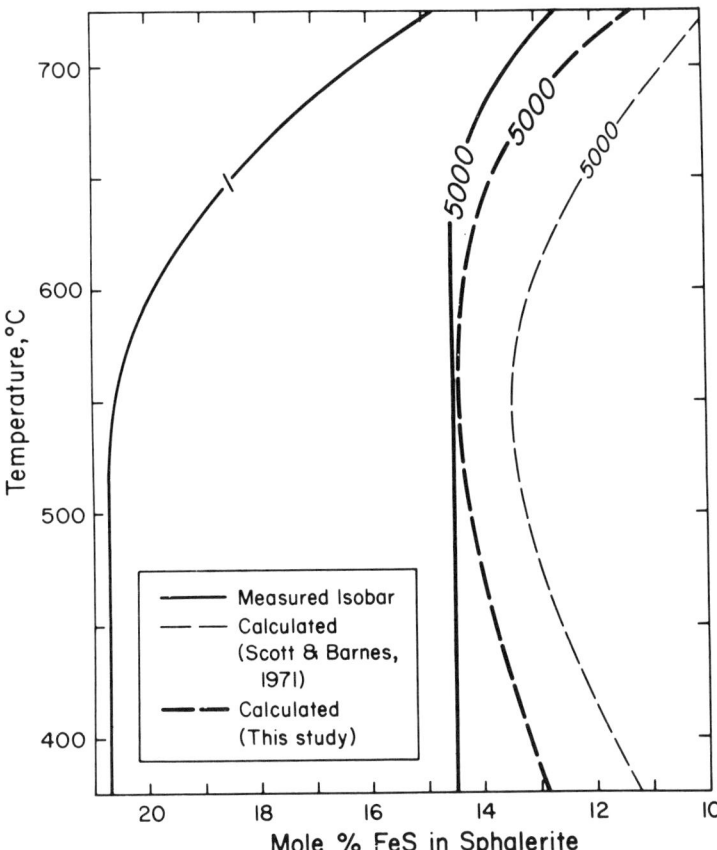

Fig. CS-21. T-X projection comparing the experimentally-determined 5kb solvus for sphalerite + pyrite + hexagonal pyrrhotite with the calculated solvi of Scott and Barnes (1971) and Scott (1973). (From Scott, 1973, *Econ. Geol.* *68*, 472).

γ_{FeS}^{sp} and d $\ln\gamma_{FeS}^{sp}$/dP were obtained from equation (cs-10) and da_{FeS}/dP from equation (cs-8). The volume data for evaluating equations (cs-10) and (cs-8) are in Tables CS-2, and 5 and their corrections for thermal expansion and compression are in Tables CS-3 and CS-4.

Other elements in the system besides Zn, Fe, and S are unlikely to have much of an effect on the phase relations in Figure CS-20. Only rarely will Co and Ni be in high enough concentration in the Fe sulfides to cause a significant decrease in a_{FeS} (Barton and Toulmin, 1966). Concentrations of Cd and Mn, common impurities in sphalerite, up to several wt. % beyond that normally found, do not affect $\bar{V}_{FeS}^{sp}$ as was shown earlier, and so will not alter the location of the isobars. Of the remaining elements commonly found in sphalerite, only Cu is problematical.

The Cu content of most sphalerites is usually less than 1 or 2 wt. %, but the common occurrence of chalcopyrite exsolution blebs indicate that solid solution was much greater at high P and T. The question arises as to whether or not the Fe in the dissolved and exsolved chalcopyrite should be included in sphalerite analyses when applying Figure CS-20 to geobarometry. There are very few data on phase relations in the Cu-Fe-Zn-S system, but indications are that the join $CuFeS_2$-ZnS is truly binary (Fujii, 1970); i.e., the substitution is $CuFe \rightleftharpoons 2Zn$. From this viewpoint, $CuFeS_2$ and FeS are separate components in sphalerite solid solutions and the iron in the exsolved chalcopyrite or dissolved $CuFeS_2$ component should not be included in the estimation of pressure.

The assemblage sphalerite + pyrite + pyrrhotite is quite common from a variety of geological environments so the sphalerite geobarometer has a potential for wide applicability. Its use requires that the three phases must have been in equilibrium during deposition or metamorphism even though it is known that the equilibrium state is not preserved in hexagonal pyrrhotite, as was discussed earlier. When one or both of the Fe sulfides are absent, the sphalerite composition may be quite variable. For example, sphalerites analyzed from a single drill hole in granulite facies metamorphic rocks at Broken Hill, Australia, in which pyrrhotite is the only Fe sulfide present, range in composition from 18.0 to 22.2 mole % FeS over a distance of 74 feet. On the other hand, sphalerites which are in intimate contact with both pyrite and pyrrhotite in ores from highly metamorphosed terrains often show much less spread in Fe content, reflecting the influence of the a_{FeS} buffer.

Applications of the sphalerite geobarometer are discussed by Scott and Barnes (1971) and Scott (1974).

Phase Relations Below 300°C

Equilibrium involving sphalerite is difficult to attain at low temperatures in the laboratory but, armed with a few experimental points, analyses of naturally-occurring sphalerites from low-temperature assemblages and knowledge of the Fe-S system, Scott and Kissin (1973) have pieced together a T-X projection shown in Figure CS-22.

The general location and slope of the univariant boundaries in Figure CS-22 are seen to be consistent with the stoichiometry of Figure CS-3 by making use of the relationships among Fe sulfide X, T, and a_{FeS} (Fig. CS-8 and CS-9) and between a_{FeS} and sphalerite composition (Fig. CS-16). In general the FeS content of sphalerite is a function only of a_{FeS} which, at a given temperature, is inversely related to the sulfur activity of the coexisting Fe sulfide assemblage. For example, Figure CS-3 shows a rapid increase in the sulfur content and hence sulfur activity of the pyrrhotite phase in equilibrium with pyrite between 262° and 254°C. Therefore, the assemblage monoclinic pyrrhotite + pyrite which is stable below 254°C must lie at lower a_{FeS} than ordered "hexagonal" pyrrhotite (NA type) +

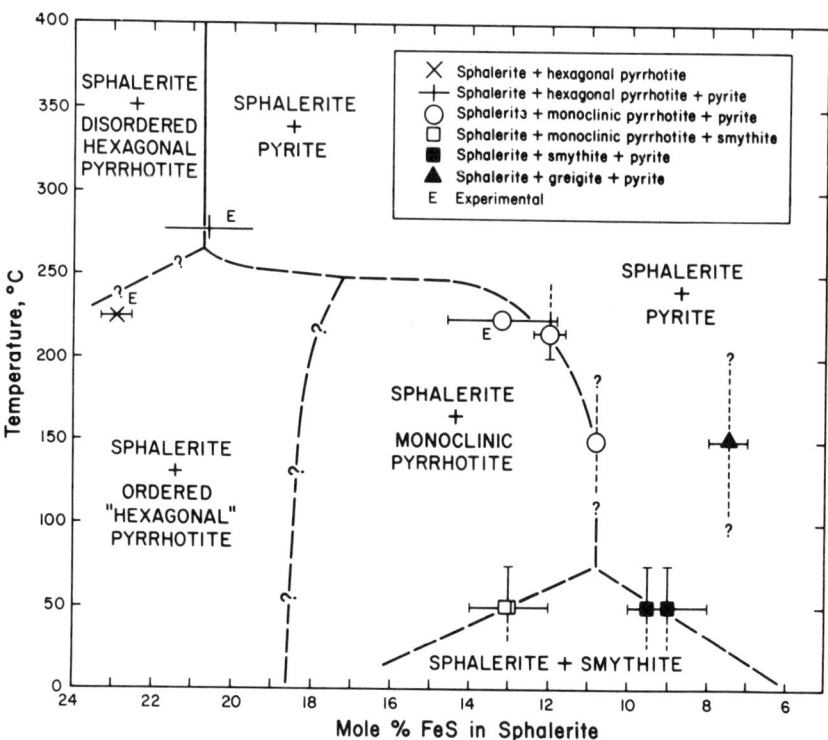

Fig. CS-22. T-X projection of a portion of the system FeS-ZnS-S at low temperatures. The solid line is the sphalerite + hexagonal pyrrhotite + pyrite solvus from Figure CS-14. Dashed lines are univariant boundaries estimated from experiments, natural assemblages and the distribution of univariant equilibria in Figure CS-3. Some of the complications involving ordered pyrrhotites between 262°C and 308°C are omitted. The invariant point at 248°C is misplaced and should be at 254°C. (From Scott and Kissin, 1973, *Econ. Geol. 68, 478*).

Pyrite at slightly higher temperatures; from Figure CS-22, sphalerite is correspondingly less FeS-rich in the assemblage monoclinic pyrrhotite + pyrite than it is in the other assemblage. Similarly, sphalerite coexisting with smythite + pyrite contains less FeS than that coexisting with monoclinic pyrrhotite + pyrite but is more FeS-rich than that coexisting with metastable (?) greigite + pyrite, the most sulfur-rich assemblage in the portion of the Fe-S system under consideration. The slope and location of the univariant boundaries sphalerite + ordered "hexagonal" (NA-type) pyrrhotite + monoclinic pyrrhotite and sphalerite + disordered hexagonal pyrrhotite (1C type) + ordered "hexagonal" pyrrhotite (NA-type), for which there are no data, cannot be determined at present and their positions in Figure CS-22 are speculative. However, they must emanate from invariant points at the temperatures shown in Figure CS-3 and the

curve for sphalerite + disordered hexagonal pyrrhotite + ordered "hexagonal"
pyrrhotite must lie at higher temperature than the experimental point at 223°C
which produced an ordered "hexagonal" N$\underline{A}$ + N$\underline{C}$ pyrrhotite.

Figure CS-22 also shows that at 1 bar the sphalerite + hexagonal pyrrhotite
+ pyrite solvus remains vertical to approximately 262°C whereupon there is an
abrupt change to a negative slope consistent with the reversal in slope of the
ordered "hexagonal" pyrrhotite (N$\underline{A}$ type) + pyrite solvus in Figure CS-3.

Partial Molar Volume of FeS in Sphalerite ($\bar{V}^{sp}_{FeS}$)

A partial molar quantity of a component (i) in a phase is obtained by drawing
at a specified composition (N_i) a tangent to the curve of the molar quantity against
mole fraction and extrapolating that tangent to $N_i = 1$. For example, consider
the partial molar volume of FeS in sphalerite which is needed to calculate the
effect of pressure on sphalerite equilibria. The raw data for molar volumes are
unit cell volumes determined by x-ray diffraction. In the case of a cubic mineral
such as sphalerite, a plot of the unit cell edge vs $100N^{sp}_{FeS}$ (mole % FeS in sphalerite)
as in Figure S-9 can be used directly to obtain the necessary information. A
tangent to the curve in Figure S-9 can be drawn graphically at a specified sphalerite
composition and extrapolated to 100% FeS to obtain a "partial molar cell edge"
for FeS from which $\bar{V}^{sp}_{FeS}$ can be calculated, but there is a more elegant way using
simple calculus. The equation of the tangent will be of the form

$$\underline{a}'' - \underline{a}' = m \ (100 - 100 \ N^{sp}_{FeS}) \tag{cs-13}$$

where $\underline{a}'$ is the cell edge in Å of sphalerite at the specified composition, $100N^{sp}_{FeS}$,
and $\underline{a}''$ is the extrapolated cell edge at 100% FeS; m is the slope of the tangent and
is obtained by differentiating the equation of the curve for cell edge vs
composition in Figure S-9 as follows:

$$m = \frac{d\underline{a}'}{d \ (100N^{sp}_{FeS})} = -0.000008214 \ (100N^{sp}_{FeS}) + 0.0005637 \tag{cs-14}$$

Substituting equation (cs-14) into (cs-13) and rearranging gives the extrapolated
cell edge

$$\underline{a}'' = \underline{a}' + [0.0005637 - 0.000008214(100N^{sp}_{FeS})] \ (100 - 100N^{sp}_{FeS}).$$

The partial molar volume (in cm^3) of FeS for the specified sphalerite composition
is given by:

$$\bar{V}^{sp}_{FeS} = \frac{(\underline{a}'')^3 \ x \ 10^{-24} \ x \ Avogadro's \ no.}{no. \ of \ formulas/unit \ cell \ (=4)}$$

Calculated values for sphalerite are in Table CS-5. Corrections for thermal
expansion and compression are given in Tables CS-3 and CS-4 under the entry for
sphalerite, i.e., $\Delta\bar{V}^{sp}_{FeS} = \Delta V^{sp}$ with T and P.

CS-56

Table CS-5. Partial Molar Volumes of FeS in Sphalerite at 25°C, 1 bar.

$100N_{FeS}^{sp}$	$\bar{V}_{FeS}^{sp}$, cc
0	24.593
10	24.488
20	24.394
30	24.311
40	24.240
50	24.179
55	24.153

The complexities of the Cu-S system, like those of the Fe-S system, have only
recently become known. Prior to 1940 most mineralogists recognized but two copper
sulfide minerals, chalcocite (Cu_2S) and covellite (CuS). Although chalcocite
was believed to exhibit solid solution as far as $Cu_{1.8}S$, digenite (the mineral
with that composition) was not confirmed as a separate phase until 1942 (Buerger,
1942). Since then three additional minerals have been identified -- djurleite
($Cu_{1.97}S$), anilite (Cu_7S_5), and "blue-remaining" covellite ($Cu_{1+x}S$). Experimental
studies of phase equilibria have produced these minerals and several nonmineralogic
polymorphs (Table CS-6).

The phase relationships above 500°C (Fig. CS-22) are straight forward;
there are but two intermediate phases, a cubic solid-solution series which
ranges in composition from chalcocite to digenite, and covellite which is stable
in the presence of its autogenous vapor up to 507°C. The low-temperature
relationships (Fig. CS-23)*are not so well established as those at high temperature.
Although digenite has always been considered a true copper-sulfide, Morimoto and
Koto (1970) have argued "that digenite is stable at room temperature only when
it contains a small amount (∿1%) of iron."

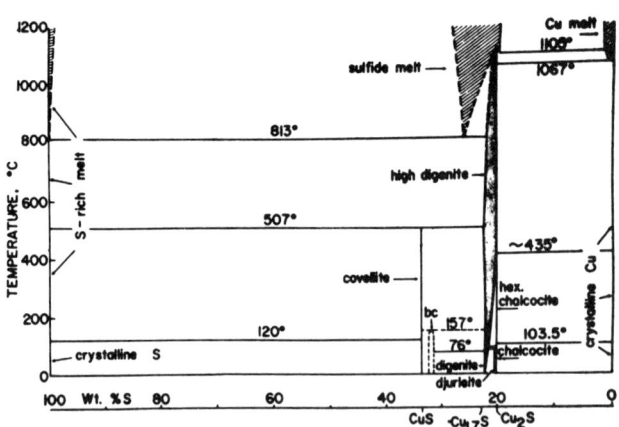

Fig. CS-22. Phase relations in the Cu-S system as known in 1966. Note the absence
of anilite. (From Roseboom, 1966, Econ. Geol. 61, 648.)

* R.W. Potter (1974) has redetermined the low-temperature equilibria among
copper sulfides (Geol. Soc. Amer. Abstracts, 1974).

Table CS-6. Minerals and Phases of the Cu-S System*

Mineral Name	Composition	Thermal Stability,°C Maximum	Minimum	Structure type (cell edges in Å)	Remarks
Chalcocite	Cu_2S	103 (1)*	--	Monoclinic (15) $P2/c$, $\underline{a}$ = 15.22 $\underline{b}$ = 11.88, $\underline{c}$ = 13.48 $\underline{\beta}$ = 116.35°	Inverts to hexag. form
--	Cu_2S	∿435 (1)	103	Hexagonal (2,3) $P6_3/mmc$ $\underline{a}$ = 3.89, $\underline{c}$ = 6.68	Inverts to cubic form
--	Cu_2S	1129 (8)	∿435	Cubic $Fm3m$ (4) $\underline{a}$ = 27.85	Complete s.s. with Cu_9S_5
--	Cu_2S	500 (5)	--	Tetragonal (7) $P422$, $\underline{a}$ = 3.40, $\underline{c}$ = 11.29	Stable only at P > 1kb
Djurleite	$Cu_{1.97}S$	93 (1)	--	Orthorhombic (10) $P*n*$, $\underline{a}$ = 26.95, $\underline{b}$ = 15.71, $\underline{c}$ = 13.56	Possibly $P2_1/n$ with β=90°
Digenite	Cu_9S_5	83 (1)	--	Cubic $Fd3m$ (4) $\underline{a}$ = 5.56	Stabilized by Fe? See text.
--	$Cu_{9+x}S_5$	1129 (8)	83	Cubic $Fm3m$ (4) $\underline{a}$ ∿27.8	Complete s.s. with Cu_2S
Anilite	Cu_7S_5	70 (12)	--	Orthorhombic (11) $Pnma$, $\underline{a}$ = 7.89 $\underline{b}$ = 7.84, $\underline{c}$ = 11.01	
"Blaubleibender" Covellite	$Cu_{1+x}S$	157 (6)	--	Hexagonal (18) $P6_3/mmc?$ $\underline{a}$ = 3.79, $\underline{c}$ = 16.35	Thermodynamically stable?
Covellite	CuS	507 (13)	--	Hexagonal (14) $P6_3/mmc$ $\underline{a}$ = 3.79, $\underline{c}$ = 16.34	
--	CuS_2	550 (9)	--	Cubic $Pa3$ $\underline{a}$ = 5.79	High-P synthesis. Pseudo cubic? (17) (16)

*Numbers in brackets are references:

(1) Roseboom (1966)
(2) Buerger & Buerger (1944)
(3) Wuensch & Buerger (1963)
(4) Morimoto & Kullerud (1963)
(5) Skinner (1970)
(6) Moh (1964)
(7) Janosi (1964)
(8) Jensen (1947)
(9) Munson (1966)

(10) Morimoto (1962)
(11) Morimoto et al (1969)
(12) Morimoto & Koto (1970)
(13) Kullerud (1965)
(14) Berry (1954)
(15) Evans (1968)
(16) Taylor & Kullerud (1971)
(17) Taylor & Kullerud (1972)
(18) Rickard (1972)

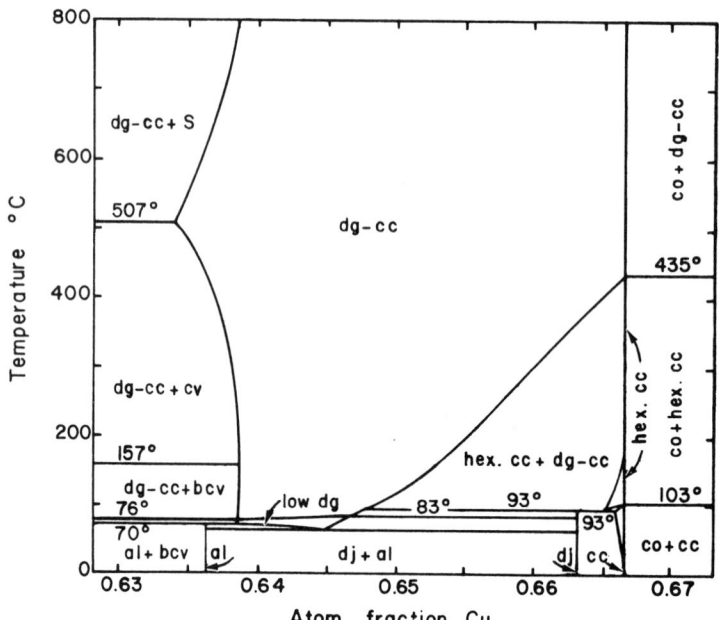

Fig. CS-23. Phase relations in a portion of the Cu-S system. Note the absence of digenite below ∿70°C. (After Barton, 1973)

Chalcocite

    Chalcocite (Cu_2S) is a common mineral in both hypogene and supergene environments. Its symmetry, though commonly listed as orthorhombic, has been reported by Evans (1968) to be monoclinic. It is stable to 103°C, above which it inverts to a hexagonal form which, in turn, is stable to ∿435°C. Although neither the low-temperature form, the only natural polymorph, nor the hexagonal form exhibit any detectable deviation from Cu:S = 2:1, the high-temperature cubic form exhibits a solid-solution field (cc-dg) which extends to Cu:S ∿ 1.75:1.

    Djurle (1958) synthesized what he believed to be a tetragonal form of djurleite. Subsequent examination of this phase (Roseboom, 1966; Luquet *et al* 1972) confirmed that it is a metastable low pressure form, possibly with considerable compositional variation of the high-pressure polymorph of Cu_2S described by Skinner *et al* (1964). Tetragonal Cu_2S is only stable at pressures above ∿0.8kb (Fig. CS-24). It is quenchable and has a specific gravity of 5.93 which is significantly greater than that of chalcocite (5.78) (Skinner, 1970).

    Although the tetragonal form of Cu_2S is readily quenched in the laboratory, Skinner (1970) reports that it inverts to chalcocite when stored at room temperature for a few weeks. However, Serebryanaya (1966) found no change in his synthetic

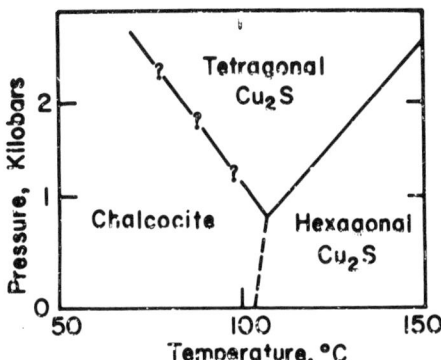

Fig. CS-24. A suggested phase diagram for Cu_2S. (From Skinner, 1970, *Econ. Geol.* <u>65</u>, 726.)

tetragonal Cu_2S after 8 months. The only natural occurrence of this phase is pseudo-mineralogic--a small amount was precipitated in the sulfides deposited by effluent brines from the Salton Sea geothermal fields (Skinner *et al*, 1967). This material inverted to chalcocite in a short time.

Digenite

Digenite, generally reported as Cu_9S_5, has recently been considered by Morimoto and Koto (1970) to be stable only when it contains a small amount of iron ($\sim 1\%$). Above about 70°C the low-temperature cubic form of digenite becomes a stable phase on the Cu-S join. At slightly higher temperature, 76° to 83°C--depending on composition, digenite inverts to a high-temperature cubic form which is isostructural with high-temperature chalcocite. Solid solution between these phases (Fig. CS-22,23) becomes complete above about 435°C.

Djurleite

Djurleite, $Cu_{1.96}S$, was simultaneously described by Roseboom (1962) and Morimoto (1962) after its synthetic analog had been found by Djurle (1958). The maximum thermal stability of djurleite has now been established as approximately 93°C, above which it decomposes to the hexagonal form of chalcocite and the cubic cc-dg phase. Its usual mode of occurrence as intimate intergrowths among other copper sulfides and its similarity to these minerals results in its commonly being overlooked in sulfide ores.

Anilite

The most recently discovered copper sulfide is anilite, Cu_7S_5, which was described in 1969 by Morimoto *et al*. Subsequently its maximum stability was found to be ~ 70°C, above which it decomposes to digenite and covellite (Morimoto and Koto, 1970). Anilite, like djurleite, closely resembles digenite and is difficult to recognize without careful x-ray examination.

Covellite and "Blue-remaining" Covellite

Covellite, CuS, is among the most striking of sulfides, displaying brilliant blue interference colors in plane light, and purple-red pleochroism and blood-red

CS-61

anisotropism under crossed nicols. It does not deviate measureably from CuS and is hexagonal up to its decomposition temperature of 507°C (Kullerud, 1965). Frenzel (1959) applied the name *"blaubleibender"* or "blue-remaining covellite" to a phase which closely resembled covellite except for its lack of the strong purple-red pleochroism of ordinary covellite. He also found that it has a distinct but similar x-ray pattern. Moh (1964) found that *"blaubleibender"* covellite could be synthesized below 157°C, but its thermodynamic stability remains in doubt.

CuS_2

Munson (1966) reported synthesis of a copper disulfide with the pyrite-type structure. Subsequent work by Taylor and Kullerud (1972) has defined an invariant point related to this phase and has suggested that its symmetry may be only pseudocubic.

Thermochemistry

Thermochemical aspects of the Cu-S system have recently been summarized by Barton (1973) whose data are reproduced below. The Cu-S system as a function of T and a_{S_2} is presented in Fig. CS-25. It is based upon:

(1) $4Cu + S_2 = 2Cu_2S$ (435-700°C) $\Delta G° = 62,840 + 14.71T$

(2) $4Cu + S_2 = 2Cu_2S$ (103-435°C) $\Delta G° = 74,306 + 173.63T -$
                                                    $44.998/T - 52.58T \log T +$
                                                    $0.01028T^2$

(3) $S_2 = 2S(liq)$ (200-800°C) $\Delta G = -26,810 + 30.25T$ (200-800°C)

(4) H_2S/H_2 ratios over high digenite compositions (Rau, 1967).

(5) graphical Gibbs-Duhem integrations to determine the activity of Cu_2S in the high digenite solid solution (Fig. CS-26).

(6) $2Cu_2S$ (in cc-dg) $+ S_2 = 4CuS$ $\log K = -2 \log a_{Cu_2S} - a_{S_2}$
                                                    $\log K = -10.897 + 9680.7T$
                                                    $\Delta G = -28,780 + 16.14T$ (cal/mole CuS)

The reader is referred to Barton's excellent paper for a detailed discussion. Schneeberg (1973) has measured the fugacity of diatomic sulfur vapor over the covellite-high digenite assemblage (the same as (6) above) by means of electrochemical cells. His two equations determined in separate experiments are:

$\log f_{S_2} = 24.467 - 25669 / T + 4844700 / T^2$,

$\log f_{S_2} = 24.193 - 24097 / T + 4211000 / T^2$.

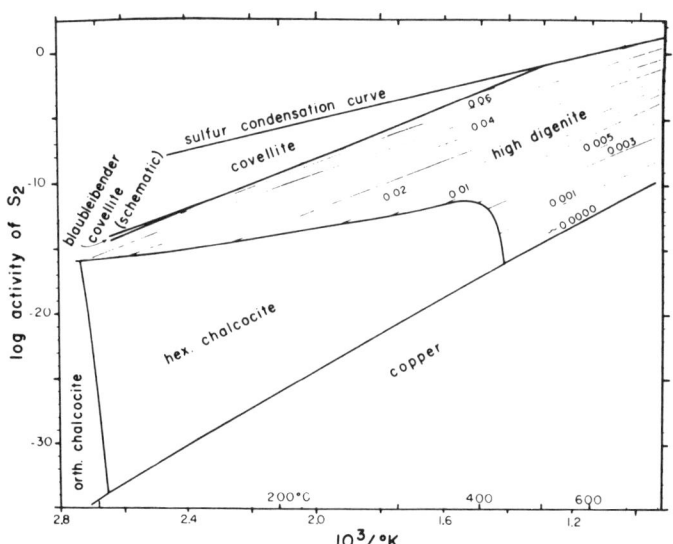

Fig. CS-25. Activity of S_2-temperature diagrams for the Cu-S system. The composition of high digenite is shown in light weight contours; the numbers indicate the ratio S_2/Cu_2S in the solid solution. (From Barton, 1973, *Econ. Geol.* *68*, 457.)

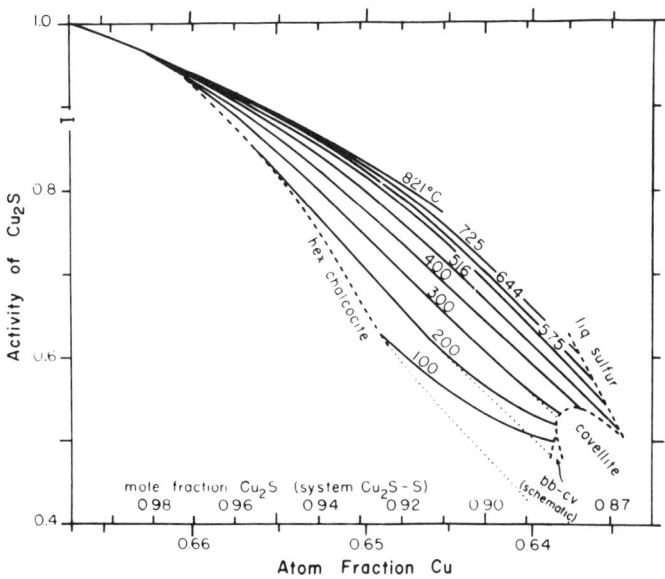

Fig. CS-26. Activity of Cu_2S composition diagram for the high digenite field. The dotted curves represent Barton's unammended extrapolations of Rau's (1967) data. The short bar at mole fraction of Cu_2S = 1.00 represents the total range of non-stoichiometry reported by Cook (1972) for "Cu_2S" in equilibrium with copper. (From Barton, 1973, *Econ. Geol.* *68*, 458.)

More time and effort have been expended in the determination of phase equilibria and mineralogical relationships among the Cu-Fe sulfides than any other ternary sulfide system. In spite of this many relationships remain enigmatic, obscured by the presence of extensive solid solutions, unquenchable phases, and metastability. The general form of phase relationships has emerged from the work of Merwin and Lombard (1937), Schlegel and Schiller (1952), Hiller and Probsthain (1956), Yund and Kullerud (1966), Kullerud et al (1969), Mukaiyama and Izawa (1970), Cabri (1973), and Barton (1973). The minerals and phases reported in the Cu-Fe-S system are listed in Table CS-7; their compositions are illustrated in Fig. CS-27.

The nature of the high-temperature relationships are now well established; a 600°C isothermal section is shown in Fig. CS-28. It is characterized by three extensive solid solutions: (1) chalcocite-digenite-bornite (cc-dg); (2) intermediate solid solution (iss); (3) pyrrhotite (po). The centrally located iss phase, which has the sphalerite-type (fcc) structure, includes a large compositional area which is slightly sulfur deficient. Cabri (1973) has noted that the iss field may be divided into three zones each characterized by a different quenching behavior (Fig. CS-29). Compositions in the first zone, which includes the S-rich portion of the iss from Cu:Fe=1 to the Fe-rich extremity, quench from 600°C to give chalcopyrite + iss. The second zone, which includes the S-deficient region on the Fe-rich end, quenches to a phase exhibiting a primitive cubic cell. The third region which separates zones 1 and 2 and which includes all of the central and Cu-rich portions of the iss quenches to give the primitive phase plus either chalcopyrite or mooihoekite. These complications but herald the complexities which appear at low temperature. The bornite-bearing solid solution extends from the Cu-S join where it is continuous with the high chalcocite - high digenite solid solution to a composition considerably more Fe- and sulfur-rich than stoichiometric bornite, Cu_5FeS_4. The pyrrhotite solid solution accepts up to about 4.5 wt. % Cu into its structure.

As temperature decreases below 600°C the simple, well understood phase equilibria gradually give way to less well understood and, in some areas, quite conjectural relationships as additional phases become stable. By far the most important change is the appearance of chalcopyrite, $CuFeS_2$, as a stable phase below 557°C. It forms in the iss-pyrite field and remains isolated from all other Cu-Fe sulfides until temperature is further decreased. Other phase appearances of consequence are covellite, CuS, at 507°C and idaite, $Cu_{5+x}FeS_{6+x}$, at 501°C.

At 400°C the phase relationships appear as shown in Figure CS-30 and at 300°C they probably look something like those in Figure CS-31. Although the general configuration of the Cu-Fe-S system probably changes little at lower temperature, there remain many questions regarding the appearance and phase relations of the chalcopyrite-like phases, talnakhite, mooihoekite, and haycockite. Cabri

Table CS-7   Minerals and Phases of Cu-Fe-S System*

Mineral Name	Composition	Thermal stability,°C Maximum	Thermal stability,°C Minimum	Structure type (cell edges in Å)	Remarks
Digenite	$(Cu,Fe)_9S_5$	83	--	Cubic $Fd3m$ $\underline{a}$ = 5.56	see Table CS-6
---	$\sim(Cu,Fe)_9S_5$	1129	83	Cubic $Fm3m$ $\underline{a}$ = 27.85 (17)	See Table CS-6
Bornite	Cu_5FeS_4	228 (13)	--	Tetragonal $P42_1c$ $\underline{a}$ = 10.44, $\underline{c}$ = 21.88 (13)	
---	Cu_5FeS_4	--		Cubic Fd** (13) $\underline{a}$ in multiples of 5.5	several metastable forms (13)
---	Cu_5FeS_4	$\sim$1100 (16)	228	Cubic $Fm3m$ (13) $\underline{a}$ = 5.5	
x-bornite	$Cu_5FeS_{4.05}$	125 (12)	--	Tetragonal (19) $\underline{a}$ = 16.5, $\underline{c}$ = 11.0	
Idaite	$Cu_{5.5}FeS_{6.5}$	501 (12)	--	Hexagonal (14) $\underline{a}$ = 3.77, $\underline{c}$ = 11.18	gives (15) $\underline{a}$ = 3.90, $\underline{c}$ = 16.95Å
Fukuchilite	Cu_3FeS_8	$\sim$ 200 (7)	--	Cubic $Pa3$ (8) $\underline{a}$ $\simeq$ 5.6	Considerable s.s. along CuS_2-FeS_2 join (8)
Chalcopyrite	$CuFeS_2$	557	--	Tetragonal $I\bar{4}2d$ $\underline{a}$ = 5.28, (2) $\underline{c}$ = 10.40	Transforms to py + iss (4, 11)
Cubanite	$CuFe_2S_3$	200-(6) 210	--	Orthorhombic $Pcmn$ $\underline{a}$ = 6.46, (5) $\underline{b}$ = 11.12, $\underline{c}$ = 6.23	Transforms to iss
Intermediate solid solution	Wide range	960 (16)	20-200	Cubic $F\bar{4}3m$ (4) $\underline{a}$ $\simeq$ 5.36	Quenchable only near $CuFe_2S_3$ comp.
Talnakhite	$Cu_9Fe_8S_{16}(3)$	$\sim$ 186 (9)	--	Cubic $I\bar{4}3m$ (1) $\underline{a}$ = 10.59	Transforms to interm. I
Intermediate phase I	$Cu_9Fe_8S_{16}(?)$	230 (9)	186	?	Transforms to interm. II
Intermediate phase II	$Cu_9Fe_8S_{16}(?)$	520 (9)	230	?	Transforms to iss
Mooihoekite	$Cu_9Fe_9S_{16}$	$\sim$ 167 (9)		Tetragonal $P\bar{4}2m$ $\underline{a}$ = 10.58, (2) $\underline{c}$ = 5.37	Transforms to interm. A

– – – – – – – (continued)

Mineral Name	Composition	Thermal stability,°C Maximum	Minimum	Structure type (cell edges in Å)	Remarks
Intermediate phase A	$Cu_9Fe_9S_{16}$	236	167	?	Transforms to iss
Haycockite	$Cu_4Fe_5S_8$	?	--	Orthorhombic $P222$ or $P2_12_12$ $\underline{a} \approx \underline{b} = 10.71$, $\underline{c} = 31.56$	Stable only at low T
Primitive cubic phase (pc)	Wide range	20-(9) 200	20- 200	Cubic ? (9)	Transforms to iss
---	$Cu_{0.12}Fe_{0.94}S_{1.00}$ (18)	?	--	?	Needs confirmation

*Numbers in brackets are references:

(1) Hall & Gabe (1972)
(2) Cabri & Hall (1972)
(3) Cabri & Harris (1971)
(4) Szymanski et al (1973)
(5) Buerger (1947)
(6) Cabri et al (1973)
(7) Kajiwara (1969)
(8) Shimazaki & Clark (1970)
(9) Cabri (1973)
(10) MacLean et al (1972)
(11) Barton (1973)
(12) Yund & Kullerud (1966)
(13) Morimoto & Kullerud (1966)
(14) Yund (1963)
(15) Frenzel (1959)
(16) Kullerud et al (1969)
(17) Morimoto & Kullerud (1963)
(18) Clark (1970)
(19) Morimoto (1970)

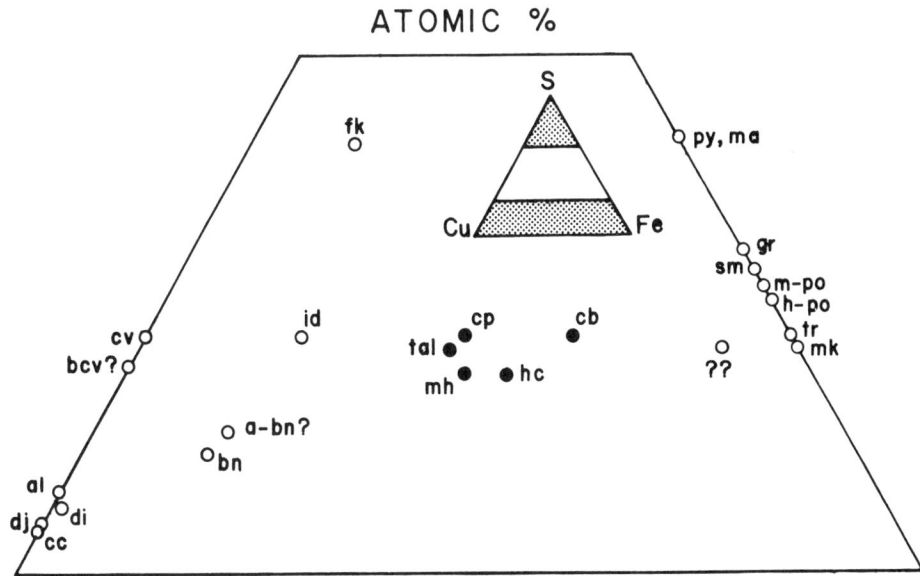

Fig. CS-27. Minerals reported within the Cu-Fe-S system. Abbreviations: cc-chal-
cocite, dj-djurleite, di-digenite, al-anilite, bcv-blue-remaining covellite, cv-
covellite, bn-bornite, a-bn-anomalous bornite, id-idaite, fk-fukuchilite,
tal-talnakhite, cp-chalcopyrite, mh-mooihoekite, hc-haycockite, cb-cubanite,
py-pyrite, mc-marcasite, gr-greigite, sm-smythite, m-po-monoclinic pyrrhotite,
h-po-hexagonal pyrrhotite, tr-troilite, mk-mackinawite, ??-Cu-mackinawite?

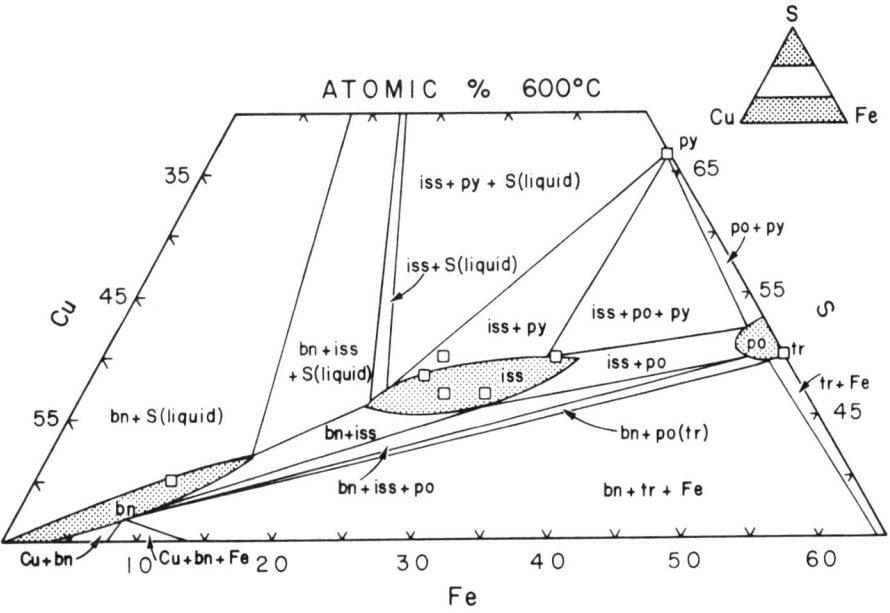

Fig. CS-28. Phase relations in the central portion of the Cu-Fe-S system at 600°C
(After Cabri, 1973).

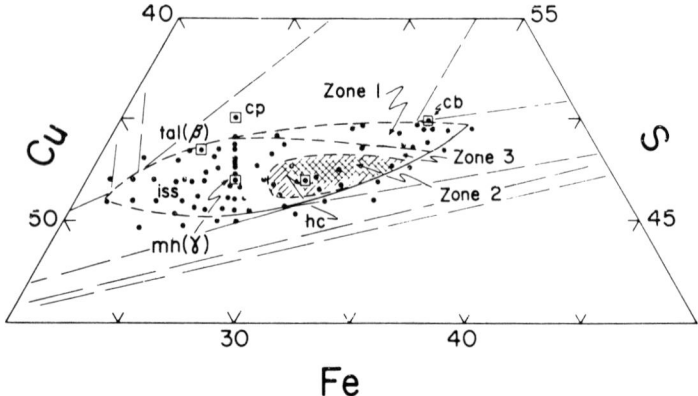

Fig. CS-29. An enlarged portion from the center of Figure CS-28 showing the three quench zones of the iss. See discussion in text (From Cabri, 1973, *Econ. Geol.* *68*, 447.)

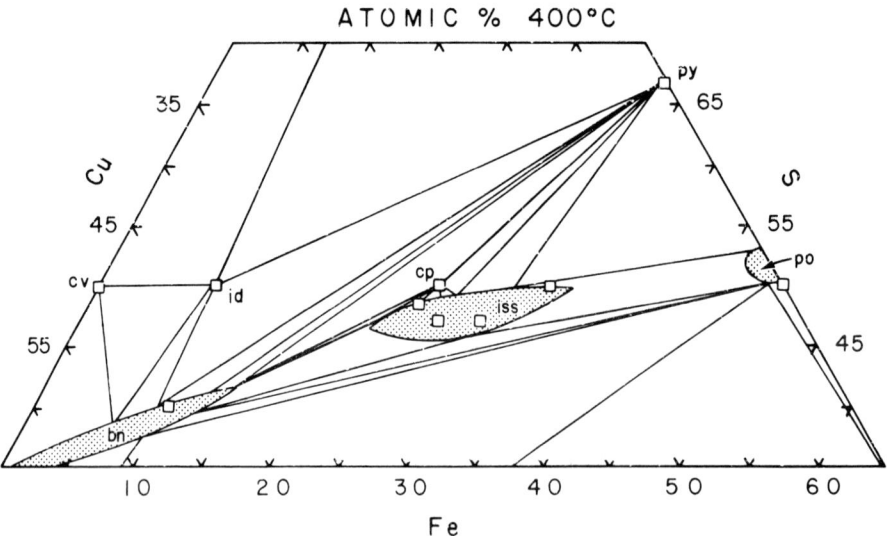

Fig. CS-30. Schematic phase relations in the central portion of the Cu-Fe-S system at 400°C. Based on data and discussions of several workers.

and his associates have done much to elucidate the nature of these phases, but apparent polymorphism and the problems of attaining low temperature equilibrium leave much work to be done.

CS-68

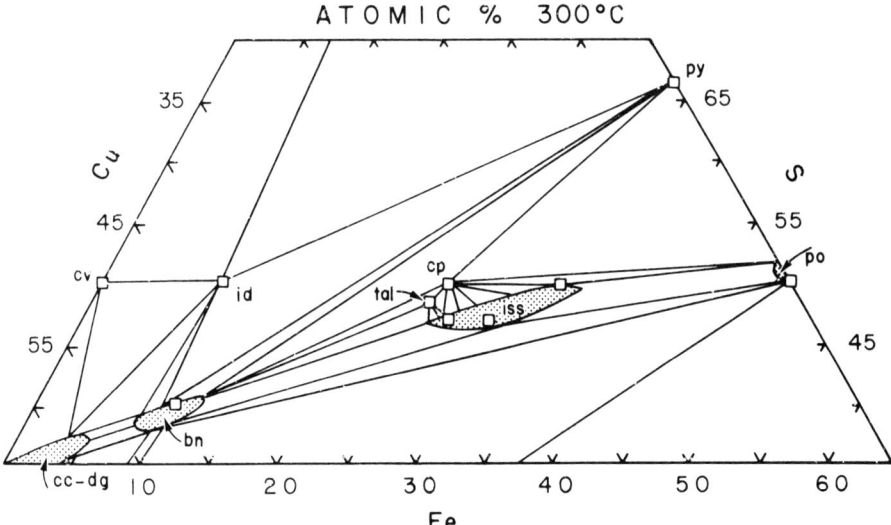

Fig. CS-31. Schematic phase relations in the central portion of the Cu-Fe-S
system at 300°C. Based on data and discussions of several workers.

Low-temperature stability relationships are uncertain; however, based on
natural occurrences and numerous other workers' contributions they likely
approximate Figure CS-32. The stable assemblages shown range from the very common
and unquestioned chalcopyrite + pyrite to the completely speculative bornite +
mooihoekite. Quite likely the phases talnakhite, mooihoekite and haycockite are
far more common than realized but have generally been overlooked because of their
strong similarity to chalcopyrite.

"Blue-remaining" covellite and x-bornite have been omitted from the low
temperature diagram because their status as stable phases is in doubt. Even among
the common Cu-Fe sulfides there remains confusion. Kullerud and Yund (1966)
suggested that pyrite and chalcocite never stably coexist and that the bornite-
pyrite assemblage is unstable below 228°C. Barton and Skinner (1967), however,
maintain that the pyrite-chalcocite tie line is a strong possibility. Bornite-
pyrite assemblages supposed by Kullerud and Yund (1966) to be unstable below 228°C
are quite common in many ore bodies, including Magma Mine, Arizona, where the
elevated rock temperatures (at least as high as 50°C) have provided us with
extraordinary long term experiments (measured in millions of years instead of
hours or days). On the other hand breakdown of anomalous bornites at temperatures
near 100°C results in exsolution of chalcopyrite and chalcocite.

It is apparent that a great deal remains to be learned of the phase equilibria
in the Cu-Fe-S system, especially below 400°C.

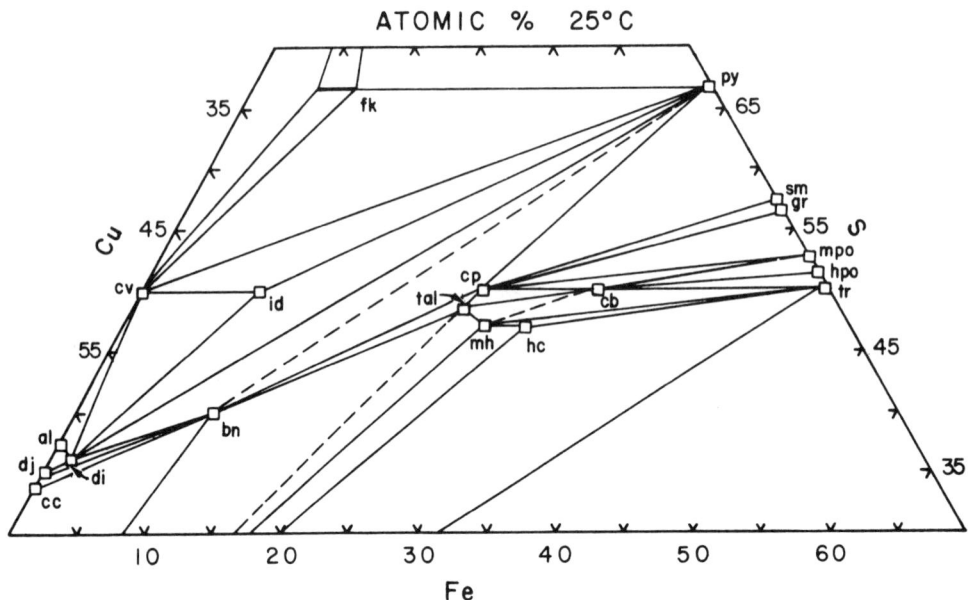

Fig. CS-32. Possible phase relations in the central portion of the Cu-Fe-S system at 25°C. The most speculative tie lines are dashed; see text for discussion.

Digenite

Digenite, Cu_9S_5, long considered to be a member of the copper-sulfide family, has recently been redesignated as a Cu-Fe sulfide by Morimoto and Kato (1970) who found that natural digenites contained small amounts (∿1%) of iron. They further found that low-temperature (<70°C) annealing experiments in the pure Cu-S system did not yield digenite but mixtures of anilite and chalcocite. As temperature is raised above 25°C, digenite exists over an increasing compositional range and becomes a stable phase in the Cu-S system at 70°C. At higher temperatures the solid solution expands to include compositions which are more S- and Fe-rich. Above ∿335°C the digenite solid solution merges with the bonite solid solution thus forming a single phase which spans from the Cu-S join to compositions containing as much as 15 at. % Fe. Jensen (1942) found that the maximum melting temperature (1129°C) of the digenite solid solution lies on the Cu-S join at a composition slightly copper deficient of Cu_2S.

Bornite

Bornite, one of the most readily recognized of sulfides, is known to exist in several polymorphic forms, all based upon a subcell of 5.5Å (Morimoto and Kullerud, 1961). The low-temperature form that is encountered in ores is tetragonal with $\underline{a}$ = 10.94, $\underline{c}$ = 21.88 Å. On heating to 228°C it inverts to a cubic form with $\underline{a} \sim 5.5$ Å. Morimoto and Kullerud (1966) found that rapid cooling of high-temperature bornite results in the development of a metastable cubic "$\underline{2a}$" structure with $\underline{a} \sim 10.94$ Å. Bornite, like digenite, exhibits extensive solid solution within the Cu-Fe-S system, especially with regard to Cu/Fe ratio. Solid solution with the digenite phase becomes complete above about 335°C. Quenching of intermediate compositions along the bornite-digenite join results in development of metastable polymorphs whose cell edges are integral multiples (2,3,4, or 5) or, in some instances, non-integral (4.7 and 5.7) repeats (Morimoto, 1970) of the basic 5.5 Å subcell, depending upon composition (Morimoto and Kullerud, 1966). The high-temperature bornite phase also exhibits solid solubility toward chalcopyrite; breakdown of these compositions yield remarkable exsolution textures commonly found in ores. Brett (1964) in a study of exsolution on the bornite-digenite, bornite-chalcocite and bornite-chalcopyrite joins found, however, that formation of such textures was neither diagnostic of initial temperature of formation nor cooling rate.

X-, Anomalous, or Sulfur-rich Bornite

Heating experiments conducted on natural bornites by a variety of workers (summarized in Brett and Yund, 1964) indicate that some bornites, instead of merely inverting to a high-temperature polymorph, exsolve both chalcopyrite and chalcocite. This has led several observers to consider these bornites as metastable S-rich bornites. Brett and Yund (1964) concluded that some bornites which formed at temperatures below 75°C have sulfur to metal ratios greater than that indicated by the stoichiometric formula Cu_5FeS_4. Yund and Kullerud (1966) extended their work and found that x-bornite, as they called it, could be synthesized in a field which possessed a slight excess (0.5 wt. %) of sulfur relative to normal bornite at temperatures below 125°C (Fig. CS-33). Whether or not this phase is ever thermodynamically stable is unknown, but its natural occurrence, especially in red-bed copper deposits, is now well established. However, the absence of x-bornite in most hypogene ores has been interpreted as indicating that it does not form during cooling through solid-state reactions but that it probably forms directly from low-temperature solutions (Brett and Yund, 1964).

Cubanite

Cubanite, $CuFe_2S_3$, is often intimately associated with chalcopyrite in which it occurs as laths which are highly anisotropic. The orthorhombic, naturally-occurring form is only stable to 200-210°C (Cabri *et al*, 1973); above this

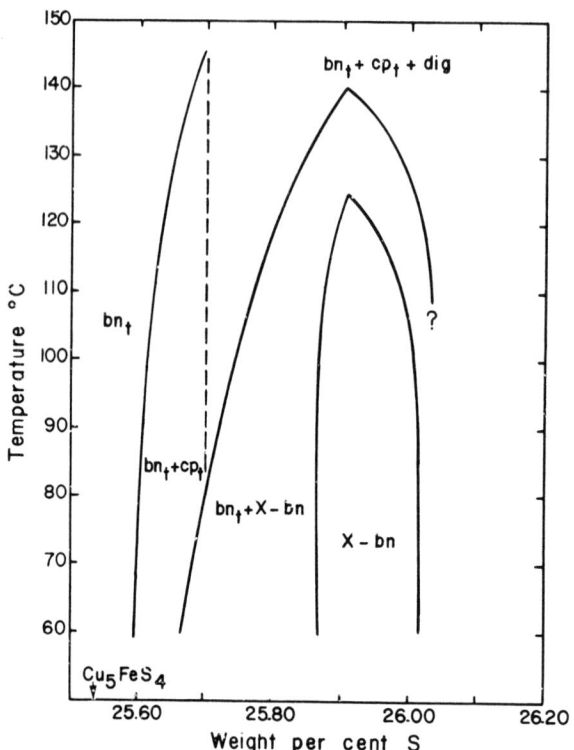

Fig. CS-33. Phase relations in a portion of the Cu-Fe-S system along the join Cu$_5$Fe-S illustrating the field of X-bornite synthesis. (After Yund and Kullerud, 1966).

temperature it transforms to cubic iss. On heating natural cubanites Cabri *et al* found that twinning normal to [111] was common; they believe that this twinning accounts for Vaasjoki's (1971) report of a hexagonal high-temperature form of cubanite. They also note that exsolution of chalcopyrite from iss near cubanite compositions yields a tetragonal x-ray pattern which probably accounts for the previously reported tetragonal form of cubanite.

Intermediate Solid Solution (iss) and the Primitive Cubic Phase (pc)

At temperatures above 500°C (Figs. CS-28,29), the iss represents the dominant ternary phase of the Cu-Fe-S system. Long believed to be a high temperature polymorph of chalcopyrite, the iss is now known to represent a distinct, though crystallographically similar, phase. It has a disordered sphalerite-type (fcc) structure and varies in composition over a range sufficiently wide (Fig. CS-29) to include the compositions of talnakhite, mooihoekite, haycockite, and cubanite. Much of the previous misinterpretation apparently resulted from the nonquenchability of large portions of the compositional range. The low-temperature decomposition of the iss is not well understood but it appears that zone 2 of Fig. CS-29 inverts at some temperature below 600°C to a primitive cubic phase which persists down to 20-200°C. Natural intergrowths of chalcopyrite and bornite and chalcopyrite and cubanite and natural assemblages of talnakhite, mooihoekite and haycockite likely form as decomposition products of initially deposited iss.

Idaite

Although a synthetic phase of composition Cu$_{5-x}$Fe$_x$S$_{6+x}$ was first reported by Merwin and Lombard (1937), no natural analog was found until 1959 (Frenzel) when the name idaite was applied. The similarity of this phase with bornite (except for a

slightly orange tint and a strong anisotropism apparently led earlier workers to refer
to idaite as "orange bornite." Idaite is stable below 501°C; its unit cell was
described by Frenzel (1959) as hexagonal $\underline{a}$ = 3.90, $\underline{c}$ = 16.95Å, but Yund (1963)
reported a hexagonal cell with $\underline{a}$ = 3.77, $\underline{c}$ = 11.18Å (*contra:* Frenzel, 1963).

Fukuchilite

Among the recent mineralogic additions to the Cu-Fe-S system is fukuchilite,
Cu_3FeS_8, a pyrite-like structure (Kajiwara, 1969). Subsequent experimental study
by means of hydrothermal synthesis in the 100-275°C range (Shimazaki and Clark, 1970)
revealed that it is possible to form a pyrite-type compound on the FeS_2-Cu_2S join
from FeS_2 to approximately $FeS_2:CuS_2$ = 3:7. Shimazaki and Clark conclude, however,
that FeS_2-CuS_2 solid solutions are thermodynamically unstable even though they form
and persist metastably in nature.

Chalcopyrite

Chalcopyrite, $CuFeS_2$, the most common of the ternary copper-iron-sulfides,
has an ordered tetragonal structure which is stable up to 557°C. It deviates very
little from ideal $CuFeS_2$, being slightly metal-rich at high temperature (Barton, 1973).
Above 557° (Fig. CS-28) chalcopyrite decomposes to pyrite + iss; it does not merely
invert to a high temperature polymorph. There is, however, a close relationship
between the disordered cubic structure of the iss ($\underline{a}$ ∿ 5.36Å) and the ordered tetragonal
cell of the chalcopyrite ($\underline{a}$ = 5.28; $\underline{c}$ = 10.40Å).

Talnakhite, Intermediate Phase I, Intermediate Phase II

Talnakhite, $Cu_9Fe_8S_{16}$, and its high temperature polymorphs I and II are examples
of sulfides which were synthesized long before their recognition in the mineral realm.
First observed by Hiller and Probsthain (1956) as their β-phase, talnakhite when
found naturally was simple referred to as "cubic chalcopyrite" (Bud'ko and Kulagov,
1963). Subsequently in an unprecedented move, the International Mineralogical
Association refused one proposed name (Cabri, 1967) and assigned the name talnakhite.
Synthetic talnakhite goes through three transitions on heating: at 186°C to phase I,
at 230°C to phase II, and between 520 and 525°C to iss. Talnakhite is cubic
($\underline{a}$ ∿ 10.59Å) but the structures of the intermediate phase are unknown.

Mooihoekite and Intermediate Phase A

Mooihoekite, $Cu_9Fe_9S_{16}$, represents a second chalcopyrite-like phase, and,
like talnakhite, was found (Cabri and Hall, 1972) after the analogous synthetic
phase had been described (Hiller and Probsthain, 1956). Like the other chalcopyrite-
like phases it has been found only rarely, probably because of its similarity to

chalcopyrite. Structurally mooihoekite, like talnakhite, is closely related to
iss, being composed of sphalerite-like 5.3Å cells. On heating, mooihoekite undergoes
a transition at 167°C to an intermediate phase A of unknown structure; phase A,
in turn, is transformed to iss on heating above 236°C.

Haycockite

Haycockite, $Cu_4Fe_5S_8$, the third of the chalcopyrite-like minerals, was described
by Cabri and Hall (1972). Although orthorhombic, its structure is also composed
of sphalerite-like cells, as are talnakhite, mooihoekite, chalcopyrite and iss.
Little is known of its stability and it has not been synthesized. It does appear,
however, that it transforms to the iss at high temperature.

$Cu_{0.12}Fe_{0.94}S_{1.00}$ Phase

Clark (1970) reported the occurrence of a new, unnamed sulfide mineral, of
composition $Cu_{0.12}Fe_{0.94}S_{1.00}$. Aside from its association with chalcopyrite and
cubanite, little is known about it. Clark hypothesizes that it may represent a
new mineral or that it may merely be a member of a cuprian mackinawite solid-
solution series.

Thermochemistry

Barton (1973) recently summarized all previous data and extended our knowledge
of the thermochemistry of the Cu-Fe sulfides, with particular emphasis on the
(CuFe)-S join. These data in the form of a log a_{S_2} - 1/T plot are given in
Fig. CS-34. Extrapolation of these data to any other portion of the system is,
however, extremely tenuous because the three solid solutions of the Cu-Fe-S
system (pyrrhotite, digenite-bornite, and the iss) "... are now known to exhibit
very strong negative departures from ideal mixing. Such non-ideal behavior seems
associated with gross departures from stoichiometry [and] the magnitude of effects
of changes in the Cu/Fe ratio is not known" (Barton, 1973, p. 464). The basis for
much of this is Barton's calculation of the variation of activity of $CuFeS_2$ in
the iss (Fig. CS-35); the reader is advised to read Barton's excellent paper.

There is little other data yet available on the thermochemistry of the Cu-Fe
sulfides, although Barton and Toulmin (1964) have determined the sulfidation
equation

$5CuFeS_2 + S_2 = Cu_5FeS_4 + 4FeS_2$ as $\Delta G = -50,730 + 56.95T$ over the temperature
range 230-500°C by means of electrum tarnish experiments. Schneeberg (1963)
redetermined the sulfidation equation for that reaction in the temperature range
of 219 to 445°C by means of an electrochemical cell as log f_{S_2} = 12.560 - 11067/T.
He also determined the sulfur fugacities over idaite + covellite + pyrite
log f_{S_2} = 18.561 - 16012/T + 1500300/T^2 (218-394°C), and bornite + pyrite + idaite
log f_{S_2} = 19.909 - 18838/T + 2571800/T^2 (218-394°C).

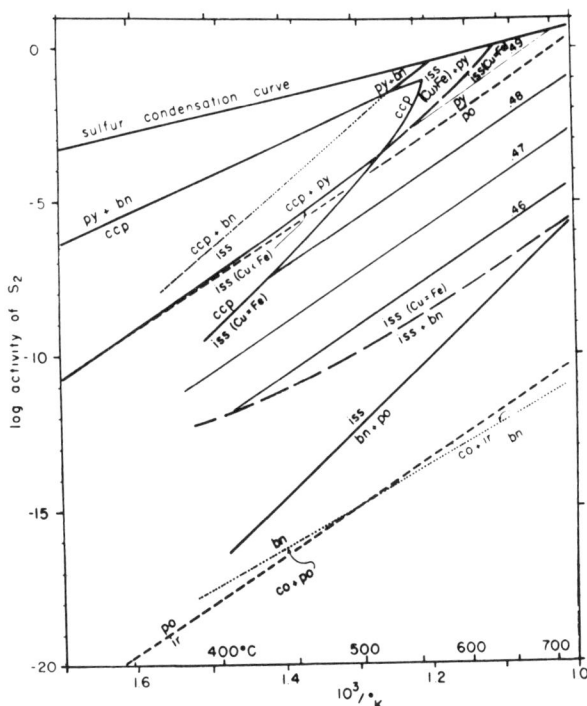

Fig. CS-34. Activity of S_2-temperature diagram for the CuFe-S join. The light lines are contours of atom fraction S in the Cu = Fe intermediate solid solution. Those univariant curves whose positions are speculative are shown by dots. Field boundaries intersecting the CuFe-S join are shown by long dashes. Reference curves not lying on the CuFe-S join are shown by short dashes. Reactions involving covellite and idaite or orange bornite are omitted (From Barton, 1973, *Econ. Geol.* <u>68</u>, 461).

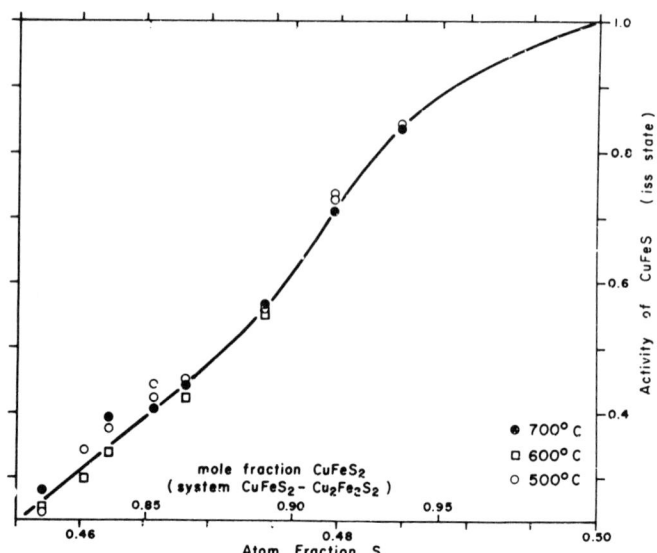

Fig. CS-35. Activity of the CuFeS$_2$ component for the system CuFeS$_2$-Cu$_2$Fe$_2$S$_2$. The reference state for CuFeS$_2$ is the hypothetical stoichiometric iss of CuFeS$_2$ composition. (From Barton, 1973, *Econ. Geol.* <u>68</u>, 463.)

The Ni-S system appears to be one of the few polymineralic binary sulfide systems in which the known mineralogy and phase equilibria are both understood and compatible. Aside from the elements, five minerals have been reported within the system: heazlewoodite (Ni_3S_2), godlevskite (Ni_7S_6), millerite (NiS), polydymite (Ni_3S_4), and vaesite (NiS_2). Three high-temperature polymorphs, one each for Ni_3S_2, Ni_7S_6, and NiS have been reported. Table CS-8 is a compilation of data for the Ni-S system. The phase equilibrium diagram determined by Kullerud and Yund (1962) using silica-tube techniques is in Fig. CS-36. The syntectic nature of the melting of NiS_2 has been revised in accordance with the more recent work of Arnold and Malik (1974).

Table CS-8. Minerals and Phases in the Ni-S System*

Mineral Name	Composition	Thermal stability,°C Maximum	Minimum	Structure type (cell edges in Å)	Remarks
Heazlewoodite	Ni_3S_2	556 (1)*	--	Hexagonal $R32$ (2) $a = 5.74$, $c = 7.14$	
---	$Ni_{3+x}S_2$	806 (1)	524		Lower stability comp. dependent
Godlevskite	$\alpha-Ni_7S_6$	400 (1)	--	Orthorhombic (3) $Bmmb$ $a = 3.27$, $b = 16.16$, $c = 11.36$	
---	Ni_7S_6	573 (1)	400		
Millerite	NiS	379 (1)	--	Hexagonal $R3m$ (5,9) $a = 9.61$, $c = 3.14$	
---	$\alpha-Ni_{1-x}S$	999 (10)	282	Hexagonal $P6_3/mmc$ (4) $a = 3.43$, $c = 5.31$	Lower stability comp. dependent
Polydymite	Ni_3S_4	356 (1)	--	Cubic $Fd3m$ (6,8) $a = 9.48$	
Vaesite	Ni_8S_2	1007 (10)	--	Cubic $Pa3$ $a = 5.67$ (7)	

*Numbers in brackets are references:

(1) Kullerud & Yund (1962)	(6) Menzer (1926)
(2) Westgren (1938)	(7) Kerr (1945)
(3) Fleet (1972)	(8) Lundqvist (1947)
(4) Alsen (1925)	(9) Rajamani & Prewitt (1974)
(5) Grice & Ferguson (1974)	(10) Arnold & Malik (1974)

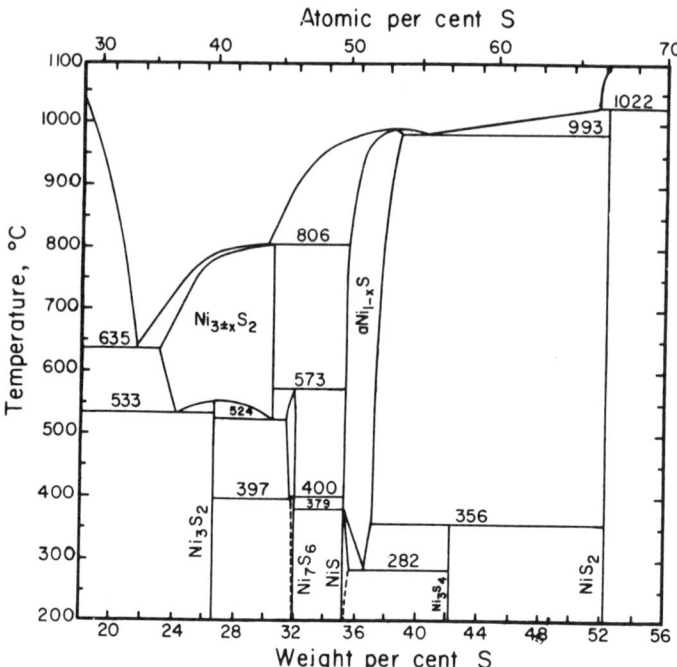

Fig. CS-36. Phase relations in the Ni-S system above 200°C from 18 to 56 wt. %
sulfur in the presence of an equilibrium vapor. (After Kullerud and Yund (1962)
modified in accordance with the data of Arnold and Malik, 1974.)

In contrast to the Fe-S and Cu-S systems, the Ni-S system appears to have no
phases with low thermal stabilities or multiple superstructures. Phase equilibria
studies have revealed phases with relatively high breakdown temperatures, rapid
and reversible transitions, and relatively uncomplicated structures. The Ni-S
minerals and their analogous synthetic phases exhibit no apparent deviation from
stoichiometry, although the high-temperature polymorphs may range widely in
composition. Kullerud and Yund (1962) indicate that the composition of the high-
temperature form of Ni_3S_2 varies as much as 3 wt. % Ni on either side of Ni_3S_2,
that high-temperature Ni_7S_6 deviates 1/4 to 1/2 wt. % Ni on either side of
Ni_7S_6, but that α-NiS, like high-temperature $Fe_{1-x}S$ varies only in the direction of
metal deficiency, forming $α-Ni_{1-x}S$. In each instance of high-temperature nonstoichiometry there is a measureable effect on the inversion temperature. For
Ni_3S_2, metal excess lowers the stoichiometric inversion temperature of 556°C to
533°C whereas metal deficiency lowers it to 524°C. Slight metal excess in the

high-temperature form of Ni_7S_6 lowers its inversion temperature from 400°C to 397°C and approximately 2 wt. % Ni deficiency in α-$Ni_{1-x}S$ lowers its inversion from 379°C to 282°C. The syntectic melting (a solid melting to two liquids) of NiS_2 is the only such case known in sulfide systems.

Mineralogically the Ni-sulfides occur in a broad range of environments but are relatively uncommon.

Heazlewoodite

Heazlewoodite, Ni_3S_2, is stable in its hexagonal mineralogical polymorph below 556°C. It occurs almost always as a product of pentlandite $(Fe,Ni)_9S_8$ reduction resulting from serpentinization of mafic rocks which contained minor amounts of sulfide. Although heazlewoodite was first described by Petterd in 1896, its validity as a mineral was not confirmed until 1947 (Peacock, 1947). Above 556°C it inverts to a nonquenchable, high-temperature form of unknown structure which exhibits a considerable range of non-stoichiometry.

Godlevskite

Godlevskite, Ni_7S_6, is another mineral whose existence was predicted before its discovery. Kullerud and Yund (1962) cleared up previous confusion regarding its composition, established the existence of the two polymorphs, and predicted its occurrence as a mineral. Subsequently it was found and named by Kulagov et al (1969). It has also been found at the Texmont Mine, Ontario (Naldrett, 1972) where it is associated with pentlandite, millerite, and heazlewoodite.

Kulagov et al (1969) and Naldrett et al (1972) indexed the powder diffraction pattern of godlevskite on the basis of an orthorhombic cell with $\underline{a} \approx 9.17$, $\underline{b} \approx 11.27$, $\underline{c} \approx 9.44$Å, but Fleet (1972) refined the structure on the basis of the orthorhombic cell parameters $\underline{a} = 3.27$, $\underline{b} = 16.16$, $\underline{c} = 11.36$Å. The high-temperature form of Ni_7S_6 is stable up to 573°C; its structure is unknown.

Millerite

Millerite, NiS, is the most common of the Ni-sulfides and occurs as a late, low-temperature alteration product of serpentinized sulfides, as hypogene sulfide ores (i.e. Marbridge #3 Mine), and as fine acicular crystals on and in calcite. The name millerite, given by Haidinger in 1845, supercedes such previous names as haarkies, schwefelnickel, nickelkies, harkise, tricopyrit, gelbnickelkies and capillose (Kullerud and Yund, 1962, p. 127).

The low temperature polymorph of NiS exhibits no detectible solid solution and is hexagonal (confirmed by Grice and Ferguson, 1974 and Rajamani and Prewitt, 1974). NiS inverts at 379°C to its high-temperature form (Kullerud and Yund, 1962) which has the NiAs structure and exhibits considerable metal deficiency: $Ni_{1-x}S$ may be stable to as low as 282°C; it melts congruently at 999°C (Arnold and Malik, 1974).

Polydymite

Polydymite, Ni_3S_4, was named by Laspeyres in 1876 because he felt that polysynthetic twinning was characteristic of the mineral. Subsequent workers have not been able to find the twinning but the name has persisted. It is stable below 356°C and has the spinel structure. Natural polydymites generally contain small amounts of Fe and/or Co substituting for Ni. Kullerud (1969) indicates that above about 400 bars pressure the spinel structure of polydymite changes to a nonquenchable high-pressure NiAs structure.

Vaesite

Vaesite, NiS_2, the nickel analog of pyrite was named by Kerr (1945). It is stable without known deviation from its NiS_2 composition to 1022°C where it melts syntectically to two liquids (Arnold and Malik, 1974).

Thermochemistry

Barton and Skinner (1967) have summarized and reduced to their most useful form the thermodynamic data of Ni-sulfides as determined in the studies by Rosenqvist (1954) and Leegaard and Rosenqvist (1964). The sulfidation reactions and the analogous free energy of reaction equations given by Barton and Skinner are:

$3Ni + S = Ni_3S_2$	$\Delta G = 79,240 + 39.01T$	(25–533°C)
$2Ni_3S + S_2 = 6NiS$	$\Delta G = 76,360 + 57.21T$	(573–806°C)
$7/2\ Ni_3S_2 = S_2 - 3/2\ Ni_7S_6$	$\Delta G = 49,260 + 19.91T$	(400–524°C)
$2Ni_7S_6 + S_2 = 14\ NiS$	$\Delta G = 50,600 + 30.53T$	(400–573°C)
$2NiS + S_2 = 2NiS_2$	$\Delta G = 49,880 + 46.67T$	(356–985°C)

The $a_{S_2} - 1/T$ diagram for the Ni-S system is given in Fig. CS-37.

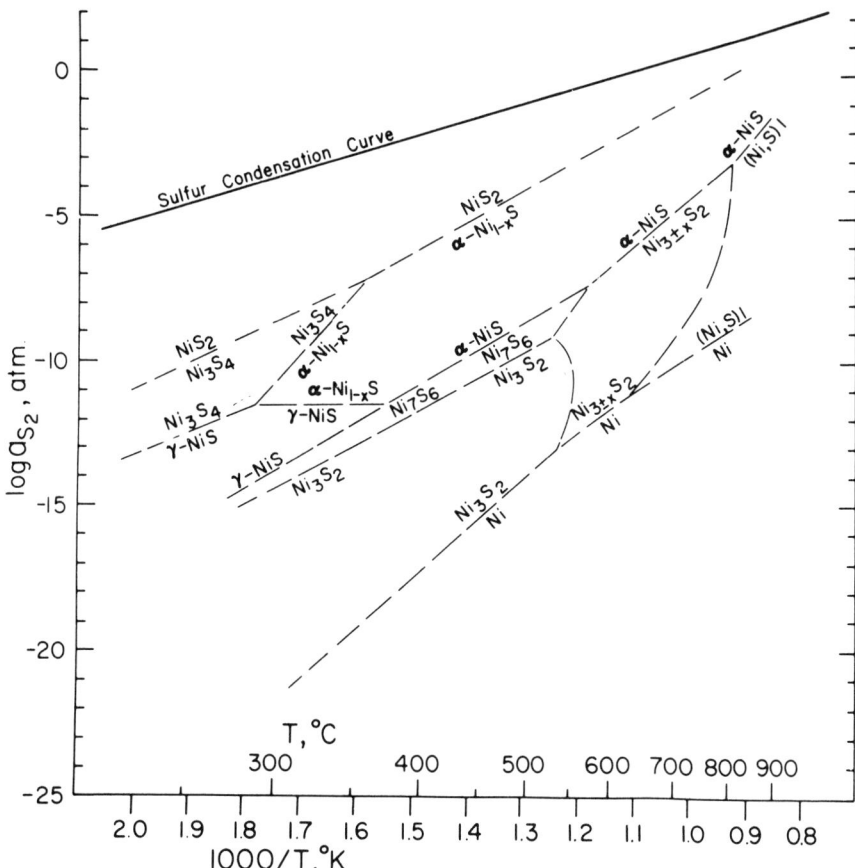

Fig. CS-37. Log a_{S_2} - 1/T plot of sulfidation reactions in the Ni-S system. (Based on data from Barton and Skinner, 1967).

The system Fe-Ni-S includes the compositions of mineral assemblages in many massive sulfide ores. It has the distinction of being the one synthetic sulfide system which is mineralogically applicable at high temperature. Analogous natural assemblages are generally viewed as having formed as immiscible sulfide melts from their parent basaltic melts at magmatic temperatures. An important consequence relative to experimental mineralogy is that high-temperature relationships, where kinetics are favorable and the phase equilibria are well understood, are directly applicable to the "real world." To be sure, there are still uncertainties in our knowledge of the low-temperature equilibria (especially < 100°C) vital to an understanding of secondary alteration effects.

The general phase relationships within the Fe-Ni-S system, first characterized by Lundqvist (1947), have been extended and refined by Kullerud (1963a), Naldrett, Craig and Kullerud (1967), Craig, Naldrett, and Kullerud (1968), Kullerud, Yund, and Moh (1969), Shewman and Clark (1970), Craig and Naldrett (1971), Malik and Arnold (1971), Craig (1971), Misra and Fleet (1973a), Craig (1973), and Misra and Fleet (1974). The known minerals and synthetic phases of the Fe-Ni-S system are listed in Table CS-9.

Table CS-9. Minerals and Phases in the Fe-Ni-S System.*

Mineral Name	Composition	Thermal stability, °C Maximum Minimum		Structure type (cell edges in Å)	Remarks
Pentlandite	$(Fe,Ni)_9S_8$	610 (1)*	--	Cubic *Fm3m* (2,3,4) $a = 10.04$	Fe/Ni variable
(mss)	$(Fe,Ni)_{1-x}S$	1192– 992 (5,6)	--	Hexagonal $a \simeq 3.45$ $c \simeq 5.6$	solid solution from $Fe_{1-x}S$ to $Ni_{1-x}S^x$
Violarite	$FeNi_2S_4$	461 (7)	--	Cubic *Fd3m* $a = 9.46$	Complete solid solution with Ni_3S_4
Bravoite	$(Fe,Ni)S_2$	137 (8)	--	Cubic *Pa3* $a = 5.51$	(9,10) suggest bv is always metastable

*Numbers in brackets are references:

(1) Kullerud (1963)
(2) Lindqvist *et al* (1936)
(3) Pearson & Buerger (1956)
(4) Rajamani & Prewitt (1973)
(5) Jensen (1942)

(6) Kullerud & Yund (1966)
(7) Craig (1971)
(8) Clark & Kullerud (1963)
(9) Springer *et al* (1964)
(10) Shimazaki (1971)

The high- temperature equilibria are dominated by the $(Fe,Ni)_{1-x}S$ solid
solution (mss) which spans the system from a temperature of nearly 1000°C
(Fig. CS-38) to below 300°C. The bulk compositions of many large Ni-bearing
sulfide ores--those of the "Sudbury-type" which are interpreted as forming as
sulfide melts (Skinner and Barton, 1973)--and of sulfide droplets in basaltic
flows (Skinner and Peck, 1969) indicate that the mss is the primary crystalline
sulfide phase and commonly the only sulfide phase to appear until the ores have
cooled several hundred degrees centigrade. The solubility of sulfur in basaltic
melts is controlled strongly by the activity of FeS which in turn is controlled
by the activities of sulfur and oxygen and to a lesser degree by CaO. The mss
phase is a metal deficient NiAs type structure which has $Fe_{1-x}S$ and $Ni_{1-x}S$ as
its end-members and is known to accomodate considerable amounts of Cu and Co into
its structure. Although over most of its compositional range mss is readily
preserved in the laboratory by rapid cooling, it is rarely, if ever, preserved in
nature except where present in subaerial or submarine basic flows. Naldrett,

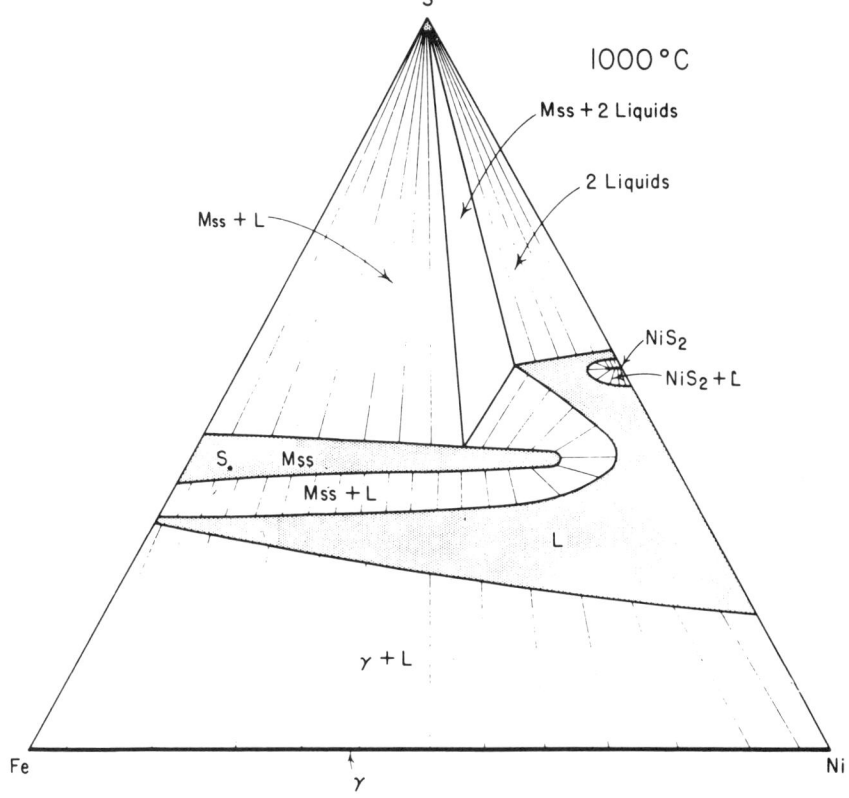

Fig. CS-38. Phase relations in the Fe-Ni-S system at 1000°C in the presence of an
equilibrium vapor (From Kullerud, Yund, and Moh, 1969, *Econ. Geol. Monograph* **4,** 336.)

Craig, and Kullerud (1967) delineated regions of the mss wherein superstructures formed regardless of the rate of quenching. Misra and Fleet (1973b) indicate that they observed superstructures throughout the mss but Francis and Craig (unpublished), using single-crystal techniques found no evidence for superstructures in some portions of the central part of the mss field. [A note of caution: superstructures are not reliably identified by means of powder diffractometry; single-crystal techniques should be employed.]

Decrease in temperature from 1000°C results in the appearance of several binary phases and the development of a variety of 2- and 3-phase regions as shown in Fig. CS-39. In this temperature range relatively little change is noted in the compositional limits of the mss. Pentlandite, $(Fe,Ni)_9S_8$, the principal ore mineral of nickel, becomes stable at 610°C (Kullerud, 1963b), forming through reaction of a nearly pure FeS composition of mss and $(Ni,Fe)_{3+x}S_2$.

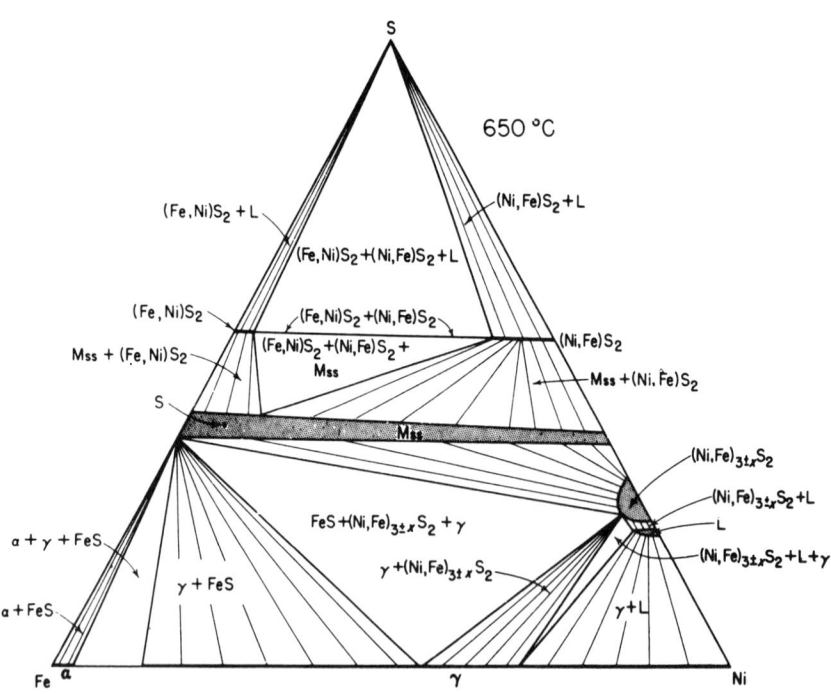

Fig. CS-39. Phase relations in the Fe-Ni-S system at 650°C in the presence of an equilibrium vapor. (From Kullerud, Yund, and Moh, 1969, *Econ. Geol. Monograph* **4**, p. 339).

The common occurrence of pentlandite as "flames" or lamallae within pyrrhotite, long considered to be exsolution features, led to an intensive examination of the central portion of the Fe-Ni-S system by Naldrett, Craig, and Kullerud (1967). They found that the compositional boundaries of the mss field are temperature sensitive and that cooling of typical massive sulfide-ore bulk compositions (initially mss) below about 400°C (a 400°C isotherm is shown in Figure CS-40) would result in exsolution of pentlandite from mss as its S-poor boundary retreated to more S-rich values with decreasing temperature. At least in its first stages, exsolution is strongly controlled crystallographically with the pentlandite lamellae forming such that the (111) plane of pentlandite (pn) is parallel to (001) in the host mss and (110) in pn is parallel to (100) in the mss. This relationship, surmised by Ehrenberg (1932) and reasoned on the basis of atomic population of planes by Gruner (1929), whose argument was correct even if the presumed structures were not, has been confirmed by x-ray single-crystal studies by Francis, Craig, and Gibbs (1974).

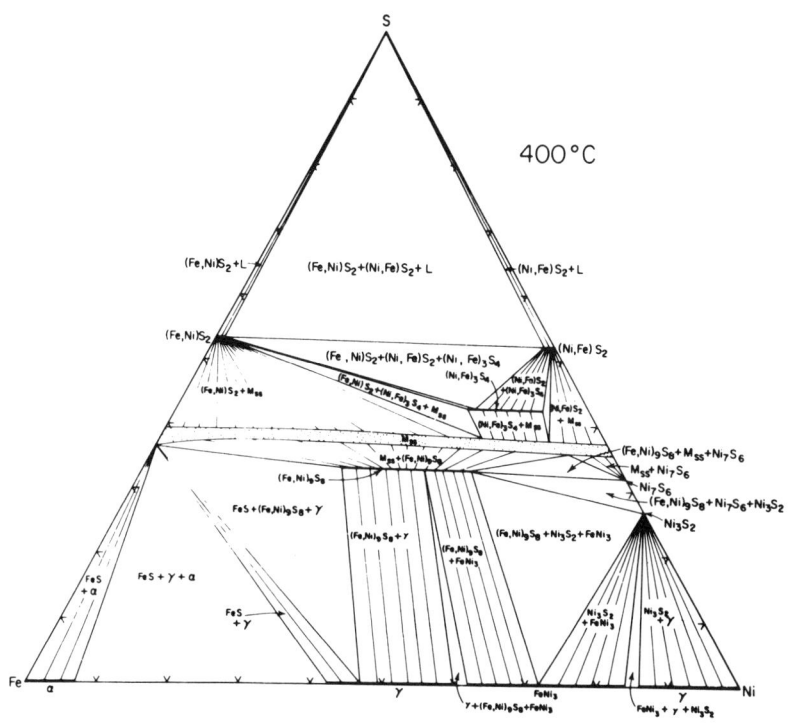

Fig. CS-40. Phase relations in the Fe-Ni-S system at 400°C in the presence of an equilibrium vapor (From Craig, Naldrett, and Kullerud, 1968, *Carnegie Inst. Wash. Yearbook* 66, p. 441).

Pentlandite of varying composition can exsolve along nearly the entire S-poor boundary of the mss field as noted by the broad expanse of the pn–mss 2-phase region in Fig. CS-40. The compositional variation of natural pentlandite in various assemblages has been examined by Nickel (1973), Shewman and Clark (1970), Graterol and Naldrett (1971), and Misra and Fleet (1973a). Not unexpectedly the compositions of pentlandite in pyrrhotite- and/or pyrite-bearing assemblages are more Fe-rich than those of pentlandites in violarite- and/or millerite-bearing assemblages (see Fig. CS-41).

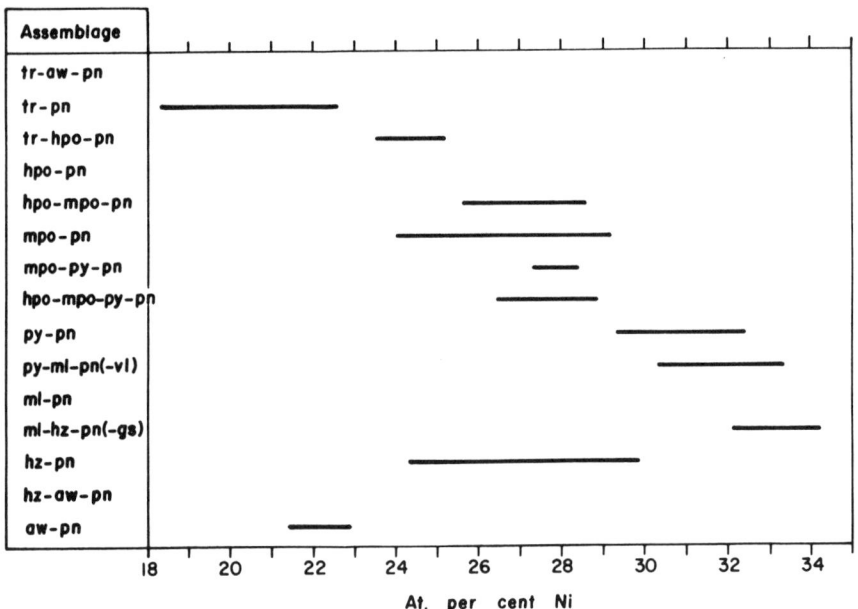

Fig. CS-41. Compositions of natural pentlandites in Fe-Ni sulfide assemblages. Data not available for assemblages tr-aw-pn, hop-pn, ml-pn, hz-aw-pn. Abbreviations are: py – pyrite, vs – vaesite, vl – violarite, pdy – polydymite, ml – millerite, gs – godlevskite, hz – heazlewoodite, pn – pentlandite, gr – greigite, sm – smythite, mpo – monoclinic pyrrhotite, hpo – hexagonal pyrrhotite, tr – troilite, aw – awarwite. (From Misra and Fleet, 1973, *Econ. Geol.* _68_, 530).

The composition of pentlandite in these assemblages raises the question of low temperature relationships in the Fe-Ni-S system. This has been the subject of several studies, most notably Misra and Fleet (1973) and Craig (1973), but remains somewhat enigmatic.

The principal question is whether or not pentlandite and violarite stably coexist, (Fig. CS-42) as suggested by Kullerud, Yund, and Moh (1969) and Misra and Fleet (1974), or are merely in metastable juxtaposition, as suggested by Graterol and Naldrett (1971), Misra and Fleet (1973), and Craig (1973). Interestingly, Misra and Fleet (1974) have noted that in a partially violaritized pentlandite grain ". . . violarite has the same crystallographic orientation as the host pentlandite, and the alteration can be effected by removal of excess metal atoms from pentlandite and redistribution of the remainder." They further suggest that ". . . violarite forms readily from pentlandite because of the ease with which the pentlandite structure can be converted to that of violarite."

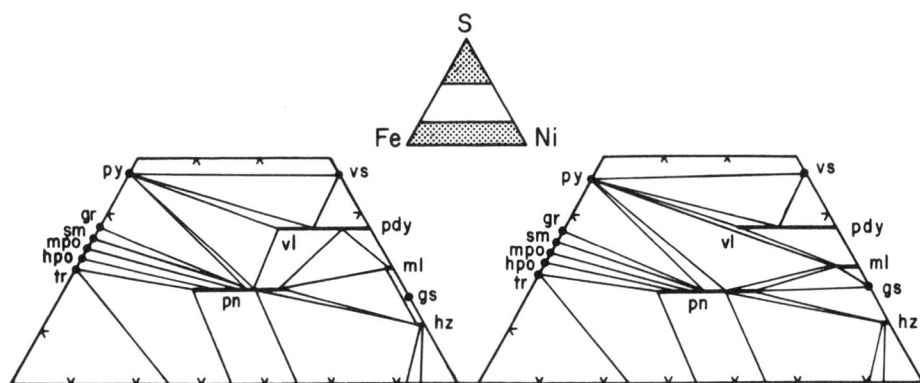

Fig. CS-42. Highly schematic presentations of the two alternative interpretations of low-temperature equilibria in the Fe-Ni-S system. Abbreviations are as in Fig. CS-41.

Pentlandite

Pentlandite, $(Fe,Ni)_9S_8$, is a common mineral in both massive and disseminated sulfides which are associated with mafic rocks. It is stable up to 610°C in the pure Fe-Ni-S system (Kullerud, 1963). However, natural pentlandites frequently contain small amounts of Co substituting for Fe and Ni. Vaasjoki, Hakli, and Toritti (1974) have shown that the presence of Co raises the thermal stability of pentlandite; e.g., the maximum thermal stability of a specimen with 7.5 wt. % Co is 630°C and that of a specimen with 40.8 wt. % Co is 746°C. Pentlandite is cubic (*Fm3m*) and its structure is based on a "cube cluster" of tetrahedral cations (Rajamarni & Prewitt, 1973).

The open nature of the pentlandite structure is responsible for two interesting phenomena. Bell *et al* (1964) found that confining pressure reduces the thermal stability of pentlandite to 425°C at 25kb, and Morimoto and Kullerud (1965), reconfirmed by Rajamani and Prewitt (1974), found that pentlandite has an unusually large coefficient of thermal expansion. This latter phenomenon apparently accounts for the common "shrinkage cracks" observed in veinlets of pentlandite.

Violarite

Violarite, $FeNi_2S_4$, is a thiospinel and occurs commonly as an alternation product of pentlandite. Although Craig (1971) found that violarite is stable up to 461°C, there is agreement that most violarites have formed at very low temperatures through meteoric weathering. At temperatures below 356°C complete solid solution exists between violarite and polydymite (Ni_3S_4). Increase in pressure reduces the maximum thermal stability of the violarite-polydymite series (Wiggins and Craig, unpublished).

In spite of the fact that Craig (1971) was unable to synthesize violarite compositions with Fe/Ni ratios greater than 1/2, natural violarites have been found by many authors with Fe/Ni > 1 (see Fig. CS-43). Natural violarites also have commonly been found to be sulfur deficient (or metal-rich) relative to the ideal metal/sulfur ratio of 3/4 (*cf*. Desborough and Czamanske, 1973).

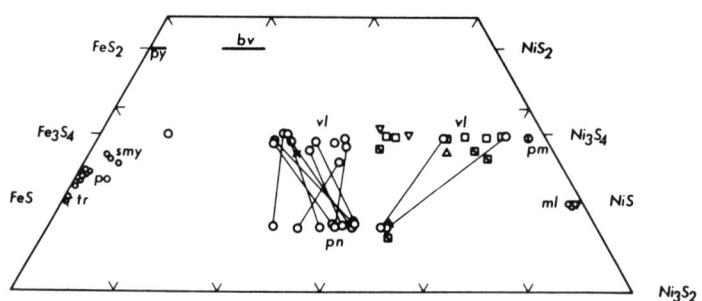

Fig. CS-43. Compositions of natural violarites in the Fe-Ni sulfide system. The central portion of the Fe-Ni-S triangle shows compositions of violarite and coexisting phases. Misra and Fleet (1974): large open circles, coexisting violarite (vl) and pentlandite (pn); open squares, vl not related to pn; small open circles, troilite (tr), pyrrhotite (po), smythite (smy), millerite (ml); open circle with bar, polydymite (pm); solid lines, pyrite (py) and bravoite (bv). Data from literature: open triangles, Graterol and Naldrett (1971); inverted open triangles, Short and Shannon (1930); open squares with bar, Buchan and Blowes (1968); cross, Desborough and Czamanske (1973. (From Misra & Fleet, 1974, *Econ. Geol.* <u>69</u>, 396.)

Violarite, like pentlandite, commonly contains minor amounts of cobalt. Phase equilibria studies have shown complete solid solution between the violarite-polydymite series and Co_3S_4 (Craig and Higgins, 1974). Desborough and Czamanske (1973), in an examination of sulfides from a kimberlite pipe, reported coexisting Ni-bearing pyrite, a new $(Fe,Ni)_9S_{11}$ phase (Ni ∿ 3%), and a new "violarite like $(Fe,Ni)_9S_{11}$ phase" (Ni ∿ 21-30%). It now appears that the new violarite-like phase is merely an iron- and sulfur-deficient violarite like that reported from several localities and that the new low Ni-$(Fe,Ni)_9S_{11}$ phase is a nickel-ferrous symthite of the type discussed by Taylor and Williams (1972).

Bravoite

Bravoite, $(Fe,Ni)S_2$, represents an intermediate phase with the pyrite structure on the pyrite-vaesite join. In spite of its occurrence in several mineralized areas, bravoite has been considered by some to be metstable and analogous to metastable disulfide phases synthesized by Springer (1964) on the FeS_2-CoS_2-NiS_2 plane. Natural bravoites typically exhibit distinct growth zoning, with variable Fe,Ni, and Co contents in adjacent zones (Vaughan, 1969).

$(Fe,Ni)_{1-x}S$ Monosulfide Solid Solution (mss)

The $(Fe,Ni)_{1-x}S$ mss is the principal phase of the Fe-Ni-S system at elevated temperatures and appears to have been the precurser of most ore mineral assemblages in Sudbury-type ores. It spans the Fe-Ni-S system from $Fe_{1-x}S$ to $Ni_{1-x}S$ from nearly 1000°C to below 300°C (see Fig. CS-38,39,40). At temperatures above 300°C mss possesses the NiAs structure; however, on quenching, superstructures analogous to those reported in quenched pyrrhotites develop in many portions of the mss (Misra and Fleet, 1973). Cooling of the mss from elevated temperatures can result in exsolution of pyrite (below 743°C), vaesite (below ∿ 900°C), and/or violarite (below 461°C) from compositions along the sulfur-rich boundary, and pentlandite (below 610°C) along the sulfur-poor boundary.

Below 300°C two miscibility gaps develop in the mss, one on each side of Fe:Ni = 1:1 (Misra and Fleet, 1973; Craig, 1973). Final decomposition of the three resulting segments apparently occurs below 100-200°C.

Thermochemistry

Although there have been several studies of the thermochemistry of the Fe-sulfides and the Ni-sulfides, little systematic investigation has been carried out within the Fe-Ni-S system. Figure CS-44 illustrates some of the sulfidation curves within the Fe-Ni-S system.

Scott, Naldrett and Gasparrini (1974) have investigated the variation of a_{FeS} within the mss at 930°C and found that $Fe_{1-x}S$ and $Ni_{1-x}S$ mix ideally for a given mole % S.

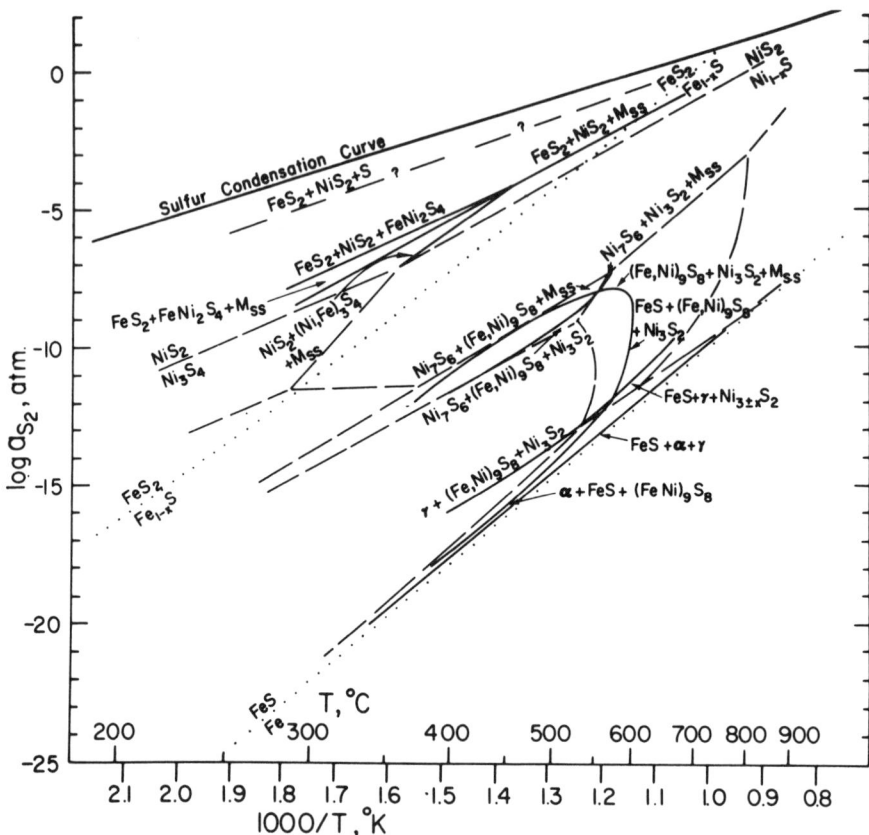

Fig. CS-44. Log a_{S_2} - 1/T plot of sulfidation reactions in the Fe-Ni-S system. Constructed from the data of Barton and Skinner (1967) and Craig and Naldrett (1971).

The sulfosalts comprise an interesting but frustratingly diverse group of minerals. Though widely dispersed in the mineralogic realm, they are inadequately understood and commonly considered merely as curiosities. The term "sulfosalt" has in itself been a point of confusion. Hellner (1958), Nowacki (1969) and Takéuchi and Sadanaga (1969) have defined sulfosalts in terms of structural units, *i.e.* "the presence of the TS_3 (T = As, Bi, Sb) pyramids in the structure is the feature that distinguishes a sulfosalt from a sulfide." On the other hand, Berry (1965) and Staples (1972) have used a chemical classification based upon the presence of one of the semimetals, As, Bi, or Sb, as an essential constituent in the mineral. Neither classification is free of ambiguity; the structural scheme has difficulty with certain transition metal bearing sulfosalts (*i.e.* samsonite, $MnAg_4Sb_2S_6$) whereas the chemical scheme would erroneously include arsenopyrite, FeAsS, and enargite, Cu_3AsS_4, among the sulfosalts.

As noted by Takéuchi (1970) and Craig and Barton (1973), most sulfosalts can be regarded as intermediate phases on joins between simple sulfides (*i.e.* the silver sulfbismulthinides lie on the $Ag_2S-Bi_2S_3$ join); accordingly "the chemical compositions of sulfosalts are in general stoichiometric, the formulae being consistent with the normal valences of the elements involved."

The sulfosalts, like the normal sulfides, exhibit extensive solid solubilities of both their metal and their semimetal constituents, especially at elevated temperatures. Thus, the silver bismuth sulfosalt, matildite, $AgBiS_2$, exhibits solubility in its high-temperature, disordered cubic polymorph both toward Ag_2S and toward Bi_2S_3 (Fig. CS-45). It also exhibits complete miscibility with galena, a solid solution in which the coupled substitution of Ag^{1+} and Bi^{3+} for $2Pb^{2+}$ occurs (Fig. CS-46).

A good example of semimetal substitution is offered by Springer (1969) who found variation from Bi_2S_3 to $(Bi_{0.45}Sb_{0.55})_2S_3$ in natural Bi-Sb sulfides, and by the experimental work of Springer and LaFlamme (1971) (Fig. CS-47) which demonstrates complete solubility of Bi_2S_3 in Sb_2S_3 above 200°C. Although extensive solid miscibility apparently remains possible among some sulfosalts even to low temperatures, many other high-temperature solid solutions decompose on cooling. This often results in formation of lovely and intricate lamellar to myrmekitic intergrowths of sulfosalts (see Fig. CS-48). The optical similarity of many sulfosalts necessitates careful x-ray and microprobe analysis in their identification.

Although there is considerable similarity in the geochemical behavior of As, Sb, and Bi, the stoichiometry of the sulfosalts formed by combination of any given metal with them may be quite different. Compare, for instance, the lead sulfarsenides (Fig. CS-49a), the lead sulfantimonides (Fig. CS-49b), and the lead sulfbismathinides (Fig. CS-49c). At lower temperatures than are shown in Fig. CS-49

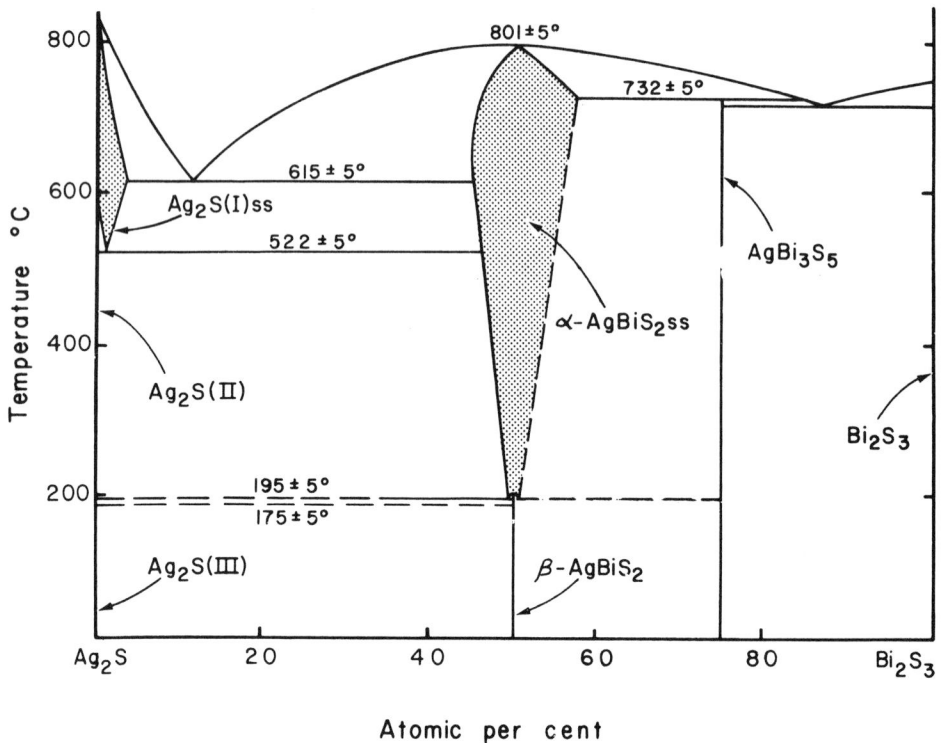

Fig. CS-45. Phase relations on the Ag_2S-Bi_2S_3 join in the presence of equilibrium vapor (after Van Hook, 1960).

the nature of phase relations is much less clear because of the indeterminate status of a large number of sulfosalts which occur naturally but which have not been synthesized. Some of the problem no doubt lies in the stabilizing effects of minor amounts of "impurities" in the complex sulfosalts. Additional complications may arise from the formation and preservation (natural and/or synthetic) of meta-stable sulfosalts because of the small stabilizing energies (discussed below in the section on thermochemistry).

Although many sulfosalts lie along metal-sulfide - semimetal-sulfide joins and may be characterized by relatively simple stoichiometries, there have been identified in recent years an increasingly large number of phases which neither lie along mineral joins nor exhibit simple stoichiometries. For example, Skinner, Luce, and Makovicky (1972) found $CuSbS_2$ (chalcostibite) and Cu_3SbS_3 (later appropriately named skinnerite) along the Cu_2S-Sb_2S_3 join, Cu_3SbS_4 (famatinite) along the hypothetical join CuS-SbS, and $Cu_{12+x}Sb_{4+y}S_{13}$, where $0 \le x < 1.92$ and $-0.02 < y < 0.27$, (tetrahedrite) as a phase without simple stoichiometry (Fig. CS-50).

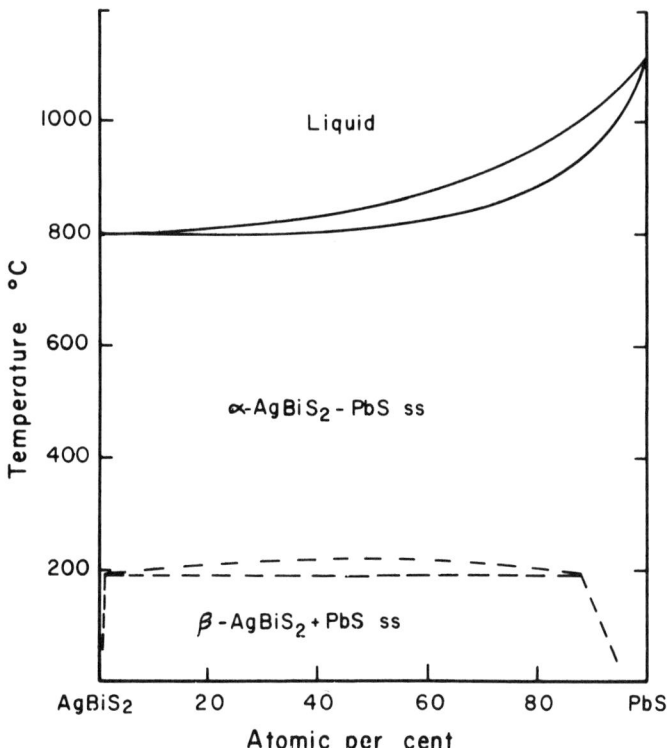

Fig. CS-46. Phase relations on the PbS-AgBiS$_2$ join in the presence of equilibrium vapor (after Van Hook, 1960).

There presently exists no good single reference on the vast realm of sulfo-salt mineralogy; however, Staples (1972) has prepared a listing of sulfosalts and Goodell (pers. comm., 1974) has assembled a compilation of information regarding sulfosalts in the Cu$_2$S-Ag$_2$S-PbS-Ag$_2$S$_3$-Sb$_2$S$_3$-Bi$_2$S$_3$ system which will probably be published in 1975. Nowacki (1962) presents an extensive compilation of the sulfo-salts and their crystallographic parameters. The current state of knowledge concerning most sulfosalts containing more than three components is meagre at best. For what information is available, the reader is referred to the references given in the listing of sulfide systems presented previously in these notes and to the chapter by B.J. Wuensch (this volume).

Thermochemistry

The stoichiometry of many sulfosalts, their relatively rapid rates of equilib-ration at temperatures of 400°C and above, and their structures characterized by subunits similar to the component simple sulfides have made the sulfosalts good candidates for thermochemical studies. Schenck and co-workers (1933, 1938, 1939a,b)

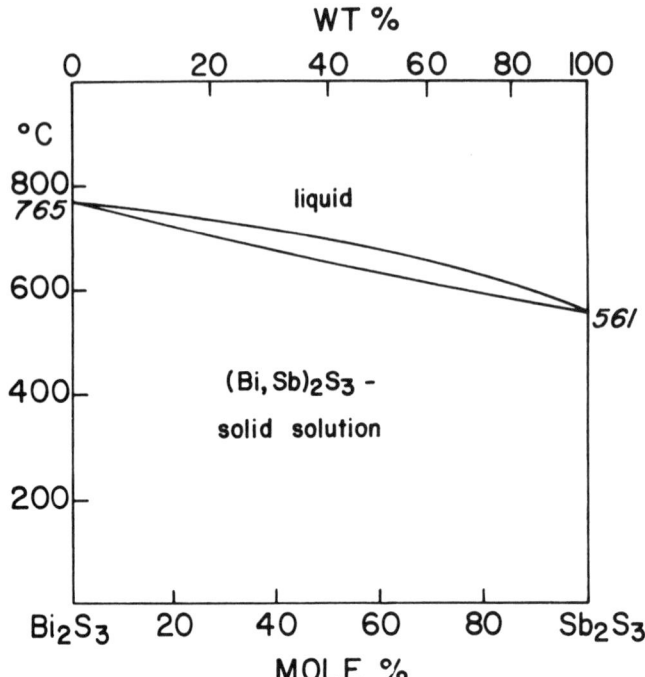

Fig. CS-47. Phase relations on the Bi_2S_3-Sb_2S_3 join in the presence of equilibrium vapor (after Springer and LaFlamme, 1971).

provided the background information (H_2S/H_2 ratios over univariant assemblages) used by Verduch and Wagner (1957), Craig and Lees (1972), and Craig and Barton (1973) in calculating thermochemical parameters for a large number of sulfosalts in the systems Ag-Bi-S, Ag-Sb-S, Pb-Bi-S, Pb-Sb-S, Cu-Bi-S, Cu-Sb-S, and Sr-Sb-S. Barton (1971) has provided additional data from studies in the Fe-Sb-S system.

As noted by Craig and Barton (1973), "any stable intermediate [sulfosalt] must be more stable than the combined end-members, i.e. free energy of the compound must be more negative than the appropriately weighted free energies of the end members." The free energy of reaction, ΔG_m, that stabilizes the sulfosalts is relatively small (Fig. CS-51), hence Barton's (1970) conclusion that "a given sulfosalt is not a great deal more stable than any of several alternative configurations representing the same bulk composition."

The calculation of the stabilizing energy is straight forward but requires knowledge of (1) the relevant phase equilibria, (2) the free energies of formation of the component simple sulfides, and (3) the activity of S_2 gas in equilibrium with the univariant assemblage in question. For example, consider matildite,

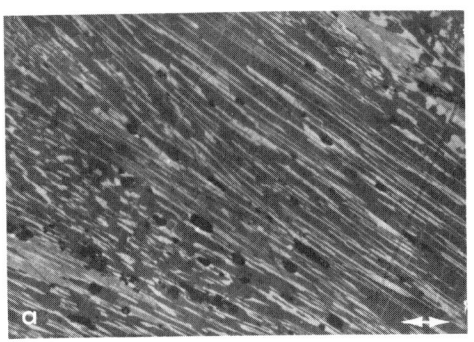

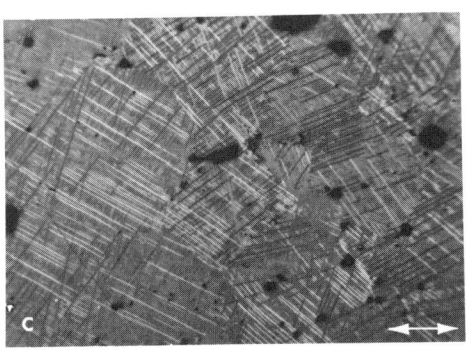

Fig. CS-48. All photomicrographs are from Craig (1965) and were taken using crossed-nicols. (a) Myrmekitic intergrowth of galenobismuthite (dark) and bismuthinite (light) resulting from entectoidal breakdown of phase V of the Pb-Bi-S system. Sample was initially synthesized at 700°C and then annealed at 575°C for 7 days (b) Mymekitic intergrowths of cosalite (black and white) and galena (gray) from Boliden, Sweden (c) Lamellar intergrowths of matildite (white and black) in galena (gray) resulting from decomposition of an initially homogeneous phase. Specimen from Lead, Colorado. The arrows are 20 microns long.

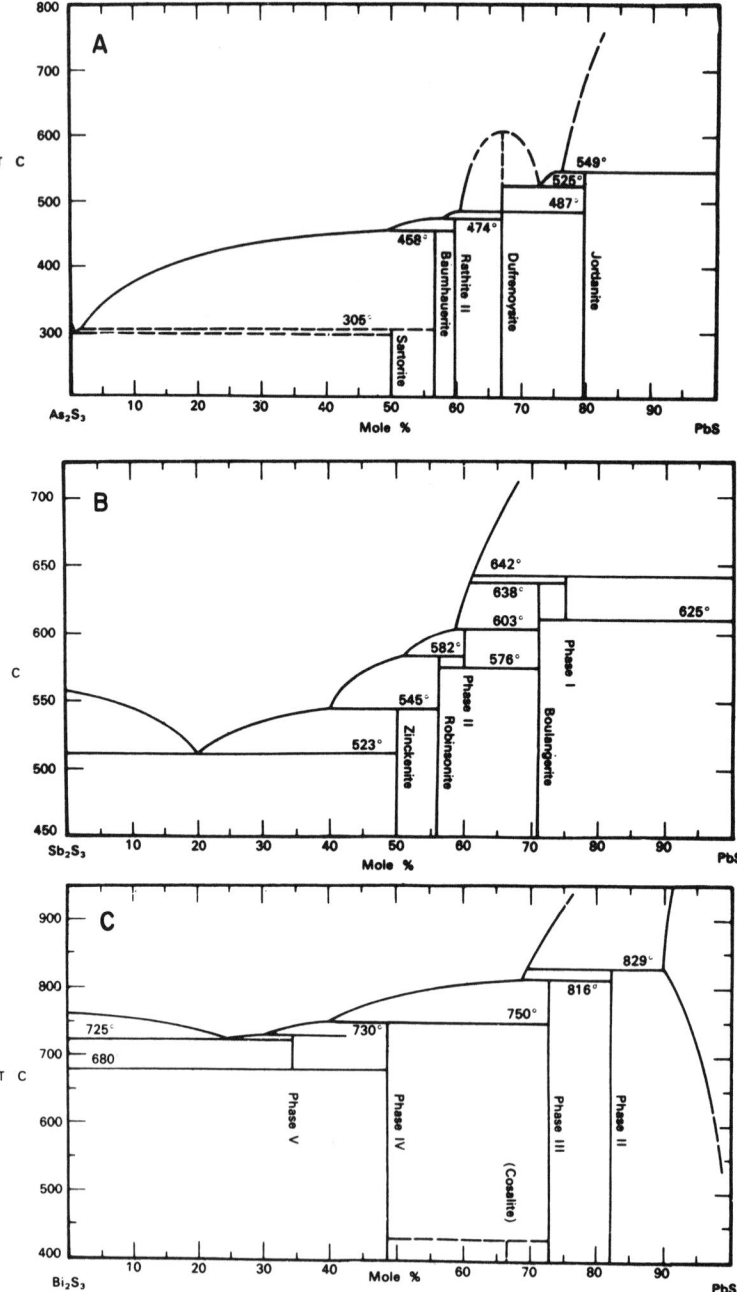

Fig. CS-49. Phase relations in the system (A) PbS-As$_2$S$_3$ (from Chang and Bever, 1973, after Kutoglu, 1969), (B) PbS-Sb$_2$S$_3$ (from Chang and Bever, 1973, after Craig, Chang and Lees, 1973), (C) PbS-Bi$_2$S$_3$ (From Chang and Bever, 1973, after Craig, 1967).

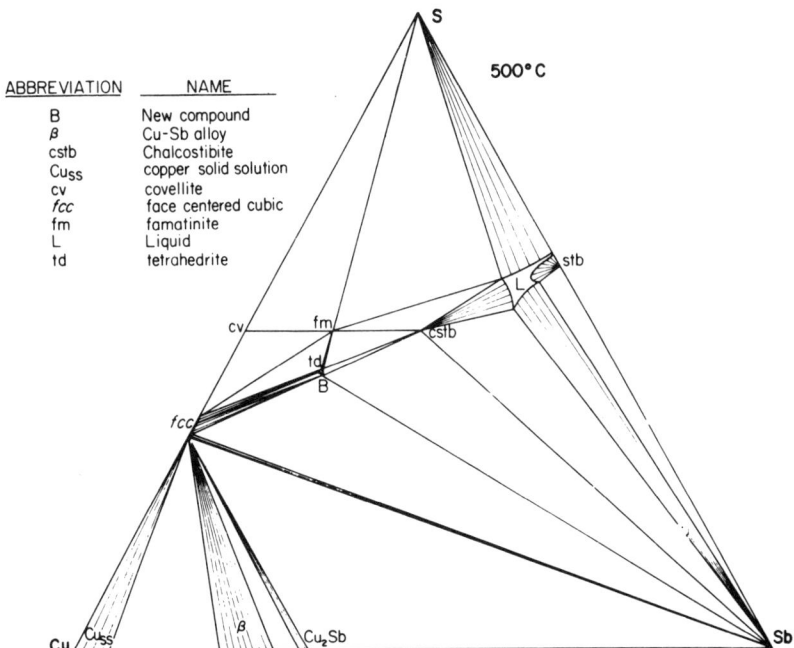

ABBREVIATION	NAME
B	New compound
β	Cu-Sb alloy
cstb	Chalcostibite
Cu$_{ss}$	copper solid solution
cv	covellite
fcc	face centered cubic
fm	famatinite
L	Liquid
td	tetrahedrite

Fig. CS-50. Phase relations in the system Cu-Sb-S at 500°C (element concentrations
in atomic percent). Symbols: B-skinnerite, CS+b-chalcostibite, fm-famatinite,
td-tetrahedrite, L-liquid, stb-stibnite, cv-covellite, fcc-cholcocite-digenite
solid solution (from Skinner, Luce, and Makovicky, 1973, *Econ. Geol.* *67, 932*).

$AgBiS_2$, at 510°C. We know from the work of Craig (1967) that there exists a
univariant assemblage which contains Ag_2S + $AgBiS_2$ + Bi-liquid + S_2 vapor (we
make the assumption that the effects solid solution of Ag_2S in $AgBiS_2$ and the
solubility of Ag and S in Bi-liquid are negligible--this is not quite true but
the errors introduced are small) and that the equilibrium S_2 activity is $10^{-6.73}$
atm. (Schenck *et al* 1939). This assemblage is equivalent to the reaction

$$2/3\ Ag_2S + 4/3\ Bi + S_2 = 4/3\ AgBiS_2.$$

If we treat the solids and Bi-liq as invariant compositions (so that their activities
are unity), the change in free energy of the reaction may be written as $\Delta G = -RT \ln \dfrac{1}{aS_2}$
and $\Delta G = -24097$ cal. The free energy change for $4/3\ Bi + S_2 = 2/3\ Bi_2S_3$ at 510°C
is only -21798 cal; the difference between these two values represents the free
energy which makes $4/3$ mole $AgBiS_2$ more stable than a simple mixture of $2/3$ mole
of Ag_2S and $2/3$ mole of Bi_2S_3. The free energy of formation of the phase $AgBiS_2$
from the elements at this same temperature is then merely the sum of the free energies
of formation of $1/2\ Ag_2S + 1/2\ Bi_2S_3$ + stabilizing energy.

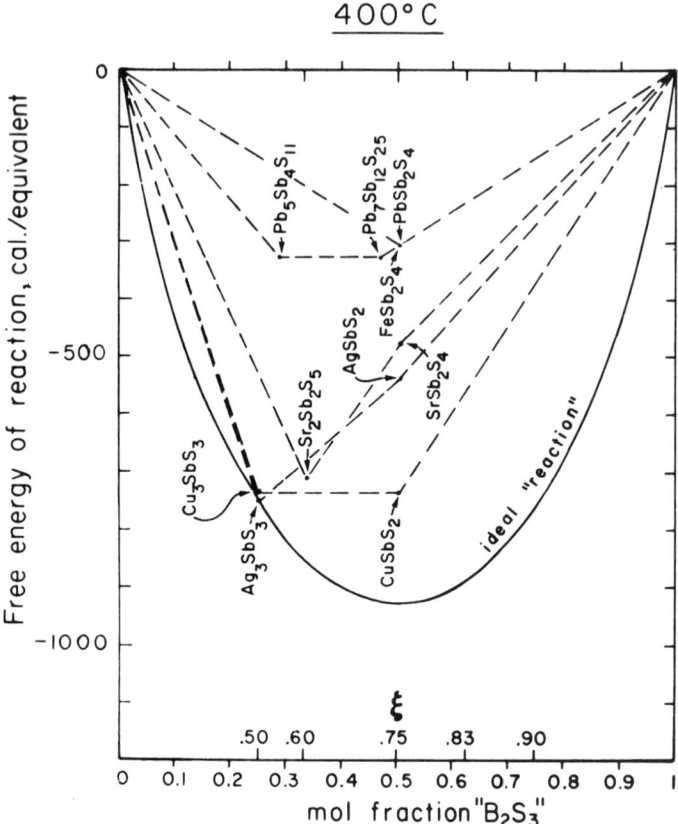

Fig. CS-51. Free energies of reaction (= stabilization) of several sulfosalts at 400°C. Mole fractions are in terms of complete sulfide formula unit (*i.e.*, Ag_2S and Sb_2S_3).

Craig and Barton (1973) converted the stabilizing energy to a temperature-independent form (= an entropy change) which permitted development of temperature-independent equations for sulfidation reactions

$$2/3 \ Ag_2S + 4/3 \ Bi + S_2 = 4/3 \ AgBiS_2 \quad \Delta G = -52891 + 34.34T \ (271-343°C)$$

The reader is referred to the papers by Craig and Lees (1972) and Craig and Barton (1973) for the detailed discussion of these calculations and their application to a large number of sulfosalts. Such calculations are equally applicable to many sulfides and permit formulation of a wide variety of phase diagrams.

It has long been known that the Law of Definite Proportions and the Law of Combining Weights do not strictly apply to crystalline solids but that the proportions of their elements can be variable. This nonstoichiometry can be a very small fraction of an atomic percent as in pyrite or very large as in $Fe_{1-x}S$ and $Ni_{1-x}S$. In either case it manifests itself in the variable electrical, optical, and chemical properties of the minerals. For example, the range of nonstoichiometry in galena is quite small, about 0.1 atomic % (Bloem and Kröger, 1956), yet it gives rise to a wide variation in hardness, reflectivity and conductivity. Fig. CS-52 shows that galena is stable over a range of f_{S_2} values from the Pb+PbS boundary to the sulfur condensation boundary. At high f_{S_2} galena has Pb-vacancies resulting in conductivity by holes (p-type) whereas at low f_{S_2} it has S-vacancies resulting in conductivity by electrons (n-type). At intermediate sulfur fugacities, galena is stoichiometric as shown by the coincidence of the n and p contours.

Some of the kinds of point defects in sulfides and their terminology are in Table CS-10 , using PbS as an example. In this section, we are primarily interested in those defect which give rise to nonstoichiometry (i.e., Pb:S≠1) and their influence on phase equilibria, particularly on polymorphic transitions. The crystal chemical aspects of nonstoichiometry are discussed further in the chapter by Prewitt and Rajamani. The comprehensive review of Kröger (1964) on defects in crystals and the paper by Scanlon (1963) relating chemical and physical properties of sulfides are useful sources of further information.

The rigid definition of the term "polymorphism" requires that the high- and low-temperature forms have identical compositions. Because of nonstoichiometry this requirement is often not met by phases which are compositionally closely-related to one another. At the transition temperature, the closely-related phases **must**, with one exception, have different compositions as required by the phase rule. This objection to the definition of polymorphism is illustrated by Fig. CS-53 which shows three types of two-phase loops which can separate high and low temperature modifications. Fig. CS-53 demonstrates first of all, that the composition limits of the two-phase loop cannot be the same for the high- and low-temperature phases. Secondly, the only exception to the requirement that the phases coexisting at a transition temperature must have different compositions, however small that difference may be, is at a minimum or maximum in the two-phase loop (Fig. CS-53B and C). It is for these reasons that Scott and Barnes (1972) recommended that the definition of polymorphism (and polytypism) be relaxed to accommodate phases which are obviously related by composition but which exhibit a nonstoichiometric proportion of elements of one or two atomic percent.

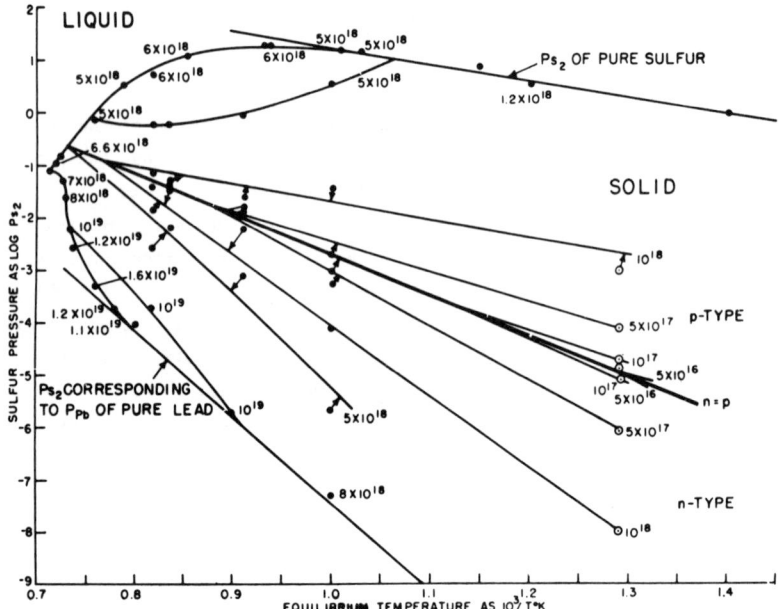

Fig. CS-52. Log f_{S_2} - 1/T diagram for lead sulfide showing contour lines of equal deviation from stoichiometry in atoms/cc. The solid line labelled n=p represents stoichiometric PbS. Above this line there are Pb-vacancies and below it, S-vacancies. Note that the temperature axis is reversed from normal partice. (From Scanlon, 1963, *Mineral. Soc. Amer. Spec. Paper 1, 135-143*).

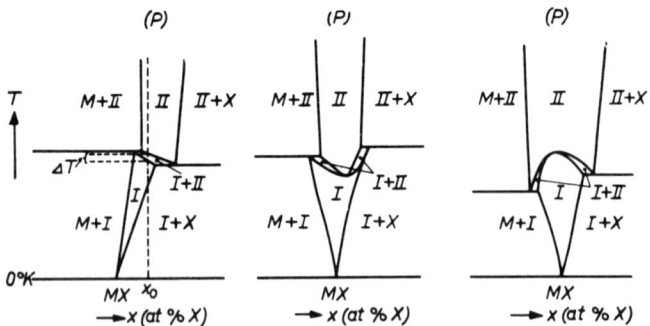

Fig. CS-53. Schematic representation of a partial T-X section of a binary phase diagram with one compound, MX, having a small existence region and a first-order transition between phases I and II. Three possibilities for the phase transition are shown. In the left diagram phases I and II will coexist over the temperature interval ΔT for the bulk composition x_0. (From Albers, 1967, *Physics and Chemistry of II-IV Compounds, p. 165-222*).

CS-100

Table CS-10 Notation for Point Defects According to Two Schemes

Using PbS as an Example (From Simkovich, 1966)

Kröger-Vink[a]	Schottky[a]	Effective Charge[*]	Description
PbS	null		Undisturbed, regular PbS lattice
V_{Pb}	$Pb_{\square x}$	0	Lead ion vacancy that has captured two electron holes, i.e., a lead *atom* vacancy in PbS
$V_{Pb'}$	$Pb_{\square'}$	−1	Lead ion vacancy that has captured one electron hole
$V_{Pb''}$	$Pb_{\square''}$	−2	Lead ion vacancy
V_S	$S_{\square x}$	0	Sulfur ion vacancy that has captured two electrons, i.e., a sulfur *atom* vacancy in PbS
$V_{S·}$	$S_{\square·}$	+1	Sulfur ion vacancy that has captured one electron
$V_{S··}$	$S_{\square··}$	+2	Sulfur ion vacancy
Pb_i	$Pb_{\square x}$	0	Lead ion interstitial that has captured two electrons, i.e., a lead *atom* interstitial in PbS.
$Pb_{i·}$	$Pb_{\square·}$	+1	Lead ion interstitial that has captured one electron
$Pb_{i··}$	$Pb_{\square··}$	+2	Lead ion interstitial
$Ag_{Pb'}$	$Ag_{\square(Pb)}$	−1	A monovalent silver ion substituted for a lead ion
$Bi_{Pb·}$	$Bi_{\square(Pb)}$	+1	A trivalent bismuth ion substituted for a lead ion
$Cl_{S·}$	$Cl_{\square(S)}$	+1	A monovalent chlorine ion substituted for a sulfur ion
n	⊖	−1	An electron
p	⊕	+1	An electron hole, e.g., Pb^{3+} substituted for Pb^{2+}

* The effective charge is the charge deviation from the normal charge located at the site designated by the symbol. E.g., if a +4 cation substitutes for a +2 ion at a regular +2 ion site, the +4 ion carries an excess of 2 plus charges.

From Fig. CS-53 it can be seen that the polymorphic inversion temperature for the metal-saturated phase will be different from that of the sulfur-saturated phase. Also, the sulfur fugacity at the metal-rich limit will be lower than at the sulfur-rich limit. Therefore, variation in polymorphic inversion temperature with f_{S_2} is, by itself, sufficient evidence of nonstoichiometry. Barnes (1973) has calculated that for a polymorphic inversion near 600°C, $f_{S_2} = 10^{-32}$ atm and an assumed ΔH of transition of 1 kcal/mole, a composition range as small as 0.004 atomic percent in a nonstoichiometric compound will produce an easily detected 2°C range in the inversion temperature. The resolution declines with decreasing temperature and increasing sulfur fugacity to 0.133 atomic percent for a 2°C temperature change near 300°C and $f_{S_2} = 10^{-2}$ atm.

The effects of nonstoichiometry on polymorphic inversions are described below for several common sulfides. Particular emphasis is placed on zinc sulfide which shows an unusually wide range in inversion temperature with f_{S_2}.

Silver Sulfide

There are three polymorphs of $Ag_{2\pm x}S$. In the pure system, the transitions cannot be quenched so only the low-temperature modification, acanthite, is found although it is often pseudomorphous after a higher-temperature modification, argentite.

Kracek (1946) measured the variation with composition in inversion temperatures for the polymorphs by DTA (Fig. CS-54). His results show, as is usually the case for sulfides, that the amount of nonstoichiometry is greater at high temperatures than it is at low temperatures. As a result, the transition temperature between acanthite (III) and argentite (II) has a very small range from 177.8°C at the S-saturated limit to 176.3°C at the Ag-saturated limit. The higher temperature transition between unnamed phase I and argentite has a much wider spread from 622°C to 586°C at the S- and Ag-saturated boundaries, respectively.

Molybdenite

Molybdenite (MoS_2) occurs in two polytypic modifications, hexagonal (2H) and rhombohedral (3R).* In his study of molybdenite from the Panasqueira deposit of northern Portugal, Clark (1970) found that the 3R polytype occurred only with pyrrhotite and the 2H polytype with pyrite that had formed by sulfidation of pyrrhotite. The fact that at a given temperature pyrrhotite is stable at lower values of f_{S_2} than is pyrite suggested to Clark that the same was true of 3R

---

* The Ramsdell notation by which polytypes are named is discussed in the chapter by Wuensch.

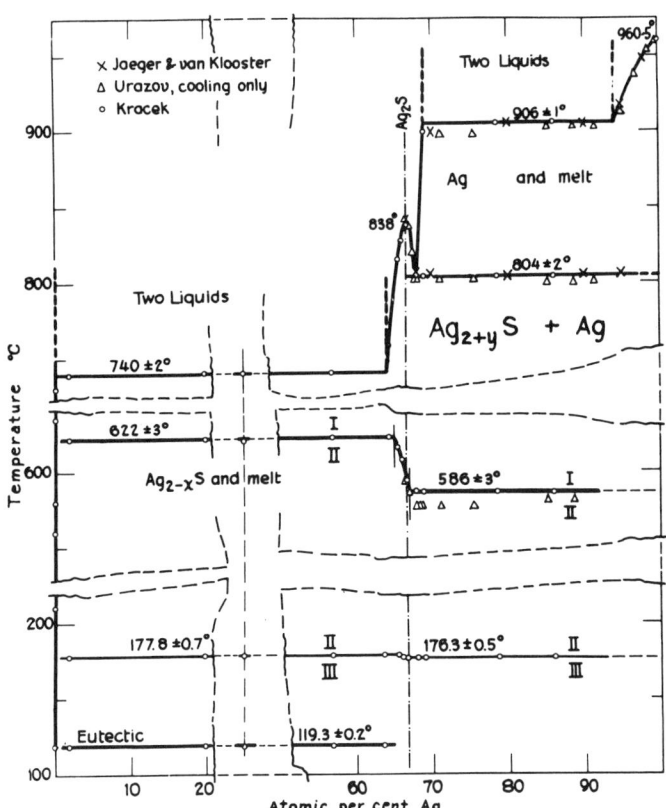

Fig. CS-54. T-X section for the system Ag-S showing the polymorphic inversions of $Ag_{2\pm x}S$. (From Kracek, 1946, *Trans. Am. Geophys. Union, 27*, p. 364-374).

and 2H molybdenite in which case 3R molybdenite should be sulfur-deficient (or Mo-rich) relative to 2H. This view was strengthened by experiments in which 3R molybdenite was converted to 2H by heating with excess sulfur at 400°C. Heating without excess sulfur at 400°C did not produce any change in the 3R polytype although it did invert to 2H at 600°C. Microprobe analyses of natural molybdenites by Clark (1970) indicated slightly more sulfur in 2H than in 3R and more molybdenum in 3R than in 2H but the data are not unequivocal.

Mercury Sulfide

Mercury sulfide occurs naturally in two polymorphic forms, cinnabar and meta-cinnabar, which until recently were regarded as stoichiometric HgS with an inversion temperature of 344°C (Dickson and Tunell, 1959). However, after an extensive experimental investigation in which f_{S_2} was carefully controlled, Potter and Barnes

(1971) found that, in comparison to most sulfides, cinnabar and metacinnabar exhibit a wide range of composition and, consequently, their inversion temperature varies considerable with f_{S_2}. In addition, they found a third polymorph tentatively named "hypocinnabar" which is stable at higher temperatures (R. W. Potter, personal communication, 1972; see also, Barnes, 1973). The maximum composition range for $Hg_{1-x}S$ found by Potter and Barnes (1971) is 50.1 ± 0.1 to 46.35 ± 0.15 atomic % Hg. The polymorphic inversion is a single loop as in Fig. CS-53 (left diagram) extending from stoichiometric HgS at 345 ± 5°C and $f_{S_2} = 10^{-4.57\pm0.07}$ atm. to an Hg deficiency of 3.63 atomic % at 316 ± 3°C and $f_{S_2} = 10^{-3.29\pm0.03}$ atm.* The maximum width of the loop is 1.65 atomic % Hg and occurs at 328 ± 3°C.

Zinc Sulfide

Sphalerite and wurtzite are commonly regarded as the low-temperature cubic and high-temperature hexagonal polymorphs, respectively, of stoichiometric ZnS, with an inversion temperature near 1020°C at 1 atm (Allen and Crenshaw, 1912). Wurtzite which formed at very much lower temperatures in nature is generally thought to be metastable. There are now reasons to reject these assumptions , however. Zinc sulfide is known to be nonstoichiometric and, although the composition range is small, it was shown by Scott and Barnes (1972) to have an unusually large effect on the temperature of the polymorphic transition between sphalerite and wurtzite.

Nomenclature. The terms polymorphism and polytypism are ambiguous as applied to zinc sulfide and require clarification. Mineralogists prefer to call "wurtzite" those polytypes of zinc sulfide whose net symmetry is rhombohedral (nR) or hexagonal (nH) and reserve, as a polymorph, the name "sphalerite" for the cubic (3C) modification (Smith, 1955). Inasmuch as more than 100 different stacking sequences have been observed for zinc sulfide, this terminology is arbitrary. From a structural viewpoint there appears to be nothing special about the three-layer periodicity of sphalerite relative to the stacking sequence of any other polytype. Thus, there is no meaningful difference for zinc sulfide between the terms polymorph and polytype, as pointed out by Smith (1955). Nevertheless, such a distinction between zinc sulfide phases is convenient because of the widespread occurrence of sphalerite in nature and the ease with which it can be distinguished from the wurtzite polytypes by gross optical properties and morphology. We will use this terminology for our present purposes to discuss general relationships without regard for the particular wurtzite polytypes which coexist with sphalerite.

---

*Potter (personal communication, 1974) now believes that the values of f_{S_2} given by Potter and Barnes (1971) may be in error as a result of revised data for speciation of sulfur gas (Haas and Potter, 1974).

Nonstoichiometry of Zinc Sulfide. As early as 1934, Buerger, after concluding that pyrite and marcasite had slightly different Fe/Zn ratios, stated that a similar relationship held for sphalerite - "...the high-sulfur form of ZnS...," and wurtzite - "...the low-sulfur form of ZnS...," (Buerger, 1934, p. 61). However, he did not publish the details of this study.

There are many supporting data in the literature showing that "ZnS" deviates from equal proportions of zinc and sulfur. Most of the information comes from studies in solid state physics and is summarized, in part, by Nickel (1965). Deviations from stoichiometry have been observed by means of electrical properties, luminescence, color, and chemical analyses. The data presented here are by no means complete but are illustrative of the kinds of information that are available on nonstoichiometry from the enormous literature on the properties of this compound.

"ZnS" is an intrinsic semiconductor displaying both p- and n-type conductivity. This means that in the pure (i.e., undoped) compound either Zn vacancies (p-type) or sulfur vacancies (n-type) can exist. When heated above 900°C in a vacuum, flowing nitrogen, or zinc vapor, "ZnS" becomes increasingly n-type, as measured at room temperature by the pulsed Dember effect (Morehead and Fowler, 1962), due to the creation of sulfur vacancies (Baba, 1963; Morehead, 1963). On the other hand, when heated in sulfur vapor, "ZnS" becomes p-type. At constant temperature the increase in the magnitude of the Dember pulse was proportional to the square-root of P_{S_2} as expected for the formation of zinc vacancies (Morehead, 1963). Similarly, Shalimova and Morozova (1965) have examined phase changes in thin films consisting of a sphalerite and wurtzite mixture which were annealed first in zinc and then in sulfur vapor at 940°C and approximately 1.3 atm vapor pressure. After heating in zinc vapor, the proportion of wurtzite was greatly increased, in some cases to 100%. Annealing in sulfur vapor produced the reverse effect; the hexagonal phase changed to sphalerite. Optical absorption studies confirmed that this hexagonal phase contained a deficiency of S and that this deficiency was removed in the cubic phase. Shalimova and Morozova concluded that (p. 471) "...disruption of the stoichiometric composition of zinc sulfide has an effect on its crystal structure. When the crystals grow under high zinc pressure and zinc penetrates the lattice, the hexagonal form results. When the composition...approaches stoichiometric composition, the lattice passes over into the cubic form ...." In a later paper, Morozova et al (1969) reached a similar conclusion from a study of point defects in zinc sulfide powders. Similar results are shown by Bansagi et al (1968) in their figure 5 summarizing the effects of zinc and sulfur pressure on the kinetics of inversion.

The relationship between stoichiometry and conditions of heat treatment has also been derived from luminescence studies. When "ZnS" is heated in zinc vapor a green fluorescence develops due to the incorporation of excess zinc (i.e., S-deficient) above the stoichiometric composition (Kröger and Vink, 1954). Uchida (1964) was able to show that this treatment as well as firing in an argon + 20%H$_2$

mixture produced an increasing concentration of sulfur vacancies with increasing temperature. Firing in sulfur vapor produced a change in the emission spectra which Uchida interpreted as being due to interstitial sulfur atoms. However, in view of Morehead's (1963) conductivity data it is more likely that zinc vacancies were produced.

Color may be another sensitive indicator of nonstoichiometry. Scott and Barnes (1972) found that honey-yellow Joplin sphalerite, when heated for several weeks in aqueous sulfide solutions below 200°C and from 1 to 28 atm H_2S pressure, became dark brown to black, depending on the duration of the experiment. The few tenths % Fe in the sphalerite remained constant and, therefore, was not responsible for the darkening. From an examination of many analyzed natural sphalerites, Togari (1961) concluded that color (chromaticity) was controlled by excess of sulfur over total metals and not just by iron. Iron was found only to enhance the brilliance (intensity) of the color. Similarly, Roedder and Dwornik (1968) could not correlate color of natural sphalerite from Pine Point with iron content. Therefore, it is not necessarily true that dark colored sphalerite is rich in iron (Platonov and Marfunin, 1968). Rather it may indicate that the sphalerite is metal-deficient and formed under highly sulfidizing conditions.

Small differences in composition between sphalerite and wurtzite are found in careful chemical analyses of "ZnS." Nickel (1965) noted that most analyses of natural sphalerite indicate a sulfur to metal ratio greater than one. Especially precise analyses by Pankratz and King (1965) of synthetic sphalerite and wurtzite (both pure and Fe doped) give S/(Fe + Zn) mole ratios of 1.001, 1.002, 1.003, and 1.004 for sphalerite and 0.995 and 0.998 for wurtzite. These data indicate that sphalerite is Zn-deficient and wurtzite is S-deficient.

Sphalerite-Wurtzite Inversion. Scott and Barnes (1972) have examined the sphalerite-wurtzite equilibrium by means of hydrothermal recrystallization and H_2/H_2S gas-mixing experiments in which f_{S_2} was controlled in the manner described in the earlier chapter on experimental methods. Their experiments, summarized in Fig. CS- 55, showed that sphalerite and wurtzite can coexist over a range of temperatures well below 1020°C as a function of f_{S_2}. The f_{S_2}-dependence of the equilibrium demands that wurtzite is S-deficient relative to sphalerite and is consistent with available data on their stoichiometries. It is also consistent with earlier reported "anomalous" inversion temperatures from 600°C to above 1240°C which must simply reflect the differences in the experimental environments leading to the formation of Zn-deficient sphalerite at high f_{S_2} or S-deficient wurtzite at low f_{S_2}.

The effect of other elements in solid solution on the stability of wurtzite vs sphalerite is largely unknown. There is evidence from unpublished experiments at 300°C and 400°C that Cd-rich mixed crystals have the wurtzite structure at values of f_{S_2} higher than those of the univariant curve in Fig. CS-55. Iron

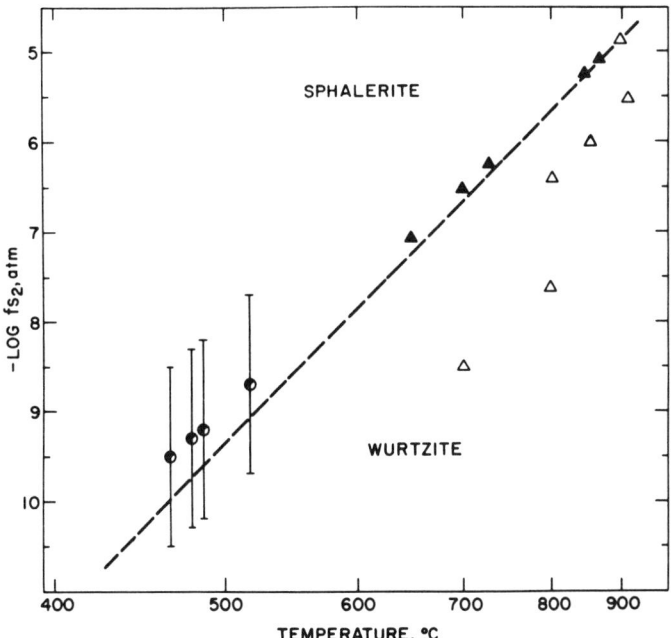

Fig. CS- 55. The univariant sphalerite-wurtzite boundary as a function of sulfur fugacity and temperature at 1 atm. Triangles represent gas mixing experiments (closed = sphalerite; open = wurtzite). Circles are calculated points for the boundary from hydrothermal recrystallization experiments in NaOH solutions; bars show the estimated accuracy of the calculated f_{S_2}. Precision in f_{S_2} in the gas mixing experiments is within each point. (From Scott and Barnes, 1972, *Geochim. Cosmochim. Acta 36, 1284.*)

substitution appears to have the opposite effect of stabilizing the sphalerite structure to very low f_{S_2} as indicated by the fact that Barton and Toulmin (1966) found iron-saturated zinc sulfide buffered by Fe + FeS to be sphalerite and not wurtzite at temperatures below 850°C. At very high temperatures, solid solution of ZnO which has the wurtzite structure is sufficient (0.9 mole % at 900°C) to cause Zn(S,O) also to have the wurtzite structure (Skinner and Barton, 1960). However, at the very much lower temperatures of wurtzite precipitation in nature, the solubility of ZnO in ZnS must be vanishingly small and is unlikely to influence the relationship in Fig. CS-55.

In nature, wurtzite of low FeS-content is normally found with pyrite or marcasite rather than with pyrrhotite, so the sphalerite-wurtzite boundary in Fig. CS-55 must, with decreasing temperature, cross the pyrite-pyrrhotite solvus and pass into the pyrite stability field. The extrapolation of the sphalerite-wurtzite boundary to lower temperatures in Fig. CS-56 is consistent with this observation and with the experimental data at higher temperatures. Of course, such an extrapolation implies that wurtzite does not have a geologically-important lower temperature limit of stability. Evidence is found in the usual occurrence of wurtzite in low-temperature sedimentary or diagenetic environments.

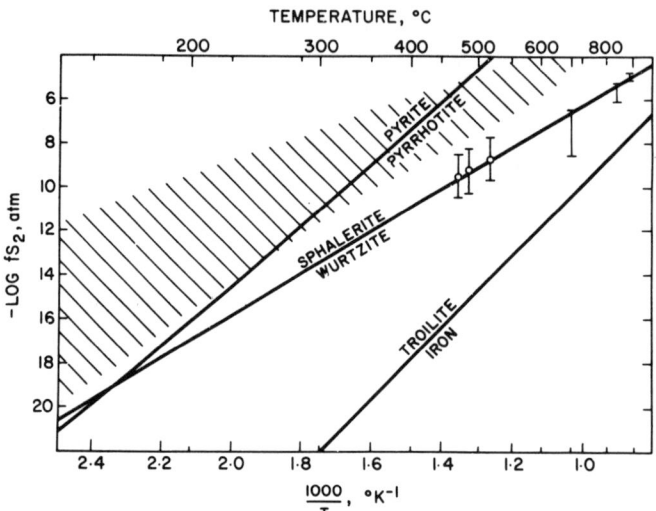

Fig. CS-56. Extrapolation of the sphalerite-wurtzite equilibrium as a function of log f_{S_2} and temperature. Data points and bracketing bars are from Fig. CS-55. The pyrite-pyrrhotite and iron-troilite equilibria are superimposed for reference and any influence of iron content on the sphalerite-wurtzite curve has been neglected. The hachured area is the "main line" ore-forming environment from Barton (1970). (From Scott and Barnes, 1972, *Geochim. Cosmochim. Acta* <u>36</u>, *1289*).

Fig. CS-56 provides a ready explanation for the preponderance of sphalerite over wurtzite in nature. At high temperatures wurtzite is stable under prohibitively low f_{S_2}'s compared to the normal sulfidation state of ore-forming environments as outlined by the hachures in Fig. CS-56. Even at low temperatures wurtzite is stable at unusually low f_{S_2}'s. Examples of such occurrences are described by Scott and Barnes (1972).

Wurtzite Polytypes. Examination of the wurtzites produced in the hydrothermal recrystallization experiments by single crystal x-ray methods revealed another univariant boundary separating the 2H and 4H polytypes (Scott and Barnes, 1967; Scott, 1968). It is not possible, for lack of thermodynamic data, to calculate actual values of f_{S_2} from these experiments but the f_{S_2}-dependence of the 2H-4H equilibrium is clear and is shown schematically in Fig. CS- 57.

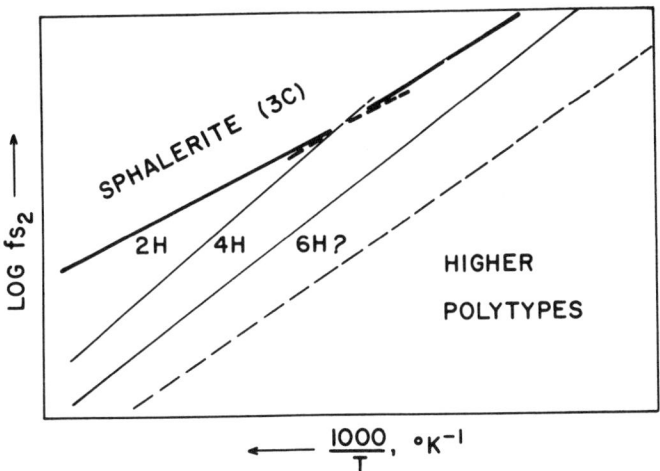

Fig. CS- 57. Schematic log f_{S_2}-temperature diagram showing possible relationships among zinc sulfide phases.

Polytypism in zinc sulfide and other compounds has been variously attributed to dislocations (Frank, 1951; Vand, 1951a,b) and to thermodynamic effects (Jagodzinski, 1954a,b; Schneer, 1955). However, there is considerable disagreement as to whether polytypes are separate phases or are continuously variable, layered structures (Verma and Krishna, 1966). The hydrothermal experiments, in which wurtzite 2H and 4H were found to coexist as a function of f_{S_2} and temperature at constant pressure, show these polytypes to be separate, thermodynamically-stable phases which undergo first-order phase changes. Furthermore, because f_{S_2} is a controlling variable, a change in composition accompanies each phase change. Wurtzite 4H must be deficient in sulfur relative to 2H which, as discussed previously, is deficient in sulfur relative to sphalerite. There are no direct measurements of the degree of nonstoichiometry exhibited by the two polytypes. However, considering that the maximum departure from stoichiometry for zinc sulfide is probably less than 1 atomic percent, the compositional difference between the two polytypes is probably on the order of a few hundredths atomic percent.

From their study, Scott and Barnes (1967) have proposed that polytypism in zinc sulfide is controlled by stoichiometry. It is unlikely that relationships between the 2H and 4H polytypes are unique in this system. Rather, all of the short-period polytypes of wurtzite are regarded as thermodynamically stable phases, each with a narrow compositional range and occupying a field in f_{S_2}-temperature space, as shown schematically in Fig. CS-57. The common occurrence of several polytypes in syntactic coalescence in both natural and vapor-grown synthetic wurtzite suggests that the isothermal f_{S_2} range over which a particular polytype is stable is quite small. Except for 2H and 4H, the relative positions that the polytypes occupy in f_{S_2}-T-X space cannot be determined without further experimental work. In the case of SiC which is isostructural with ZnS, Gomes de Mesquita (1967) has shown that the bond energy, and hence, thermodynamic stability, of some polytypes is dependent upon the percent hexagonality of the structure. There may also be a simple structural relationship, such as percent hexagonality or number of layers in a repeated unit as a function of f_{S_2} and temperature for wurtzite polytypes, but as yet there is no evidence.

Unfortunately, the present data provide no information on the long-period polytypes. It is unlikely that compositions and temperature control these more complicated structures because there are a large number of polytypes to fit into a small compositional range. It is suggested that these long-period polytypes are metastable phases produced by stacking faults in the short-period ("parent") structures in terms of the Frank or Vand dislocation theories. In support of this, wurtzite crystals from nature, where equilibrium between the crystal surface and aqueous solution are likely to prevail (Barnes and Czamanske, 1967), commonly have short repeat units. These grew under quiescent conditions at low temperatures (100°C or less) where the stresses necessary to produce dislocations are minimal. On the other hand, synthetic crystals, grown by vapor transport where thermal stresses are large, commonly contain polytype families in which the coalesced long-period forms are structurally related to and, in fact, may have been derived from shorter-period "parent" polytypes (Mardix, Alexander, Brafman and Steinberger, 1967).

Mineral. Soc. Amer. Spec. Pap. **3**, 187–198 (1970).   Reprinted by permission.

Ch. 6

SULFIDE PETROLOGY[1]

Paul B. Barton, Jr.

U. S. Geological Survey, Washington, D. C. 20242

ABSTRACT

The petrology of sulfide-rich rocks has the following special characteristics: (1) despite overall mineralogical variety, many ore deposits exhibit simple mineralogy at any single stage of mineralization; (2) the depositional pattern of many ores is such that even a single hand specimen may represent a complex series of superposed chemical systems; (3) sulfides are very susceptible to retrograde metamorphic processes. The interpretative study of sulfide-rich rocks, therefore, requires especially careful effort to reconstruct the history of the system.

Because many ore-forming systems tend to be open (in the sense of Korzhinskii, 1959) and to contian relatively simple phase assemblages for the number of components, quantitative knowledge of the compositions of phases is especially important in reducing the number of degrees of freedom so that the physical-chemical environment of ore formation may be evaluated. In general, the compositions of individual ore minerals reflect the activities of components in the environment more strongly than they do the simple effects of pressure and temperature on the minerals.

Aside from the presence of the major metals themselves, the two most important variables in determining sulfide mineral assemblages are activity of sulfur and temperature; total pressure ranks a poor third. An abundance of sulfidation-reduction reactions makes naturally occurring sulfide systems particularly amenable to thermodynamic quantification. The free-energy changes of univariant sulfidation-reduction equilibria are well represented as linear functions of temperature. The many experimentally determined invariant points help tie the grid of sulfidation reactions into a tightly knit package, whose internal consistency exceeds the accuracy of the individual measurements.

In spite of the widespread occurrence of its minerals, sulfur is a minor constituent of the crust compared to other redox-participating elements such as iron, carbon, hydrogen and oxygen; it is therefore these latter elements which ultimately exert the controlling influence on the activity of sulfur and thus on the mineralogy of the sulfides.

INTRODUCTION

The study of sulfide-rich rocks has always been properly considered as one facet of petrology; however, it is an aspect that is all too often ignored by the modern petrologist and has appeared almost solely within the realm of the economic geologist. My goal here is to broaden petrology, not to restrict economic geology.

Despite the presumptive title, this paper will not attempt to cover all aspects of sulfide petrology, for that would fill a thick volume. Instead, I shall consider briefly some of the problems and potentials in interpreting sulfide mineral assemblages. Supplementary material for this subject should include at least the topics covered in *Geochemistry of Hydrothermal Ore Deposits* edited by H. L. Barnes (1967) plus the articles by Gunnar Kullerud and Harold Helgeson in this volume. Despite excellent recent studies, such as those described in the Graton-Sales Volume (Ridge, 1968) or the work of Holland and coworkers at Providencia and Bluebell (Ohmoto and others, 1969, Sawkins, 1964), there has been, and remains, a profound deficiency in descriptive material relating to sulfide-rich rocks. In no instance of an appreciable degree of complexity are the mineral assemblages quantitatively understood. This lack is rapidly becoming the most serious barrier to additional progress.

Much of our effort in studying sulfide-rich rocks is directed toward understanding the petrogenesis of the sulfides, not merely their descriptive petrography. Therefore, we are not only concerned with what *is* but also with what *was*. We are especially interested in deducing the environmental parameters (temperature, pressure and

[1] Publication authorized by the Director, U.S. Geological Survey.

activities of all components) that prevailed when the deposit was forming, or was being modified from a still earlier form. Moreover, to evaluate the probability of various possible processes that may have operated on the rock, we would like to know the sequences of environments in time and space. Such are the goals, now let us consider some of the difficulties that have thusfar barred the way.

COMPLEXITY OF THE PROBLEM OF INTERPRETING MINERAL ASSOCIATIONS

The problem originates chiefly in the unraveling of a complex depositional pattern which may have been masked or erased by post-depositional processes. The initial goal of an investigation is then to recognize and characterize equilibrium mineral assemblages (groups of minerals that represent a current or former equilibrium state). This, however, is the most difficult and most neglected part of sulfide petrology.

First of all, with the exception of magmatic segregation, the processes of formation of most sulfide ores involve deposition from a dominantly aqueous fluid. There are many important deposits (such as those of the Mississippi Valley type, and many, if not most, of the geologically young deposits of the Cordilleran region of North and South America) which preserve a record of the mineral relationships at the time of formation. Minerals from such deposits frequently exhibit depositional detail in a bewildering array (see Fig. 1A thru D), an array marked principally by changes in chemistry but also by relatively minor changes in temperature. Sulfide ores tend to be depositories of the rarer elements; as a consequence of this and the chemical changes, the number of mineral species in a deposit as a whole can be large, sometimes much larger than would be permitted by the phase rule.

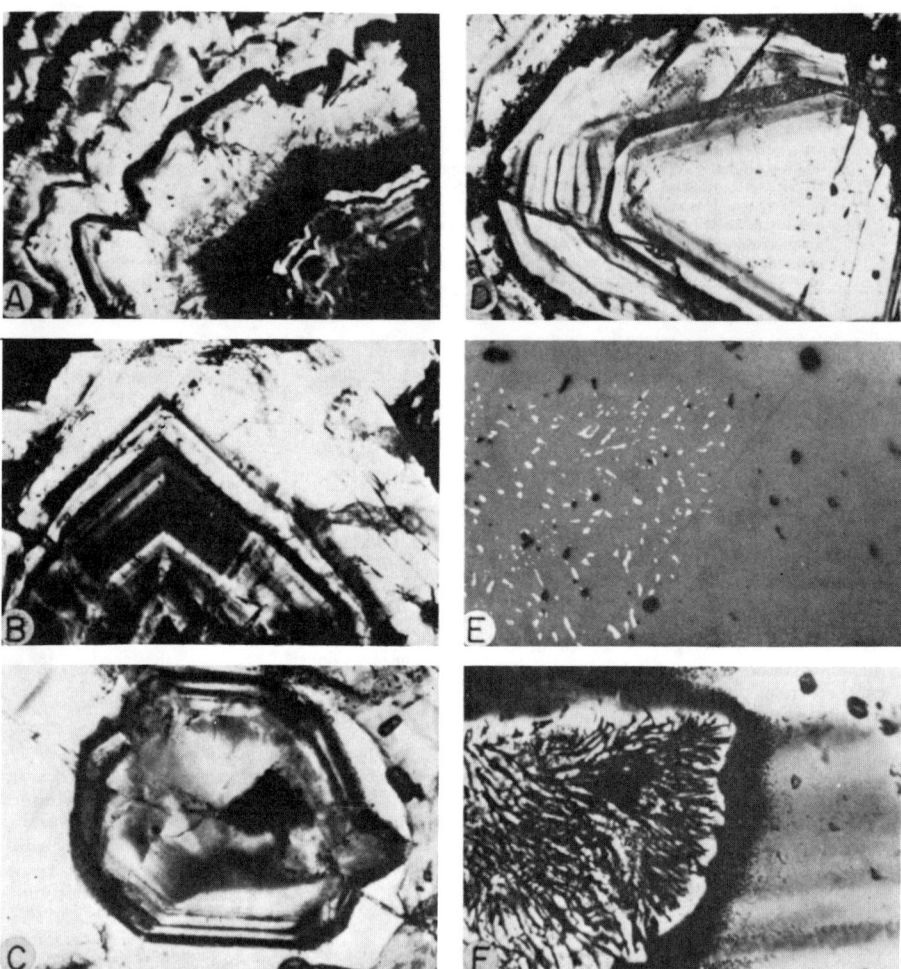

FIG. 1. Growth-zoned sphalerite from several localities. In each instance the area photographed is almost entirely sphalerite and represents less than 20 percent of the sphalerite paragenesis for that sample. The specimens are prepared as doubly polished sections from 0.05 to 0.5 mm thick. In general, color correlates with iron content ranging from 1 to 5 percent Fe in the darkest bands to <0.1 percent in the lightest ones. A. Piquette mine, Wisconsin. Width of photo is 3.3 mm. Transmitted light; B. OH vein, Creede, Colorado. Width of photo is 3.3 mm. Transmitted light; C. Leonard mine, Butte, Montana. Width of photo is 0.5 mm. Transmitted light; D. OH vein. Creede, Colorado, Width of photo is 3.3 mm. Transmitted light. E. OH vein, Creede, Colorado. Width of photo is 0.25 mm. Reflected light. Photo shows nearly monotonous sphalerite with a few "blebs" of chalcopyrite. See Fig. 1F for the same field in transmitted light. F. Same field as Fig. 1E, but in transmitted light. This shows nearly colorless sphalerite with arborescent chalcopyrite preferentially replacing iron-rich zones in growth-banded pale yellow to brown sphalerite. The diffuse dark interface between the colorless and banded sphalerite is caused by disseminated fine chalcopyrite.

However, detailed study inevitably reduces the mineralogical complexity to the extent that for any *single stage of mineralization* (representing an interval of deposition during which there was no discernable chemical or physical change) there are usually far too few, rather than too many, phases for the number of components. In the terminology of Korzhinskii (1959), the system tends to be "open" with respect to many components.

Over the years a considerable body of observational data on the textural interrelationships of sulfides has accumulated (for example, see Bastin, 1950, or Edwards, 1947, or Ramdohr, 1960.) It is certain that ore textures present much information, but it is equally certain that there are few areas of scientific endeavor that are more subject to misinterpretation than the study of ore textures. The interpretation of ore textures is the most maligned, most difficult, and most important aspect of the study of these rocks. As an example of the difficulty in recognizing features, consider the standard polished section (Fig. 1E) showing, in reflected light, a uniform field of sphalerite with a few blebs of chalcopyrite to one side. One might pass a hundred such fields without comment except, perhaps, to wonder inconclusively whether the chalcopyrite had grown with, exsolved from, replaced, or been replaced by, the sphalerite. The same field of view illuminated with transmitted light (Fig. 1F) shows colorless sphalerite with chalcopyrite replacing banded yellow sphalerite, a feature that would never have been recognized in a normal examination of the polished section. How many times have supposedly definitive studies overlooked such essential detail?

The second major aspect of the problem is the intervention of postdepositional processes.[1] For example, many sulfide deposits have been oxidized and subjected to supergene enrichment. This is a readily discernable process which, while it may mask the nature of the original deposit, does not often lead the observer astray. In contrast, there is metamorphism, a process which has been overlooked and misinterpreted by literally generations of geologists. The rates of reaction for sulfides vary widely,

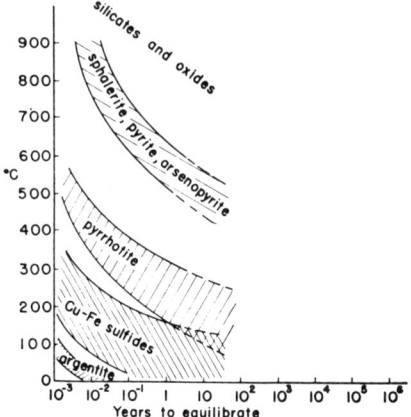

FIG. 2. Time for equilibration for various minerals as functions of temperature.

but in general they are much higher than for silicates or oxides (see Fig. 2). As a consequence, some sulfide minerals such as argentite or chalcopyrite may react internally to homogenize initial compositional zoning or reactions may occur between some sulfides while adjacent, more refractory sulfides (such as sphalerite or pyrite) or silicates remain unaltered. Such changes could occur during deposition, in which case the compositional zoning of crystals would be destroyed as fast as it tried to form, during cooling following deposition, or during subsequent events such as regional or contact metamorphism.

Sulfides as a group will be among the first minerals to exhibit very low grade metamorphic effects, and they may well react so completely as to erase all of the fine-scale record of their heritage, even when the country rocks show little evidence of metamorphism. The older literature is heavily populated with interpretations of bedded and massive sulfide deposits in terms of the nearly complete replacement of pre-existing sedimentary rocks by sulfide minerals. There are without doubt such things as true replacement deposits (mantos, sulfide-bearing contact skarns, replacement veins, etc.) yet many recent investigators, Anderson (1969), Kinkle (1966), or Stanton (1966), have concluded that the conformable portions of such deposits as Mount Isa, Broken Hill, Sullivan, Ducktown, Rio Tinto, Rammelsberg, and many others may have originated as low temperature hydrothermal or even sedimentary deposits at, or near, the sea floor. They were subsequently buried and, in most cases, isochemically metamorphosed to mineral assemblages and textures that differ from the original ones. It is also probable that metamorphosed ores cooled slowly enough from the conditions of the metamorphic maximum for extensive retrograde reactions to occur among the sulfides so that the presently observed rock is several stages removed from its

---

[1] In 1963 Barton, Bethke, and Toulmin presented an extensive review and discussion entitled "Equilibrium in Ore Deposits"; since then several additional significant studies have been made. Craig, Naldrett and Kullerud (1968) have summarized extensive experimentation in the Cu-Ni-Fe-S system at the Geophysical Laboratory concluding that the initial crystallization of Sudbury-type ores was as a copper- and nickel-bearing pyrrhotite solid solution and that a large amount of subsolidus reequilibration was required to produce the presently observed assemblages. Brett (1964) critically examined exsolution reactions in the Cu-Fe-S system. Stanton (1964) and Stanton and Gorman (1968) investigated the rate of approach toward textural equilibrium of strained sphalerite, galena and chalcopyrite; their results are most interesting and their conclusions in basic agreement with those of the author. Inexplicably, the rapid textural equilibration that they found for sphalerite is seemingly inconsistent with the sluggish rates of chemical equilibration observed by Barton and Toulmin (1966).

initial state. As an example of such reactions, Yund and Kullerud (1966) have shown that above 334°C chalcopyrite+pyrrhotite react to yield pyrite+the high temperature polymorph of cubanite. Many sulfide ores contain the low-temperature pyrrhotite+chalcopyrite assemblage, but few, if any, show cubanite+pyrite even though the associated silicate rocks may indicate temperatures well in excess of 334°C. Such retrograde phenomena are so common in sulfides that one must either find evidence[1] for the lack of prograde or retrograde metamorphic reactions or reconstruct the original state before proceeding on to the task of interpreting the mineral association in terms of the environment of initial ore deposition. For some deposits this may mean simply that the sulfide mineral assemblage can tell us nothing at all about the environment of initial ore deposition.

Having discussed some of the field-related problems concerning mineral assemblages, let us now turn to the interpretation of mineral assemblages within the context of laboratory-acquired data on mineral stabilities.

Thermodynamic Approach to Phase Diagrams

In contrast to chemical engineering, where pilot plant experiments can successfully evaluate unknown parameters, much of experimental geology is afflicted with the you-can't-have-your-cake-and-eat-it-too syndrome. Rocks represent many-component, many-condition chemical systems; but in the laboratory only one or two independent variables may be studied simultaneously. Moreover, individual experiments that last for more than a few months, or at most a few years, are frowned on by impatient experimentalists, administrators, fund providers, and even the custodial staff. Therefore, one must experiment under conditions that are not identical to the much more leisurely geologic processes, and we must extrapolate our experimental work: from simple systems to complex ones, and from high temperature to low. Fortunately, the basis for extrapolation is firmly based in thermodynamics.

We shall begin by separating possible variables into relatively unimportant ones which we shall subsequently neglect and important ones which we shall discuss. We can dispense with electrical, magnetic, and gravitational fields for, despite the fact that they may all occur in natural environments, their magnitudes are far too small to influence mineralogical equilibria measureably.

Next let us consider pressure, and it is not so easy to write off. Each cubic centimeter of volume change in a reaction corresponds to a pressure coefficient of 0.0239 cal/bar; that is, for a pressure change of 1500 bars a reaction with a ΔV of 2 cc will have its equilibrium shifted by $2 \times 1500 \times 0.0239 = 71.7$ cal. Volume changes for reactions between condensed phases are usually small but may range up to 2 or 3 cc/g atom, amounting to changes in the free energy of reaction of several tens of cal/kbar. Most

ore deposits that were formed at depths greater than 5 or 10 km,[2] and perhaps many that were formed at shallower depths, will likely be cooled so slowly that the initial record will be completely erased, thereby making any initial state calculations of moot value. Therefore, neglecting pressure will not usually result in an uncertainty of greater than about a hundred calories/g atom in the free energy for a reaction of significance of sulfide petrogenesis. This uncertainty should be compared to the 100 to 1000 cal (or more) uncertainty in standard free energies. Of course, if one is dealing with sulfides in the mantle or lower crust, pressure will be much more important.

Two further concerns in evaluating the role of pressure are thermal expansion and compressibility. Skinner's (1966) compilation of thermal expansion data for sulfides shows average volume increases of 1 or 2 percent (a few tenths of a cc) from room temperature to 400°C; moreover, thermal expansions on opposite sides of a reaction tend to compensate. Of even less importance is compressibility, for Birch's (1966) compilation shows that a pressure increase of 2 kbar achieves only 0.2 to 1 percent volume decrease.

A summary of measured univariant P-T curves where vapor is not present (Fig. 3) shows very little effect of pressure on the equilibrium temperature. In summary, phase relations of sulfides are relatively insensitive to pressures of the magnitude found in the upper crust, and most phase relations have much more promise as thermometers than barometers.[3]

The remaining variables of state are temperature and composition, both of which are very important. Composition can be an awkward variable and we, therefore, find it convenient to discuss the variation in phase assemblages for fixed compositions. Another parameter which is particu-

[2] A hydrostatic load gives about a maximum of about 10 km/kbar; lithostatic gives about 3.5 km/kbar.

[3] For possible exceptions see Scott and Barnes (1969) and Clark (1960)

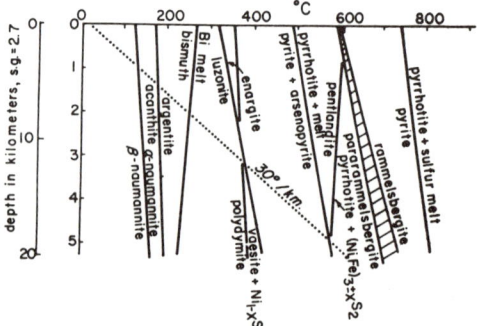

Fig. 3. Depth-temperature plot for vapor-free univariant equilibria. Depth coordinate also shows pressure in kbar. Data are from various sources as cited by Barton and Skinner (1967).

[1] Some criteria for doing this are suggested by Barton, Bethke, and Toulmin (1963).

larly useful is the activity of a component common to several phases. In the case of the sulfides we find that the activity of sulfur serves as a unifying variable with which to compare different bulk compositions.

The convenient standard state for sulfur is the ideal diatomic gas, S_2, at a fugacity of one atmosphere and at the temperature of consideration. This state is used even though it is physically unattainable (due to the condensation of solid or liquid sulfur) below 614°C (the point at which $Ps_2 = 1$ atm). The activity of S_2, as_2, is thus numerically equal to the partial pressure of S_2 in atmospheres, but bear in mind that the presence or absence of a gas phase is inconsequential. The S_2 gas standard state is convenient because curves for sulfidation reactions are not required to bend arbitrarily at the melting, transition, and boiling points of sulfur as would be the case if the standard state for sulfur were chosen, in the conventional manner, as the stable form at one atmosphere at the temperature of interest. Compilations of data in the literature frequently use the latter standard state, so some extra care in calculation is warranted.

One of the most useful ways to present sulfidation data is by plotting reactions so as to generate a metallogenetic grid, the coordinates of which are temperature and activity of S_2. As will be discussed below, there is a sensibly linear relationship between the free energy change, ΔG, and temperature for many sulfide reactions. Because $\Delta G = -RT\ln K$ where T is in degrees Kelvin and K is the equilibrium constant, and because most sulfidation reactions can be written so that all of the reactants and products except for S_2 are in their standard state, it follows that $\log as_2$ is a sensibly linear function of $1/T$. (See Barton and Toulmin, (1964) and Barton and Skinner, (1967) for further discussion). Figure 4 shows a series of such sulfidation curves for several metals. Many other such curves are compiled by Barton and Skinner (1967) and by Richardson and Jeffes (1952).

The general tendency for ΔG versus T, or $\log as_2$ versus $1/T$, curves to be linear has been pointed out by many, including Richardson and Jeffes (1948) and Kubaschewski,

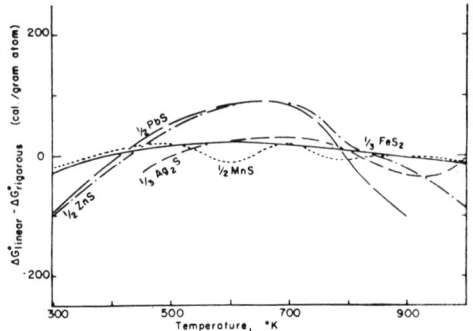

FIG. 5. Plot against temperature of differences in standard free energies of formation between the tabulated values of Robie and Waldbaum (1968) and linear equations fit to the same data.

Evans and Alcock (1967, p. 30). A linear curve implies constant values for the enthalpy and entropy changes for the reaction, a relationship that is generally recognized *not* to be rigorously true, even though it may well be valid. There is little to be gained in work with sulfides by applying a highly precise deviation from a straight line when the absolute position of the line is uncertain by at least several hundred calories (see Kubaschewski, Evans, and Alcock, 1967, p. 29). The principal source of uncertainty is the heat of reaction; this is because mineral sulfides are so refractory, and tend to yield such poorly characterized solution products that satisfactory heats of solution are difficult to obtain. Also, many sulfidation reactions are too sluggish to obtain heats of reaction directly. The best approach for sulfides appears to be to measure the equilibrium constant of formation (which gives $\Delta G°$) of a phase at high temperature and then to extrapolate to other temperatures using heat content data, *plus* an evaluation of activities if solid solutions are involved. An alternative of measuring $\log as_2$ as a function of temperature and simply extrapolating the curve is subject to error because the temperature range of measurement is often too limited to obtain an adequate control on slope; however, this method does have the advantage of automatically including at least an approximate correction for solid solution between phases in the reaction. As examples of the minimal departure from linearity for sulfidation reactions (and for solid-solid reactions as they can be constructed by adding sulfidation reactions) I have have fitted linear least squares curves to the data for several sulfides from the compilation of Robie and Waldbaum (1968) who used a rigorous computation method involving the enthalpy at 289°K and high-temperature heat content measurements. The departure of the points from the linear function is too small to show on ΔG versus T plot, and I have therefore used a plot of ΔG linear-ΔG rigorous versus T for comparison (Fig. 5). The departures are mostly less than 100 cal/g atom. In view of the uncertainty in the positions of the equilibria

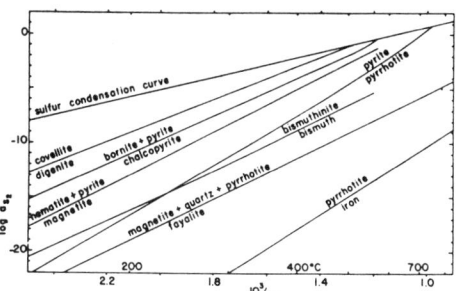

FIG. 4. Log as_2-temperature grid showing typical sulfidation reactions.

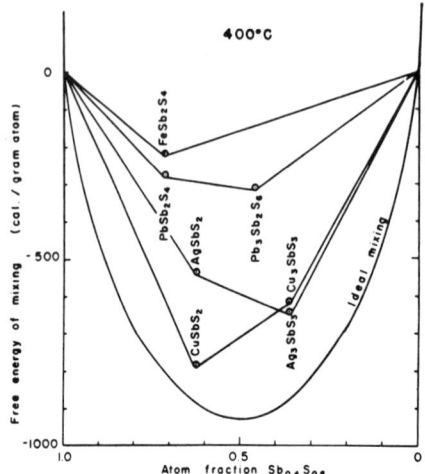

FIG. 6. Free energies of mixing for several sulfosalts. The data for the Cu, Ag, and Pb compounds are from Hall (1967) and do not necessarily represent all of the phases stable on a given join.

there appears little reason to abandon the linear curves. However, reactions involving aqueous species are usually pronounced exceptions to the above-defended linear ΔG versus T generalization.

The sulfidation reactions as written are univariant: that is, in the presence of a gas phase the number of condensed phases equals the number of components. Either temperature or pressure, but not both, may be varied arbitrarily. If we replace gas pressure by a_{S_2}, we still have a univariant curve (providing the role of total pressure is negligible, as noted earlier). The intersections of univariant curves give rise to invariant points, the positions of which can be calculated easily (see Barton and Skinner, 1967). Because the intersections of sulfidation curves tend to be very shallow, it is obvious that an error of several hundred calories (or of a few tenths of a log unit a_{S_2}) would shift individual curves and hence the calculated temperature of an invariant point by many degrees. Conversely, an experimentally determined invariant point provides a valuable triangulation station that aids in the refinement of the locations of *all* of the univariant curves emanating from the invariant point.

There are few data for sulfosalt phases, and those that are available do not always correspond to recognized minerals. Obviously a great deal of work is yet to be done before sulfosalts will be well distinguished thermodynamically. Figure 6 shows the free energy of formation of several sulfosalts *from the simple sulfide components* (data from Hall (1967) and Barton, unpubl.). Note the expanded scale. It is evident that a given sulfosalt is not a great deal more stable than any of several alternative configurations representing the same bulk composition. Therefore, one

should not be too surprised to find that relatively small changes in temperature, pressure and composition might rather strongly modify the configuration of sulfosalt fields in phase diagrams. For the same reason, calculation of sulfosalt phase diagrams from independent thermochemical information would require unreasonably precise data. Sulfosalts are similar to most silicates in that they are commonly intermediate phases along joins between simple compounds. Natural sulfosalt assemblages potentially may contain far more precise thermochemical information than experimental data now available; the analogous situation for silicate assemblages has been pointed out by Garrels (1957).

Having generated a grid of sulfidation reactions, we are able to locate certain assemblages relative to the temperature—a_{S_2} environment. For example, a hematite+galena assemblage must have formed in the upper shaded field in Figure 7, and a pyrrhotite+bismuthinite assemblage somewhere in the lower field.[1] Many more such limited assemblages might be devised, but as we noted earlier, mineral assemblages rarely, if ever, consist of a number of phases in *excess* of the number of components; therefore, univariant or invariant conditions are not often defined by the available phase assemblage. Consequently, phase assemblages cannot be expected to furnish either unique temperatures or a_{S_2} values or any other condition or state, for the conditions of mineral equilibration. In order to reduce the number of degrees of freedom further, and thereby to be able to solve for temperature, a_{S_2}, and the activities of all other components at the time of mineral equilibration, we must consider the *compositions*, not merely the identities, of the phases. This will be the topic for much of the remainder of this paper, but first we need to mention buffers and indicators, a subject also discussed by Barton and Skinner (1967).

A sulfide-bearing assemblage that constitutes the reactants and products for one of the isobarically univariant curves shown in Figure 7 can be considered as either an S_2 buffer or an S_2 indicator. In the role of a buffer the assemblage performs the function of either providing or consuming sulfur, at constant temperature, as needed to maintain a constant value for a_{S_2}. As an indicator the assemblage merely records the passing of the environment from one side of the univariant curve to the other, the mineral array being preserved only because the later assemblage effectively armors the earlier one and thus prevents its complete destruction. In ores, this phenomenon is commonly observed as partial replacement of one mineral (or mineral assemblage) by another. The distinction between buffers and indicators is one of scale rather than principle, because the indicator reaction generates its own microenvironment within which it functions as a buffer, without controlling the chemistry of the larger body.

[1] An attempt to define mineral facies in terms of a_{S_2} and temperature would be of dubious value because a_{S_2} can be highly variable, within a single hand specimen, or even within the same crystal.

Buffers and indicators may be of either the "fixed point" or the "sliding scale" type. A fixed-point buffer (or indicator) exists when the number of phases equals the number of independent components, that is, under divariant conditions. At constant temperature and pressure it defines the activity of each component at a precisely determined level. An example might be pyrite which in the unary system FeS_2 controls a_{FeS_2} at unity; to the extent that pyrite is a stoichiometric phase, the a_{S_2} and a_{FeS} values are not fixed, however, and they may vary widely, though not independently. Another example is the assemblage pyrite+magnetite+hematite which controls the activities of all possible components in the Fe-S-O system, i.e., a_{FeS} is controlled through the reaction $FeS_2 + 3Fe_3O_4 = 4Fe_2O_3 + 2FeS$. This and several other fixed-point-type reactions that control a_{FeS} in the presence of pyrite are shown in Figure 8. A sliding scale buffer or indicator also defines the activity of a component, but the level at which the activity is defined is a function of the composition of the buffer. The buffering (or, more probably, indicating) of a_{FeS} by the (Zn, Fe)S solid solution is an example, and Figure 8 shows this quantitatively within the framework of the fixed-point-type reactions. The ionic buffers commonly used in aqueous chemistry are of the sliding scale type. In contrast to the fixed buffers which are divariant (=isobarically univariant), the sliding scale buffers are at least trivariant and must have their variance reduced by specification of the composition of phases. Each time the concentration of a component in a phase of variable composition is specified, the variance of the system is reduced by one. Because a major goal is to reconstruct the environment at the time of equilibration, we wish to reduce the variance as much as possible, hence the emphasis on composition of minerals.

General Factors Influencing the Compositions of Minerals

The equation relating the composition, X, of a phase to the activity, a, of a component is

$$X = a/\gamma$$

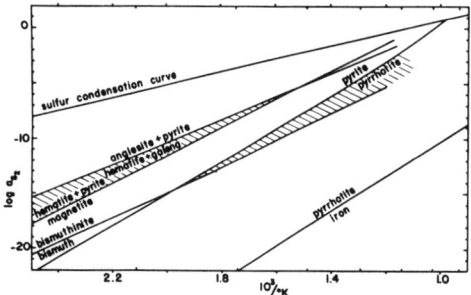

Fig. 7. Log a_{S_2}-temperature grid showing limited fields for hematite+pyrite+galena and pyrrhotite+bismuthinite assemblages.

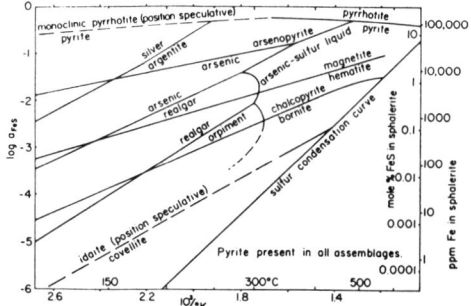

Fig. 8. Log a_{FeS}-temperature grid with superposed sulfidation equilibria of the sort $MS_x + FeS = MS_{x-1} + FeS_2$. The composition of sphalerite that would coexist with these assemblages is indicated.

where composition is expressed as a mole fraction and γ is the activity coefficient.

The relation of these quantities to chemical potential is

$$\mu - \mu° = RT \ln (X \cdot \gamma)$$

In dealing with mineral solid solutions there are two obvious choices for standard state of the solute. The first is the pure solute component *in a crystal structure identical to the solvent*. Such a state may or may not be physically attainable, but such a choice is very convenient for describing and theorizing about the properties of a solid solution. However, in comparing the uptake of a component into several different phases, each of which has a different structure, this first choice becomes awkward in that the comparable activities must be related through the chemical potentials of the individual standard states, i.e.

$$\mu°_1 - \mu°_2 = RT \ln \frac{a_2}{a_1}$$

The second choice is to use a single standard state for a given component regardless of the nature of the solid solution. This alternative will be used in this paper because it makes clearer the discussion of the distribution of trace elements between coexisting phases.

Choice of components. The proper recognition of components is essential when dealing with solid solutions in sulfides. Although we commonly refer to the "cadmium" or "iron" content of a sphalerite, the solid solutions for our purpose lie on the ZnS-CdS and ZnS-FeS joins, and it is a_{CdS} and a_{FeS}, not a_{Cd} or a_{Fe}, that is important here. However, such simple one-for-one substitution does not create either conceptual or operational difficulties.

A particularly troublesome situation in minerals is presented by coupled substitution, the condition in which a pair of ions of different charge, e.g., Ag^+ and Sb^{3+}, substitute for two ions in the host phase, e.g. 2 Pb^{2+}, in such a way as to maintain the charge balance. Because minerals are so often compositionally complex at the trace level, the activities of components participating in coupled substitution are so involved in the total array of multiple-charge substitution that quantification is virtually worthless. As an example, what kind of components could usefully be extracted from a galena composition in which the substitution is of the form

(Ag, Cu, Tl) (As, Sb, Bi, Ga, In) S$_2$ for 2 PbS?

Factors controlling the activity and activity coefficient. In considering the uptake of a minor component by a growing crystal it is convenient to separate the two factors of the equation,

$$X = a/\gamma$$

The activity, a, deals with the chemical environment imposed on the growing crystal by its surroundings, most specifically by the fluid phase from which crystallization may be occurring. In order of importance, the activity is a function of the composition, temperature and pressure of the environment. All of these factors are external to anything going on within the crystal. The activity can be locally buffered, as described previously; or it may be controlled remotely, as a_{NiS} might be controlled through the interaction of H$_2$S-bearing fluids with nickel-bearing silicates far removed from the site of deposition. Whether or not a component is buffered locally determines whether it is termed "inert" or "perfectly mobile" following the terminology of Korzhinskii (1959).

The activity coefficient, γ, is determined solely by the temperature, composition, and pressure of the growing crystal itself. The solvent in a dilute solid solution obeys Raoult's Law in that the activity of the solvent component is equal to its mole fraction. The compositional range over which "dilute" behavior is maintained varies from system to system, but in general, the greater the degree of solid solution, the greater the range of "dilute" behavior. The compositional range of Raoult's Law also tends to be much wider for simple substitutional solid solutions, such as (Zn, Fe)S, than for omission type solid solutions such as Fe$_{1-x}$S or Cu$_{1+x}$Fe$_{1+x}$S$_2$.

The Gibbs-Duhem relationship requires that so long as the solvent obeys Raoult's Law, the activity of the solute is proportional to its mole fraction. However, the proportionality constant (=activity coefficient) is generally not unity. Previous arguments regarding the minimal role of pressure apply here also and we shall probably be safe in assuming that the influence of pressure is negligibly small. The role of composition is not so minor, but it appears to be small so long as coupled substitution is excluded. The effect of temperature on activity coefficients is variable, but not of large magnitude.

Summarizing available data, variations in composition and temperature can produce effects of up to one order of magnitude, and rarely more, on the *activity coefficient*. In contrast, the *activities* of many components may vary by not just one log unit, but by many! For example, Figure 8 shows the variation of a_{FeS} over a range of 6 log units

while being in equilibrium with either pyrite or pyrrhotite. The series of mineral assemblages along the univariant curves superposed on the diagram show that natural environments do indeed span most of this range. The figure also shows how the composition of sphalerite will vary over the a_{FeS}-temperature range covered by the diagram.

For components such as SnS, MnS, or In$_2$S$_3$ that seldom appear as major constituents of ore minerals the variation in activity may be even greater than that for FeS.

Thus the range in variability of activity is drastically greater than that for the activity coefficient, and it is obvious that the concentration of a nonessential constituent in a mineral is influenced far more strongly by a than by γ. It is, therefore, futile to try to use the trace component composition of a single phase (such as the silver content of galena or the mercury content of sphalerite) to try to define some parameter such as the temperature of mineral deposition unless the activity of that component is somehow fixed.

It is possible that some geochemical reason might prevail to limit variability in a. For example, there are no feasible geochemical processes for effectively separating Zn from Cd; therefore, the Cd/Zn ratio is relatively uniform in base metal deposits and the a_{CdS} in sphalerite-depositing environments rarely varies by more than an order of magnitude (which is still far too large a variation for useful thermometry.)

Now let us consider the uptake of a component which is not observed as a separate entity in nature, and in fact, need not even have a stable existence as a pure phase.

The entrance of gold into a simple sulfide such as galena might be an example. Based on only the most preliminary sort of experimental data, let us consider the gold content of galena in equilibrium with free gold. Gold might enter as the un-ionized metal atom in interstitial positions, or it might be present as a gold sulfide component, *e.g.*, Au$_2$S, AuAuS$_2$, or Au$_2$S$_3$ none of which is known as a compound stable relative to gold plus sulfur. Very preliminary experiments (Barton, unpubl.) show that gold enters galena only in the presence of excess sulfur (the quantitative relationship is still obscure) and that silver decreases and bismuth increases the solubility of gold in galena. Therefore, the gold is probably present, at least in part, as the Au$_2$S component whose solubility is increased by the coupled substitution of AuBiS$_2$ (analogous to the enhanced solubility of argentite in galena by the substitution of AgBiS$_2$, Van Hook, 1960). The reaction 4Au+S$_2$=2Au$_2$S must lie in the metastable region beyond the reach of pure sulfur vapor as shown schematically in Figure 9. From the stoichiometry of the reaction we can contour the log a_{S_2} versus T grid in terms of a_{Au_2S}. If the activity coefficient for Au$_2$S in galena were known as a function of temperature we could contour the diagram in terms of gold content of gold-saturated galena. Granted that this part of the discussion is purely schematic, it nevertheless illustrates two points: (1) As temperature changes, the behavior of a component in a saturated solid solution is not simple; it may decrease on cooling as would be the case along the pyrite+pyrrhotite curve, or it might increase on cooling as along the sulfur condensation curve. The difference between some roasting and free-milling gold ores might well be the effective sulfur buffer system that functioned during the post-depositional history of the ore. Of course, other gold-bearing solid solutions (such as pyrite or arsenopyrite) may not behave as does galena, but the principles should be similar. (2) It is entirely possible to work satisfactorily with components which may not be seen as minerals. The copper content of pyrite might be expected to have a similar dependency on a_{S_2} provided that the copper-rich pyrite lies on the FeS_2-CuS_2 join.[1] A further extension of the discussion of the behavior of a component in solid solution is that of the distribution of a component between two or more phases as discussed below.

If we consider the equations for the same component in two different phases and then divide one expression by the other, *i.e.*,

$$\frac{X_1}{X_2} = \frac{a_1/\gamma_1}{a_2/\gamma_2} = \frac{\gamma_2}{\gamma_1} = D$$

the activity terms cancel out and the distribution coefficient, D, is equal to the inverse ratio of the activity coefficients. The activity coefficients are functions of the temperature, pressure and composition of the host phase, but we have noted already that the role of pressure is minor. Two typical isotherms, (P. M. Bethke and Barton unpubl. data), for the distribution of CdS between sphalerite and galena are shown in Figure 10. So long as we are dealing with dilute solid solutions the deviation of the activity coefficients from a constant value should be trivial, and temperature alone exerts a significant control on the

---

[1] The studies of copper-rich pyrite by Frenzel and Ottemann (1967), Einaudi (1968), and Shimazaki (1969) demonstrate the possible significance of this example.

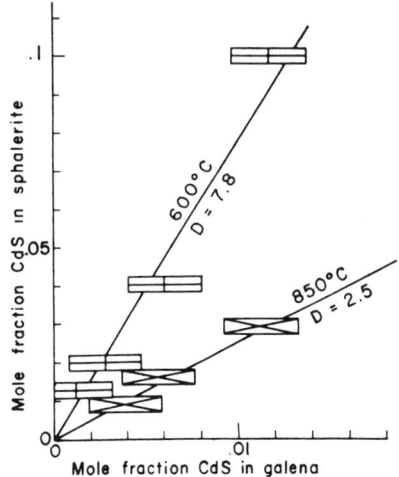

FIG. 10. Two isotherms showing the experimentally determined distribution of CdS between sphalerite and galena.

distribution coefficients. The method appears to provide promising geothermometers, but the difficulty of finding and separating for analysis samples that were deposited in mutual equilibrium presents a serious problem for successful application, as is evident from consideration of the highly complex ore textures shown in Figure 1. The distribution of sulfides of monovalent and trivalent metals between coexisting sulfides of divalent metals, for example, Ag_2S, Tl_2S, In_2S_3, or Sb_2S_3 between sphalerite and galena, will be extremely difficult to quantify in such a way as to be useful because they inherently become involved in coupled substitutions.

THE SULFIDATION STATE OF NATURAL ENVIRONMENTS

The metallogenic grid of sulfidation reactions shown in Figure 11 covers a large range of sulfur activities, and some

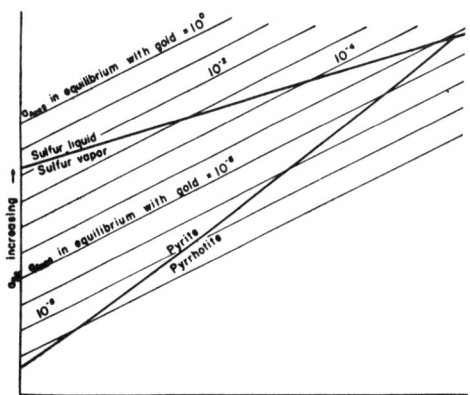

FIG. 9. Hypothetical log a_{S_2}-temperature grid suggesting the wide range of variability of a_{Au_2S} in equilibrium with native gold.

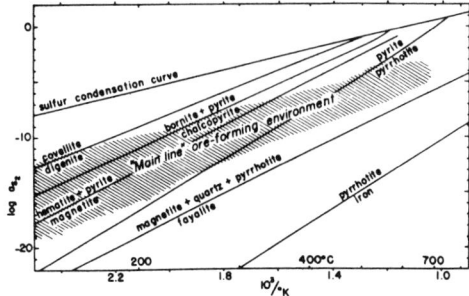

FIG. 11. Log a_{S_2}-temperature grid showing the region of principal ore-forming environments.

deposits may have sufficient mineralogical variation to be represented by one-third or more of the total a_{S_2} range. As noted earlier the sulfide-forming environments usually do not buffer themselves on a given sulfidation curve. Instead the sulfides seem to precipitate under arbitrary conditions that may either vary systematically or apparently irregularly, showing that the solid phases being precipitated do not buffer a_{S_2}, but function as indicators of a_{S_2} in the depositing solution, and that the solutions themselves are not of constant composition. Something determines the a_{S_2} of solutions; if not the local precipitates, then what? The source of ore fluids is responsible for the initial state of the fluids, and, although the specific volume of rock responsible for a given ore fluid cannot often be identified, much less examined, the following observation is pertinent. Neither igneous nor metasomatic metamorphic rocks commonly have sulfide assemblages that alone would control a_{S_2}; instead the buffer systems appear to be such as:

(1) $2FeS + 8FeSiO_3$ (in pyroxene) $+ 2Fe_3O_4$
$$= 8Fe_2SiO_4 \text{ (in olivine)} + S_2$$
(2) $6FeS + 8KAlSi_3O_8$ (in feldspar) $+ 8H_2O + 6Fe_3O_4$
$$= 8KFe_3AlSi_3O_{10}(OH)_2 \text{ (in biotite)} + 3S_2$$
(3) $4FeCO_3 + 5FeS_2 = 3Fe_3O_4 + 4C + 5S_2$

Reactions such as these involve rock-forming minerals that are available in huge quantities relative to ore deposits. The activities of several of the components in the above reactions are not fixed because they are part of solid solutions functioning as solid state buffers.

There are many geologically feasible reactions which may be written as buffers of S_2, but simply buffering S_2 is not enough. This is because the dominant sulfur bearing species is not S_2; S_2 is only a convenience in calculation. The major, low- to medium-temperature aqueous species are H_2S, $(K, Na, H)SO_4^-$, and possibly others, as discussed by Helgeson (1969). Because these species can be participants in redox reactions, the relative proportions of other components that may also participate in redox reactions, i.e., CO_2, CH_4, H_2, Fe^{2+}, etc., are very important, especially because these other components may individually or collectively be present in excess of sulfur. Even further complexities exist, for the dominant sulfur-bearing species are involved in hydrolysis and base exchange equilibria (for example, $3H_2S + KSO_4^- + 2H^+ = 2S_2 + K^+ + 4H_2O$). This means that at the source, in route, and at the site of deposition there are diverse types of reactions ranging from magmatic crystallization (or metamorphic or diagenetic recrystallization), to wallrock alteration, mixing of fluids from different sources, and ore and gangue precipitation which can play important roles in determining the activities of S_2 and of other components as well. The complexities of these processes are appreciated (see, for example, Meyer and Hemley, 1967) but not quantitatively understood as yet, although significant strides in this direction have been taken by Holland (1965) and especially Helgeson and others (1970) and Helgeson (1970).

The "main line" sulfidation state of the most ore deposits tends to run from the pyrrhotite field at high temperature well into the pyrite field at low temperatures, as suggested in Figure 11. The reason, of course, is the general position of the multiple equilibria of the sort just discussed. There is considerable variation within this trend, and an understanding of the specific reasons for a given pattern is a major goal of current research.

Low sulfidation states. Except for the near-surface, oxidizing environment where a_{S_2} is decreased through the formation of sulfates, extremely low sulfidation states (near the iron-pyrrhotite curve) are rare. Three principal types of occurrence are worth mentioning: (1) In meteorites, or perhaps in deep mantle material, the bulk chemistry is such that there is not enough oxygen and sulfur to use up all of the metals; this is not the situation with most crustal rocks. (2) Reaction with organic materials may reduce the oxygen activity and lead to the reduction of a_{S_2} through reactions such as:

$$S_2 + 2H_2O + C = CO_2 + 2H_2S$$

(3) The instability of wustite $(Fe_{1-x}O)$ relative to magnetite+iron coupled with the paucity of silica to react with FeO to form ferrous silicates under the conditions of serpentinization of some peridotite bodies yields iron which can be sufficiently abundant to use up any available sulfur:

$$80Mg_{1.8}Fe_{0.2}SiO_4 + 104H_2O + S_2$$
$$= 24Mg(OH)_2 + 2Fe + 2FeS + 4Fe_3O_4 + 40Mg_3Si_2O_5(OH)_4$$

Such chemistry might well be responsible for the generation of highly reduced, high pH solutions that would be favorable for the generation of mercury deposits.

High sulfidation states. Very high sulfidation states, approaching the native sulfur field, are also rare. These occur principally in five situations, none of which concern "normal" hydrothermal environments: (1) Volcanic sublimates frequently contain native sulfur derived from the rapid cooling of tenuous vapors containing free sulfur generated at least in part by the reaction $SO_2 + 2H_2S = 2H_2O + 3/2 S_2$; (2) The oxidation and acidification of H_2S-bearing, hot spring waters often produce free sulfur; (3) The partial oxidation of sulfide ores may yield native sulfur; (4) The biogenic reduction of sulfate often produces native sulfur and associated high sulfidation state sulfides (e.g. hauerite, MnS_2). (5) The heating to high temperatures of sulfide assemblages which had previously formed at low to moderate temperature can produce a high sulfidation state. For example, a basalt dike intruded into a pyrite vein will certainly break down the pyrite adjacent to it; whether the sulfur escapes outward to produce a more sulfur-rich halo (as by converting chalcopyrite to bornite+pyrite) or moves into the dike as it cools to sulfidize iron-bearing silicates, depends on local fracturing and other factors beyond the scope of this paper.

I wish to reemphasize that, despite the essential non-existence of a molecular sulfur species in the ore forming environment, the activity of sulfur, specifically a_{S_2}, is extremely useful because it relates different sulfide assemblages to a common variable that exerts a significant control over sulfide mineralogy.

GOALS OF CURRENT RESEARCH

Because difficulty in dealing experimentally with multi-component systems requires us to extrapolate from simple to complex systems, and because we must extrapolate downward in temperature from the conditions under which meaningful experiments can be carried out, the methods of thermodynamics are particularly attractive. In addition to the acquisition of a reservoir of data on end-member compounds, a thorough understanding of the thermodynamic behavior of solid solutions is essential. The most convenient line of thermochemical study appears to be the examination of sulfidation reactions, bearing in mind that any thermodynamic study requires a firm understanding of the phase equilibria before the results can be interpreted.

Our knowledge of specific mineral relationships is woefully inadequate. We need field studies in which the understanding of mineral paragenesis is pushed to the ultimate, far beyond the usual simple bar diagram. Finally, the most important requirement of sulfide petrology is to integrate the studies of sulfides with those of all other phases occurring in rocks and to view the sulfides in terms of the whole petrologic environment.

ACKNOWLEDGEMENTS

I wish to thank several of my colleagues, especially Priestley Toulmin, Philip Bethke, and Brian Skinner for extended discussions from which evolved many of the ideas incorporated into this article.

The manuscript has also benefited from technical reviews by Gunnar Kullerud, Gerald Czamanske, and Priestley Toulmin.

REFERENCES

ANDERSON, C. A. (1969) Massive sulfide deposits and volcanism. *Econ. Geol.* **64**, 129–146.

BARNES, H. L. ed. (1967) *Geochemistry of Hydrothermal Ore Deposits.* Holt, Rinehart and Winston, Inc., New York. 670 p.

BARTON, P. B., JR., P. M. BETHKE, AND P. TOULMIN III (1963) Equilibrium in ore deposits. *Mineral. Soc. Amer. Spec. Pap.* **1**, 171–185.

———, AND B. J. SKINNER (1967) Sulfide mineral stabilities, *In* Barnes, H. L. (ed.), *Geochemistry of Hydrothermal Ore Deposits,* Holt, Rinehart and Winston, New York.

———, AND P. TOULMIN III (1964) The electrum-tarnish method for the determination of the fugacity of sulfur in laboratory sulfide systems. *Geochim. Cosmochim. Acta* **28**, 619–640.

——— , AND ——— (1966) Phase relations involving sphalerite in the Fe-Zn-S system. *Econ. Geol.* **61**, 815–849.

BASTIN, E. S. (1950) Interpretation of ore textures. *Geol. Soc. Amer. Mem.* **45**.

BIRCH, F. (1966) Compressibility—elastic constants. *Geol. Soc. Amer. Mem.* **97**, 97–173.

BRETT, P. R. (1964) Experimental data from the system Cu-Fe-S and their bearing on exsolution textures in ores. *Econ. Geol.* **59**, 1241–1269.

CLARK, L. A. (1960) Arsenopyrite As:S ratio as a possible geobarometer (Abstr.). *Geol. Soc. Amer. Bull.* **71**, 1844.

CRAIG, J. R., A. J. NALDRETT, AND G. KULLERUD (1968) Succession of mineral assemblages in pyrrhotite-rich Ni-Cu ores. *Carnegie Inst. Wash. Year Book* **66**, 431–434.

EDWARDS, A. B. (1947) *Textures of the ore minerals and their significance.* Australasian Inst. Min. Metall., Melbourne, 185 p.

EINAUDI, M. T. (1968) Copper zoning in pyrite from Cerro de Pasco, Peru. *Amer. Mineral* **53**, 1748–1752.

FRENZEL, G., AND J. OTTEMANN (1967) Eine sulfidparagenese mit Kupferhaltigem Zonarpyrit von Nukudamu Fiji. *Mineralium Deposita* **1**, 307–316.

GARRELS, R. M. (1957) Some free energy values from geologic relations. *Amer. Mineral.* **42**, 780–791.

HALL, H. T. (1967) Application of thermochemical data to problems of ore deposition (Disc.). *Econ. Geol.* **62**, 1108.

HELGESON, H. C. (1969) Thermodynamics of hydrothermal systems at elevated temperatures and pressures. *Amer. J. Sci.* **267**, 729–804

———, T. H. BROWN, A. Nigrini, AND T. A. Jones (1970) Calculation of mass transfer in geochemical processes involving aqueous solutions. *Geochim. Cosmochim. Acta*, **34**, in press.

HOLLAND, H. D. (1965) Some applications of thermochemical data to problems of ore deposits. II, Mineral assemblages and the composition of ore-forming fluids. *Econ. Geol.* **60**, 1101–1166.

KINKLE, A. R., JR. (1966) Massive pyritic deposits related to volcanism, and possible methods of emplacement. *Econ. Geol.* **61**, 673–694.

KORZHINSKII, D. S. (1959) *Physiochemical basis of the analysis of the paragenesis of minerals.* [English transl.] Consultants Bureau, New York. 142p.

KUBASCHEWSKI, O., E. L. EVANS AND C. B. ALCOCK (1967) *Metallurgical Thermochemistry, 4th Ed.* Pergamon Press, New York. 426 p.

MEYER, C., AND J. J. HEMLEY (1967) Wall rock alteration *In* H. L. Barnes, (ed.) *Geochemistry of Hydrothermal Ore Deposits,* Holt, Rinehart and Winston, New York. 167–235.

OHMOTO, H., M. BORCSIK AND H. D. HOLLAND (1969) Chemistry and origin of hydrothermal fluids at the Bluebell mine, British Columbia. Abstr. Progr. *Geol. Soc. Amer.* Part 7, 165.

RAMDOHR, P. (1960) *Die Erzmineralien und ihre Verwachsungen.* Akademie-Verlag, Berlin. 1089 p.

RICHARDSON, R. D., AND J. H. E. JEFFES (1948) The thermodynamics of substances of interest in iron and steel making; I. Oxides. *J. Iron Steel Inst.* 160, 261–270.

———, AND ——— (1952) The thermodynamics of substances of interest in iron and steel making; III. Sulfides. *J. Iron Steel Inst.* 171, 165–175.

RIDGE, J. D., (ed.) (1968) *Ore Deposits of the United States, 1933–1967—The Graton-Sales Volume.* 2 vol. AIME, New York, 1880 p.

ROBIE, R. A., AND D. R. WALDBAUM (1968) Thermodynamic properties of minerals and related substances at 298.15°K (25.0°C) and one atmosphere (1.013 bars) pressure and at higher temperatures. *U. S. Geol. Surv. Bull.* **1259**.

SAWKINS, F. J. (1964) Lead-zinc ore deposition in the light of fluid inclusion studies, Providencia mine, Zacatecas, Mexico. *Econ. Geol.* 50, 883–919.

SCOTT, S. D., AND H. L. BARNES (1969) Sphalerite geobarometry. *Abst. Prog. Geol. Soc. Amer.* Part 7, 202–203.

SHIMAZAKI, H. (1969) Synthesis of a copper-iron disulfide phase (abstr.). *Can. Mineral.* **10**, 146.

SKINNER, B. J. (1966) Thermal expansion. *Geol. Soc. Amer. Mem.* 97, 75–96.

STANTON, R. L. (1964) Mineral interfaces in stratiform ores. *Trans. Inst. Mining Met. Sec. B,* 74, 45–79.

——— (1966) Compositions of stratiform ores as evidence of depositional processes. *Trans. Inst. Mining Met. Sec. B,* 75, 75–84.

———, AND H. GORMAN (1968) A phenomenological study of grain boundary migration in some common sulfides. *Econ. Geol.* **63**, 907–923.

VAN HOOK, H. J. (1960) The ternary system Ag$_2$S-Bi$_2$S$_3$-PbS. *Econ. Geol.* 55, 759–788.

YUND, R. A., AND G. KULLERUD (1966) Thermal stability of assemblages in the Cu-Fe-S system. *J. Petrology* 7, 454–488.

REFERENCES

Abrahams, S. C. (1955) The crystal and molecular structure of orthorhombic sulphur. Acta Crystallogr. 8, 661-671.

_____ (1961) Scale factors, form factors, and bond lengths in orthorhombic sulphur. Acta Crystallogr. 14, 311.

Addamiano, A., and M. Aven (1960) Some properties of zinc sulfide crystals grown from the melt. J. Appl. Phys. 31, 36-39.

Adiwidjaja, G., and J. Löhn (1970) Strukturver feinerung von Enargit, Cu_3AsS_4. Acta Crystallogr. B26, 1878-1879.

Albers, W. (1967) Physical chemistry of defects. In, M. Aven and J. S. Prener, Eds., Physics and Chemistry of II-VI Compounds. North Holland Publ. Co., Amsterdam, p. 165-222.

_____, and K. Schol (1961) The P-T-X phase diagram of the system Sn-S. Philips Res. Rep. 16, 329-342.

Allen, E. T. and J. L. Crenshaw (1912) The sulfides of zinc, cadmium, and mercury; their crystalline forms and genetic conditions. Am. J. Sci. Ser. 4, 34, 341-396.

Allmann, R., I. Baumann, A. Kutoglu, H. Rösch, and E. Hellner (1964) Die Kristallstruktur des Patronits $V(S_2)_2$. Naturwissenschaften, 11, 263-264.

Alsén, N. (1925) Röntgenographische Untersuchung der Kristallstrukturen von Magnetkies, Breithauptit, Pentlandit, Millerit und verwandten Verbindungen. Geol. Foren. Stockh. Forh. 45, 606-609.

Andersen, C. A., Ed. (1973a) Microprobe Analysis. John Wiley and Sons, New York, 571 p.

_____ (1973b) Analytical methods and applications of the ion microprobe mass analyzer. In, C. A. Andersen, Ed., Microprobe Analysis. John Wiley and Sons, New York, p. 531-553.

Arnold, R. G. (1958) Pyrrhotite-Pyrite Equilibrium Relations between 325 and 743°C. Ph.D. Thesis, Princeton University, New Jersey.

_____ (1962) Equilibrium relations between pyrrhotite and pyrite from 325° to 743°C. Econ. Geol. 57, 72-90.

_____ (1966) Mixtures of hexagonal and monoclinic pyrrhotite and the measurement of the metal content of pyrrhotite by X-ray diffraction. Am. Mineral. 51, 1221-1227.

_____ (1967) Range in composition and structure of 82 natural terrestrial pyrrhotites. Can. Mineral. 9, 31-50.

_____ (1969) Pyrrhotite phase relations below 304±6°C at <1 atm total pressure. Econ. Geol. 64, 405-419.

_____ (1971) Evidence for liquid immiscibility in the system FeS-S. Econ. Geol. 66, 1121-1130.

_____, and O. P. Malik (1974) The NiS-S system above 980°C - a revision. Econ. Geol. (in press).

_____, and L. E. Reicher (1962) Measurement of the metal content of naturally occurring, metal-deficient, hexagonal pyrrhotite by an X-ray spacing method. Am. Mineral. 47, 105-111.

Aurivillus, Karin Lundborg (1950) On the crystal structure of cinnabar. Acta Chem. Scand. 4, 1413-1436.

Auvray, Patrick and Francoise Genet (1973) Affinement de la structure cristalline du cinabre α-HgS. Bull. Soc. fr. Minéral. Cristallogr. 96, 218-219.

Avetisyan, K. K. and G. I. Gnatyshenko (1956) Thermal and metallographic study of the lead sulfide-zinc sulfide-iron sulfide system. *Ivest. Akad. Nauk Kazakh SSSR, Ser. Gorn. Dela, Stroimaterialov i Met.* 6, 11-25.

Avilov, A. S., R. M. Imamov, and L. A. Muradyan (1971) An electron diffraction study of some phases in the Cu-Sb-S system. *Sov. Phys. Crystallogr.* (Trans.) 15, 616-619.

Baba, H. (1963) Some effects of zinc atmospheres on zinc sulfide. *J. Electrochem. Soc.* 110, 79-81.

Bannister, F. A., and M. H. Hey (1932) Determination of minerals in platinum concentrates from the Transvaal by X-ray methods. *Mineral. Mag.* 23, 188-206.

Bansagi, T., E. A. Secc., O. K. Srivastava, and R. R. Martin (1968) Kinetics of hexagonal-cubic phase transformation of zinc sulfide *in vacuo*, in zinc vapor, and in sulfur vapor. *Can. J. Chem.* 46, 2881-2886.

Barnes, H. L. (1971) Investigations in hydrothermal sulfide systems. In, G. C. Ulmer, Ed., *Research Techniques for High Pressure and High Temperature.* Springer-Verlag, New York, p. 317-355.

_____ (1973) Polymorphism and polytypism. *Final Rep. U.S. Army Res. Office (Durham), Contract No. DAHCO4-69-C-0043*, 38 p.

_____, and G. K. Czamanske (1967) Solubilities and transport of ore minerals. In, H. L. Barnes, Ed., *Geochemistry of Hydrothermal Ore Deposits.* Holt, Rinehart, and Winston, New York, p. 334-381.

_____, and G. Kullerud (1961) Equilibria in sulfur-containing aqueous solutions, in the system Fe-S-O, and their correlation during ore deposition. *Econ. Geol.* 56, 648-688.

_____, S. B. Romberger, and M. Stemprok (1967) Ore solution chemistry, II. Solubility of HgS in sulfide solutions. *Econ. Geol.* 62, 957-982.

Barnett, D. E., R. S. Boorman, and J. K. Sutherland (1971) New data on ternary phases in the system Zn-In-S. *Phys. Stat. Solids*, 4, K49-K52.

Barstad, J. (1959) Phase relations in the system Ag-Sb-S at 400°C. *Acta Chem. Scand.* 13, 1703-1708.

Bartholome, P., F. Koteska, and J. Lopez Ruiz (1971) Cobalt zoning in microscopic pyrite from Kamoto, Republic of the Congo (Kinshasa). *Min. Deposita* 6, 167-176.

Barton, P. B., Jr. (1969) Thermochemical study of the system Fe-As-S. *Geochim. Cosmochim. Acta*, 33, 841-857.

_____ (1970) Sulfide petrology. *Mineral. Soc. Am. Spec. Pap.* 3, 187-198.

_____ (1971) The Fe-Sb-S system. *Econ. Geol.* 66, 121-132.

_____ (1973) Solid solutions in system Cu-Fe-S Part I. The Cu-S and CuFe-S join. *Econ. Geol.* 68, 455-465.

_____, P. M. Bethke, and P. Toulmin, III (1963) Equilibrium in ore deposits. *Mineral. Soc. Am. Spec. Pap.* 1, 171-185.

_____, and B. J. Skinner (1967) Sulfide mineral stabilities. In, H. L. Barnes, Ed., *Geochemistry of Hydrothermal Ore Deposits.* Holt, Rinehart and Winston, Inc., New York, p. 236-333.

_____, and P. Toulmin, III (1964a) Experimental determination of the reaction chalcopyrite + sulfur = pyrite + bornite from 350° to 500°C. *Econ. Geol.* 59, 747-752.

_____, and _____ (1964b) The electrum-tarnish method for the determination of the fugacity of sulfur in laboratory sulfide systems. *Geochim. Cosmochim. Acta*, 28, 619-640.

_____, _____ (1966) Phase relations involving sphalerite in the Fe-Zn-S system. *Econ. Geol.* 61, 815-849.

Baumann, I. H. (1964) Patronite, VS_4, und die Mineral-Paragenese der bituminosen schiefes von Minasragra, Peru. *Neues Jahrb. Mineral. Abh.* 101, 97-108.

Bayliss, P. (1968) The crystal structure of disordered gersdorffite. *Am. Mineral.* 53, 290-293.

_____ (1969a) Isomorphous substitution in synthetic cobaltite and ullmannite. *Am. Mineral.* 54, 426-430.

_____ (1969b) X-ray data, optical anisotropism and thermal stability of cobaltite, gersdorffite and ullmannite. *Mineral. Mag.* 37, 26-33.

_____, and Werner Nowacki (1972) Refinement of the crystal structure of stibnite, Sb_2S_3. *Z. Kristallogr.* 135, 308-315.

_____, and N. C. Stephenson (1967) The crystal structure of gersdorffite. *Mineral. Mag.* 36, 38-42.

_____, _____ (1968) The crystal structure of gersdorffite (III), a distorted and disordered pyrite structure. *Mineral. Mag.* 36, 940-947.

Beglaryan, M. L., and H. K. Abrikasov (1959) The $Bi_2Te_3Bi_2Se_3$ system. *Dokl. Akad. Nauk SSSR* 129, 135.

Bell, P. M., A. E. Goresy, J. L. England and B. Kullerud (1970) Pressure-temperature diagram for Cr_2FeS_4. *Carnegie Inst. Wash. Year Book,* 68, 277-278.

_____, J. L. England and G. Kullerud (1964) Pentlandite: Pressure effect on breakdown. *Carnegie Inst. Wash. Year Book* 64, 206-207.

_____, and G. Kullerud (1970) High pressure differential thermal analysis. *Carnegie Inst. Wash. Year Book* 68, 276-277.

Bell, R. E. and R. E. Herfert (1957) Preparation and characterization of a new crystalline form of molybdenum disulfide. *J. Am. Chem. Soc.* 79, 3351-3355.

Bennett, C. E. G., J. Graham, and M. R. Thornber (1972) New observations on natural pyrrhotites. Part I - Mineragraphic techniques. *Am. Mineral.* 57, 445-462.

Bente, K. (1974) Untersuchungen im pseudobinaren system Stannin - (Cu_2FeSuS_4) - Briartite (Cu_2FeGeS_4) *Neues Jahrb. Mineral. Monatsch.* 8-13.

Berkowitz, J. (1965) Molecular composition of sulfur vapor. In, C. B. Meyer, Ed., *Elemental Sulfur.* John Wiley and Sons, p. 125-159.

_____, and J. R. Marguart (1963) Equilibrium composition of sulfur vapor. *J. Chem. Phys.* 39, 275-283.

Berner, R. A. (1964) Iron sulfides formed from aqueous solution at low temperatures and atmospheric pressure. *J. Geol.* 72, 293-306.

_____ (1971) *Principles of Chemical Sedimentology.* McGraw-Hill, New York, 205 p.

Bernhardt, H. J. (1972) Untersuchungen in pseudobinaren system Stannin - Kupferkies *Neues Jahrb. Mineral. Monatsch.* 553-556.

_____, J. Y. Lee, G. Moh, Roy-Choudhury, and O. Vaasjoki (1972) Experimentelle Untersuchungen in system Cu-Fe-Zn-Su-S. *Fortschr. Mineral.* 50, 13-14.

Berry, L. G. (1954) The crystal structure of covellite, CuS, and klockmannite, CuSe. *Am. Mineral.* 39, 504-509.

_____ (1965) Recent advances in sulfide mineralogy. *Am. Mineral.* 50, 301-313.

Bertaut, E. F. (1952) La structure de la pyrrhotine. *C. R. Acad. Sci. Paris,* 234, 1295-1297.

Bethke, P. M. and P. B. Barton (1961) Unit cell dimension *versus* composition in the systems PbS-CdS, PbS-PbSe, ZnS-ZnSe, and $CuFeS_{1.90}$-$CuFeSe_{1.90}$. *U.S. Geol. Soc. Prof. Pap.* 429B, 266-270.

_____, _____ (1971a) Distribution of some minor elements between coexisting sulfide minerals. *Econ. Geol.* 66, 140-163.

Bethke, P. M. and P. B. Barton (1971b) Subsolidus relations in the system PbS–CdS. *Am. Mineral.* 56, 2034–2039.

Bither, T. A., R. J. Bouchard, W. H. Cloud, P. C. Donohue, and W. J. Siemons (1968) Transition metal pyrite dichalcogenides. High-pressure synthesis and correlation of properties. *Inorg. Chem.* 7, 2208–2220.

Blank, Z., and W. Brenner (1971) The growth of group II-VI crystals in gels. *J. Crystal Growth,* 11, 255–259.

_____, W. Brenner, and Y. Okamoto (1968) The growth of single crystals of lead sulfide in silica gels at ambient temperatures – Preliminary characterization and effect of various organic compounds as sulphide ion donors. *Mater. Res. Bull.* 3, 555–561.

Blitz, W. and F. Wiechmann (1936) Zum System Mangan/Schwefel: Abbau and Synthese des Haurits (MnS_2). *Zeit. annorg. Allg. Chem.* 228, 268–274.

Bloem, J. and F. A. Kroger (1956) The *P-T-X* phase diagram of the lead sulphur system. *Z. Phys. Chem. N.F.* 7, 1–14.

Boctor, N. Z., and G. Kullerud (1973) Distribution of Se, Fe, and Zn in some mercury ores (abstr.). *Geol. Soc. Am. Abstr. Programs,* 5, 554–555.

Boer, K. W., and N. J. Nalesnik (1969) Semiconductivity of CdS as a function of S-vapor pressure during heat treatment between 500° and 700°C. *Mat. Res. Bull.* 4, 153–160.

Boorman, R. S. (1967) Subsolidus studies in the $ZnS-FeS-FeS_2$ system. *Econ. Geol.* 62, 614–631.

_____, and J. K. Sutherland (1969) Subsolidus phase relations in the $ZnS-In_2S_3$ system 600–1080°C. *J. Mat. Sci.* 4, 658–671.

_____, _____, and L. V. Chernyshev (1971) New data on the sphalerite-pyrrhotite-pyrite solvus. *Econ. Geol.* 66, 670–673.

Born, L. and E. Hellner (1960) A structural proposal for boulangerite. *Am. Mineral.* 45, 1266–1271.

Bouchard, R. J. (1968) The preparation of pyrite solid solutions of the type $Fe_xCo_{1-x}S_2$, $Co_xNi_{1-x}S_2$ and $Cu_xNi_{1-x}S_2$. *Mat. Res. Bull.* 3, 563–570.

Bragg, W. Lawrence (1914) The analysis of crystals by the *X*-ray spectrometer. *Proc. Roy. Soc.* A89, 468–489.

Braune, H., S. Peter and V. Neveling (1951) Die Dissoziation des Schwefeldsampfes. *Z. Naturforsch.* 62, 32–37.

Brett, P. R. (1964) Experimental data from the system Cu-Fe-S and their bearing on exsolution textures in ores. *Econ. Geol.* 59, 1241–1269.

_____, and G. Kullerud (1967) The Fe-Pb-S system. *Econ. Geol.* 62, 354–369.

_____, and R. A. Yund (1964) Sulfur-rich bornites. *Am. Mineral.* 49, 1084–1098.

Brockway, L. O. (1934) The crystal structure of stannite, Cu_2FeSnS_4. *Z. Kristallogr.* 89, 434–441.

Brostigen, Gunnar, and Arne Kjekshus (1970) Bonding schemes for compounds with the pyrite, marcasite and arsenopyrite type structures. *Acta Chem. Scand.* 24, 2993–2012.

_____, _____, and C. Romming (1973) Compounds of the marcasite type crystal structure VIII. Redetermination of the prototype. *Acta Chem. Scand.* 27, 2791–2796.

Brower, W. S., H. S. Parker, and R. S. Roth (1974) Reexamination of synthetic parkerite and shandite. *Am. Mineral.* 59, 296–301.

Buchan, R. and J. H. Blowes (1968) Geology and mineralogy of a millerite nickel ore deposit. *Can. Inst. Min. Metall. Bull.* 61, 529–534.

Buerger, J. J., and M. H. Buerger (1944) Low chalcocite. *Am. Mineral.* 29, 55–65.

Buerger, Martin J. (1931) The crystal structure of marcasite. *Am. Mineral.* 16, 361-395.

_____ (1934) The pyrite-marcasite relation. *Am. Mineral.* 19, 37-61.

_____ (1936) The symmetry and crystal structure of the minerals of the arseno-pyrite group. *Z. Kristallogr.* 95, 83-113.

_____ (1939) The crystal structure of gudmundite (FeSbS) and its bearing on the existence field of the arsenopyrite structural type. *Z. Kristallogr.* 101, 290-316.

_____ (1947a) Derivative crystal structures. *J. Chem. Phys.* 15, 1-16.

_____ (1947b) The crystal structure of cubanite. *Am. Mineral.* 32, 415-425.

_____ (1960) *Crystal-Structure Analysis.* John Wiley and Sons, Inc., New York.

_____, and Theodor Hahn (1955) The crystal structure of berthierite, $FeSb_2S_4$. *Am. Mineral.* 40, 226-238.

_____, and B. J. Wuensch (1963) Distribution of atoms in high-chalcocite, Cu_2S. *Science,* 141, 276-277.

Buerger, N. W. (1934) The unmixing of chalcopyrite from sphalerite. *Am. Mineral.* 525-530.

_____ (1942) X-ray evidence of the existence of the mineral digenite, Cu_9S_5. *Am. Mineral.* 27, 712-716.

Buhlmann, E. (1965) *Untersuchungen im System Cu-Bi-S.* Ph.D. Thesis, University of Heidelberg.

_____ (1971) Untersuchungen in System Bi_2S_3-Cu_2S und geologische Schlussfol-gerungng. *Neues Jahrb. Mineral. Monatsch.* 137-141.

Bunch, T. E. and L. H. Fuchs (1969) A new mineral: brezinaite, Cr_3S_4 and the Tuscon meteorite. *Am. Mineral.* 54, 1503-1518.

Burgmann, W. Jr., G. Urbain, and M. G. Frohberg (1968) Contribution à l'étude du système fer - soufre limité au domaine du mono-sulfure de fer (pyrrho-tine). *Mém. Sci. Rev. Métall.* 65, 567-578.

Burkart-Baumann, I., J. Ottemann, and G. C. Amstutz (1966) Neue Beobachtungen au dem röntgen - amorphen sulfiden von Cerro de Pasco, Peru. *Neues Jahrb. Mineral. Monatsch.* 353-361.

_____, _____, _____ (1972) The X-ray amorphous sulfides from Cerro de Pasco, Peru and the crystalline inclusions. *Neues Jahrb. Mineral. Monatsch.* 433-446.

Burns, R. G. (1970) *Mineralogical Applications of Crystal Field Theory.* Cambridge University Press, Cambridge, England.

_____, and D. J. Vaughan (1970) Interpretation of the reflectivity behavior of ore minerals. *Am. Mineral.* 55, 1576-1586.

Buzek, Z. and K. S. Prabhala (1965) Character and type of solutions of S in the Fe-S-Ni and Fe-S-Cr system. *Sb. Ved. pr. Vys. sk. banske v. Ostrave,* 11, 563-568.

Byström, A. (1945) Monoclinic magnetic pyrites. *Ark. Kemi Mineral. Geol.* 19B, 1-8.

Cabri, J. L. (1972) The mineralogy of the platinum group elements. *Minerals Sci. Eng.* 4, 3-29.

_____ (1973) New data on phase relations in the Cu-Fe-S system. *Econ. Geol.* 68, 443-454.

_____, and S. R. Hall (1972) Mooihoekite and haycockite, two new copper-iron sulfides, and their relationship to chalcopyrite and talnakhite. *Am. Mineral.* 57, 689-708.

_____, _____, J. T. Szymanski, and J. M. Stewart (1973) On the transformation of cubanite. *Can. Mineral.* 12, 33-38.

Cabri, L. J., and D. C. Harris (1971) New compositional data for talnakhite, $Cu_{18}(Fe,Ni)_{16}S_{32}$. Econ. Geol. 66, 673-675.

_____, _____, and J. M. Stewart (1970a) Costibite (CoSbS) a new mineral from Broken Hill, New South Wales. Am. Mineral. 55, 10-17.

_____, _____, _____ (1970b) Paracostibite (CoSbS) and nisbite ($NiSb_2$) new minerals from the Red Lake area, Ontario, Canada. Can. Mineral. 10, 232-246.

Cambi, L., and M. Elli (1965) Hydrothermal processes: III. synthesis of thiosalts derived from silver and antimony sulfides. Chim. Ind. (Milan), 47, 282-290.

_____, _____, and I. Tangerini (1966) The ternary system $Ag_2S-Cu_2S-Sb_2S_3$. Chim. Ind. (Milan) 48, 567-575.

Canneri, G., and L. Fernando (1925) Contributo allo studio di alcuni minerali contenenti tallio. Analise ternuca dei sistemi $Tl_2S-As_2S_3$, Tl_2S-PbS. Reale Accad. Naz. Dei Lincei Atti 6th Ser. R. Classe Sci. Fis. Mat. Nat. 1, 671-676.

Carpenter, R. H., and G. A. Desborough, (1964) Range in solid solution and structure of naturally occurring troilite and pyrrhotite. Am. Mineral. 49, 1350-1365.

Chang, L.L.Y. (1963) Dimorphic relations in Ag_3SbS_3. Am. Mineral. 48, 429-432.

_____, and J. E. Bever (1973) Lead sulphosalt minerals: Crystal structures, stability relations, and paragenesis. Mineral. Sci. Eng. 5, 181-191.

_____, and W. R. Brice (1971) The herzonbergite - teallite series. Mineral. Mag. 38, 186-189.

Chen, T. T., and L. L. Chang (1971) Phase relations in the systems $Ag_2S-Sb_2S_3-Bi_2S_3$ and $Cu_2S-Sb_2S_3-Bi_2S_3$. Geol. Soc. Am. Abstr. Program, 3, 524.

Chenavas, J., J-C. Joubert, and M. Marezio (1971) Low-spin → high-spin state transition in high-pressure cobalt sesquioxide. Solid State Commun. 9, 1057-1060.

Chernyshev, L. V., and V. N. Anfilogov (1968) Subsolidus phase relations in the $ZnS-FeS-FeS_2$ system. Econ. Geol. 63, 841-844.

_____, _____, T. M. Pastushkova, and T. A. Suturina (1968) A study of the system Fe-Zn-S under hydrothermal conditions. Geol. Rud. Mest. 3, 50-64 [transl. Geochem. Int. 5, 196-209 (1968)].

Chevreton, M. and A. Sapet (1965) Structure de V_3S_4 et de quelques sulfures ternaires isotypes. C. R. Acad. Sci. Paris, 261, 928-930.

Cho, Seung-Am, and B. J. Wuensch (1970) Crystal chemistry of the plagionite group. Nature, 225, 444-445.

_____, _____ (1974) The crystal structure of plagionite, $Pb_5Sb_8S_{17}$, the second member in the homologous series $Pb_{3+2n}Sb_8S_{15+2n}$. Z. Kristallogr. (in press).

Clark, A. H. (1966a) Heating experiments on gudmundite. Mineral. Mag. 35, 1123-1125.

_____ (1966b) Stability field of monoclinic pyrrhotite. Inst. Min. Metal. Trans. 75B, 232-235.

_____ (1970a) An unusual copper-iron sulfide, $Cu_{0.12}Fe_{0.94}S_{1.00}$ from the Ylojarvi Copper tungsten deposit, Finland. Econ. Geol. 65, 590-591.

_____ (1970b) Compositional differences between hexagonal and rhombohedral molybdenite. Neues Jahrb. Mineral. Monatsch. 1, 33-38.

_____ (1970c) Cuprian sphalerite and a probable copper zinc sulfide, Cachiyuyo de Llampos, Copiapo, Chile. Am. Mineral. 55, 1021-1025.

_____ (1970d) Supergene metastibnite from Mina Alacrán, Pampa Larga, Copiapo, Chile. Am. Mineral. 55, 2104-2106.

Clark, A. H. (1971) Molybdenite 2H$_1$, molybdenite 3R, and jordisite from Carrizal Alto, Atacama, Chile. *Am. Mineral.* 56, 1832-1835.

———— (1974) Hypogene and supergene cobalt-copper sulfides, Carrizal Alto, Atacama, Chile. *Am. Mineral.* 59, 302-306.

————, and R. H. Sillitoe (1971) Cuprian galena solid solutions, Zapallar Mining district, Atacama, Chile. *Am. Mineral.* 56, 2142-2145.

Clark, L. A. (1990a) The Fe-As-S system: Phase relations and applications. *Econ. Geol.* 55, 1345-1381, 1631-1652.

———— (1960b) The Fe-As-S system: Variations of arsenopyrite composition as functions of T and P. *Carnegie Inst. Wash. Year Book,* 59, 127-130.

————, and G. Kullerud (1963) The sulfur-rich portion of the Fe-Ni-S system. *Econ. Geol.* 58, 853-885.

Coleman, R. G. (1959) The natural occurrence of galena-clausthalite solid solution series. *Am. Mineral.* 44, 166-175.

Cook, W. R. (1972) Phase changes in Cu$_2$S as a function of temperature. In, R. S. Roth and S. J. Shneider, Eds., *5th Mater. Res. Symp., Proc. Nat. Bur. Stand. Pub.* 364, 703-711.

Corlett, M. (1968) Low-iron polymorphs in the pyrrhotite group. *Z. Kristallogr.* 126, 124-134.

Corrl, J. A. (1964) Recovery of the high pressure phase of cadmium sulfide. *J. Appl. Phys.* 35, 3032-3033.

Cotton, F. A. (1971) *Chemical Applications of Group Theory,* 2nd Ed. John Wiley, New York.

————, and G. Wilkinson (1966) *Advanced Inorganic Chemistry,* 2nd Ed. Interscience, New York.

Cracknell, A. P. (1968) *Applied Group Theory.* Pergamon Press, Oxford.

Craig, J. R. (1967) Phase relations and mineral assemblages in the Ag-Bi-Pb-S system. *Mineral. Deposita,* 1, 278-306.

———— (1970) Livingstonite, HgSb$_4$S$_8$; synthesis and stability. *Am. Mineral.* 55, 919-924.

———— (1971) Violarite stability relations. *Am. Mineral.* 56, 1303-1311.

———— (1973) Pyrite-pentlandite and other low-temperature relations in the Fe-Ni-S system. *Am. J. Sci.* 273-A, 496-510.

————, P. B. Barton and B. H. Sepenuk (1971) Experimental Investigations in the Bi-Fe-S system (abstr.) *Geol. Soc. Am. Abstr. Programs* 3, 305.

————, ———— (1973) Thermochemical approximations for sulfosalts. *Econ. Geol.* 68, 493-506.

————, ————, ———— (1971) Experimental investigations in the Bi-Fe-S system (abstr.). *Geol. Soc. Am. Abstr. Programs* 3, 305.

————, L. L. Y. Chang, and W. R. Lees (1973) Investigations in the Pb-Sb-S system. *Can. Mineral.* 12, 199-206.

————, and J. B. Higgins (1973) Thiospinels: The carrollite-linnaeite series (abstr.). *Geol. Soc. Am. Abstr. Programs,* 5, 586.

————, ———— (1974) Thiospinels: The linnaeite-siegenite-polydymite series (abstr.). *Geol. Soc. Am. Abstr. Programs* 6 (in press).

————, and G. Kullerud (1967a)The Cu-Fe-Pb-S system. *Carnegie Inst. Wash. Year Book* 65, 344-352.

————, ———— (1967b) Sulfide melts in the Cu-Fe-Pb-S system. *Geol. Soc. Am. Abstr. Programs,*

————, ———— (1968) Phase relations and mineral assemblages in the copper-lead-sulfur system. *Am. Mineral.* 53, 145-161.

Craig, J. R., and G. Kullerud (1969) Phase relations in the Cu-Fe-Ni-S system and their application to magmatic ore deposits. In, H. D. B. Wilson, Ed., *Magmatic Ore Deposits, Econ. Geol. Monogr.* 4, 344-358.

_____, _____ (1973) The Cu-Zn-S system. *Mineral. Deposita* 8, 81-91.

_____, and A. J. Naldrett (1971) Phase relations and P_{S_2}-T variations in the Fe-Ni-S system (abstr.). *Geol. Assoc. Can. - Mineral. Assoc. Can. Sudbury,* 16-17.

_____, _____, and G. Kullerud (1968) The Fe-Ni-S system: 400°C isothermal diagram. *Carnegie Inst. Wash. Year Book* 66, 440-441.

_____, B. J. Skinner, C. A. Francis, F. D. Luce, and E. Makovicky (1974) Phase relations in the As-Sb-S system. *Trans. Am. Geophys. Union* 55, 483.

Cubicciotti, D. (1962) The bismuth-sulfur phase diagram. *J. Phys. Chem.* 66, 1205-1206.

_____ (1963) Thermodynamics of liquid bismuth and sulfur. *J. Phys. Chem.* 67, 118-123.

Curlook, W. and L. J. Pidgeon (1953) The Co-Fe-S system. *Can. Mineral. Met. Bull.* 46, 493, 297-301.

Czamanske, G. K. (1969) The stability of argentopyrite and sternbergite. *Econ. Geol.* 64, 459-461.

_____, and F. E. Goff (1973) The character of Ni^{2+} as demonstrated by solid solutions in the Ni-Fe-Zn-S system. *Econ. Geol.* 68, 258-268.

_____, and R. R. Larson (1969) The chemical identity and formula of argentopyrite and sternbergite. *Am. Mineral.* 54, 1198-1201.

Darrow, M. S., W. B. White, and R. Roy (1966) Phase relations in the system PbS-PbTe. *Trans. Metall. Soc. AIME* 236, 654-658.

Debaerdemaker, T., and A. Kutoglu (1973) The crystal and molecular structure of a new elemental sulfur, S_{18}. *Naturwissenschaften* 60, 49.

_____, E. Hellner, A. Kutoglu, M. Schmidt, and E. Wilhelm (1973) Synthese, Kristall und Molekül-Struktur von Cycloikosaschwefel, S_{20}. *Naturwissenschaften,* 60, 300.

Delafosse, D., and Can Huang Van (1962) Etude du systeme Ni-Co-S a temperature elevee. *C. R. Acad. Sci. Paris,* 254, 1286-1288.

Dell, C. I. (1972) An occurrence of greigite in Lake Superior sediments. *Am. Mineral.* 57, 1303-1304.

Demirsoy, S. (1969) Untersuchungen uber den Einfluss der chemischem Zusammensetzung auf die Spektrale Reflexionsfunktionen und Mikroeindruckharten im System FeS_2-NiS_2-CoS_2 an zonen sines naturliche Bravoit-Kristalls. *Neues Jahrb. Mineral. Monatsch.* 17, 323-333.

Desborough, G. A., and R. H. Carpenter (1965) Phase relations of pyrrhotite. *Econ. Geol.* 60, 1431-1450.

_____, and G. K. Czamanske (1973) Sulfides in eclogite nodules from a kimberlite pipe, South Africa, with comments on violarite stoichiometry. *Am. Mineral.* 58, 195-202.

_____, R. H. Hiedel, and G. K. Czamanske (1971) Improved quantitative electron microprobe analysis at low operating voltage: II. Sulfur. *Am. Mineral.* 56, 2136-2141.

Detry, D., J. Drowart, P. Goldfinger, H. Keller, and H. Rickert (1967) Zur Thermodynamik von Schwefeldampf. *Z. Phys. Chem.* 55, 314-319.

Dickerson, R. E., H. B. Gray, and G. P. Haight, Jr. (1974) *Chemical Principles,* 2nd ed. W. A. Benjamin, Menlo Park, Calif.

Dickinson, Roscoe G., and Linus Pauling (1923) The crystal structure of molybdenite. *J. Am. Chem. Soc.* 45, 1466-1471.

Dickson, F. W., A. S. Radtke, B. G. Weissberg, and C. Heropoulas (1974) Solid solutions of antimony, arsenic and gold in stibnite (Sb_2S_3), orpiment (As_2S_3), and realgar (As_2S_2). *Econ. Geol.* (in press).

_____, L. D. Shields, and G. C. Kennedy (1962) A method for the determination of equilibrium sulfur pressures of metal sulfide reactions. *Econ. Geol.* 57, 1021-1030.

_____, and G. Tunell (1959) The stability relations of cinnabar and metacinnabar. *Am. Mineral.* 44, 471-487.

Djurle, S. (1958a) An *X*-ray study on the system Cu-S. *Acta Chem. Scand.* 12, 1415-1426.

_____ (1958b) An *X*-ray study of the system Ag-Cu-S. *Acta Chem. Scand.* 12, 1427-1436.

Dolanski, J. (1974) Sulvanite from Thorpe Hills, Utah. *Am. Mineral.* 59, 307-313.

Donohue, Jerry, Aimery Caron, and Elihu Goldish (1961) The crystal and molecular structure of S_6 (sulfur-6). *J. Am. Chem. Soc.* 83, 3748-3751.

_____, Stewart H. Goodman, and Michael Crisp (1969) On the structure of fibrous sulfur. *Acta Crystallogr.* B25, 2168.

Dornberger-Schiff, K., and E. Höhne (1959) Die Kristallstruktur des Betechtinit $Pb_2(Cu,Fe)_{21}S_{15}$. *Acta Crystallogr.* 12, 646-650.

D'sugi, J., K. Shimizu, T. Nakamura, and A. Onodera (1966) High pressure transition in cadmium sulfide. *Dev. Phys. Chem. Japan* 36, 59-73.

Dunitz, J. D., and L. E. Orgel (1957) Electronic properties of transition element oxides. II. Cation distribution amongst octahedral and tetrahedral sites. *J. Phys. Chem. Solids* 3, 318-333.

DuPreez, J. W. (1945) A thermal investigation of the parkerite series. *Univ. Stellenbosch* 22, A, 97-104.

Edenharter, A., and W. Nowacki (1970) Verfeinerung der Kristallstruktur von Bournonit [$(SbS_3)_2|Cu_2^{IV}Pb^{VII}Pb^{VIII}$] und von Seligmannit [$(AsS_3)_2|Cu_2^{IV}Pb^{VII}Pb^{VIII}$]. *Z. Kristallogr.* 131, 397-417.

Ehlers, E. G. (1972) *The Interpretation of Geological Phase Diagrams*. W. H. Freeman and Co., San Francisco, 280 p.

Ehrenberg, H. (1932) Orientierte Verwachsungen von Magnetkies und Pentlandite. *Z. Kristallogr.* 82, 309-315.

Einaudi, M. T. (1968) Sphalerite-pyrrhotite-pyrite equilibria - a re-evaluation. *Econ. Geol.* 63, 832-834.

El Goresy, A., and G. Kullerud (1969) Phase relations in the system Cr-Fe-S. In, P. M. Millman, Ed., *Meteorite Research* D. Reidel Pub. Co., Dordrecht - Holland, p. 638-656.

Elliott, Norman (1960) Interatomic distances in FeS_2, CoS_2, and NiS_2. *J. Chem. Phys.* 33, 903-905.

Ellis, A. J., and W. Giggenbach (1971) Hydrogen sulfide ionization and sulfur hydrolysis in high temperature solution. *Geochim. Cosmochim. Acta* 35, 247-260.

Engel, P., and W. Nowacki (1966) Die Verfeinerung der Kristallstruktur von Proustit, Ag_3AsS_3 und Pyrargyrit, Ag_3SbS_3. *Neues. Jahrb. Mineral. Monatsch.* 6, 181-184.

_____, ._____ (1968) Die Dristallstruktur von Xanthokon, Ag_3AsS_3. *Acta Crystallogr.* B24, 77-81.

_____, _____ (1969) Die Kristallstruktur von Baumhauerit. *Z. Kristallogr.* 129, 178-202.

Engel, P., and W. Nowacki (1970) Die Kristallstruktur von Rathit - II [As₂₄S₅₆|
$Pb^{VII}_{6.5}Pb^{IX}_{12}$]. Z. Kristallogr. 131, 356-375.

Erd, R. C., H. T. Evans, Jr., and D. H. Richter, (1957) Smythite, a new iron
sulfide and associated pyrrhotite from Indiana. Am. Mineral. 42, 309-333.

Euler, Robert, and Erwin Hellner (1960) Uber komplex zusammengesetzte sulfidische
Erze Vi. Zur Kristallstruktur des Meneghinits, $CuPb_{14}Sb_7S_{24}$. Z. Kristallogr.
113, 345-372.

Evans, H. T., Jr. (1968) A crystal structure study of low chaococite (abstr.).
Geol. Soc. Am. Abstr. Programs, 92.

_____ (1970) Lunar troilite: Crystallography, Science 167, 621-623.

_____ (1971) Crystal structure of low chalcocite. Nature 232, 69-70.

_____, C. Milton, E. C. T. Chao, I. Adler, C. Mead, B. Ingram, and R. A. Berner
(1964) Valeriite and the new iron sulfide, mackinawite. U.S. Geol. Surv.
Prof. Pap. 475-D, A133, 64-69.

Feiss, P. G. (1974) Reconnaissance of the tetrahedrite-tennantite/enargite-
famatinite phase relations as a possible geothermometer: Econ. Geol.
69, 383-390.

Fleet, M. E. (1970) Refinement of the crystal structure of cubanite and polymorphism of $CuFe_2S_3$. Z. Kristallogr. 132, 276-287.

_____ (1971) The crystal structure of a pyrrhotite (Fe_7S_8). Acta Crystallogr.
B27, 1864-1867.

_____ (1972) The crystal structure of $\alpha-Ni_7S_6$. Acta Crystallogr. B28, 1237-1241.

_____ (1973a) The crystal structure and bonding of lorandite, $Tl_2As_2S_4$.
Z. Kristallogr. 138, 147-160.

_____ (1973b) The crystal structure of parkerite ($Ni_3Bi_2S_2$). Am. Mineral.
58, 435-439.

_____, and N. MacRae (1969) Two-phase hexagonal pyrrhotites. Can. Mineral.
9, 699-705.

Francis, C. A., J. R. Craig, and G. V. Gibbs (1974) Crystallographic relationships of pentlandite exsolution (abstr.). Geol. Soc. Am. Abstr. Programs
6 (in press).

Francotte, J., J. Moreau, R. Ottenburgs , and L. Levy (1965) La briartite,
$Cu_2(Fe,Zn)GeS_4$ un nouvelle espece minerale, Bull. Soc. franc. Mineral.
Crystallogr. 88, 432-437.

Frank, F. C. (1951) The growth of carborundum: dislocation and polymorphism.
Phil. Mag. 42, 1014-1021.

_____, J. B. Mullin, and H. S. Peiser, Eds. (1968). Crystal growth 1968.
Proceedings of the second international conference on crystal growth,
Birmingham, U.K., 15-19, July, 1968. J. Crystal. Growth 3/4, 1-842.

Frantz, J. D. and H. P. Eugster (1973) Acid-base buffers: Use of Ag + AgCl in the
experimental control of solution equilibria at elevated pressures and
temperatures. Am. J. Sci. 273, 268-286.

Franzen, H. F., J. Smeggil, and B. R. Conard (1967) The group IV di-transition
metal sulfides and selenides. Mat. Res. Bull. 2, 1087-1092.

Fredriksson, K., and C. A. Anderson (1964) Electron probe analyses of Cu in
meneghinite. Am. Mineral. 49, 1467-1469.

Frenzel, G. (1959) Idaite und blaubleibender Covellin. Neues Jahrb. Mineral.
Abh. 93, 87-132.

Friedrich, K. (1907) Die Schmelz-diagramme der binären systeme Bleiglenz-
magnetkies und Bleiglanz-Schwefelsilber. Metallurgie 4, 477-485.

Frondel, C.(1963) Isodimorphism of the polybasite and pearcite series: Am. Mineral.
48, 565-572.

Frondel, C. (1967) Voltzite. *Am. Mineral.* 52, 617-634.

_____, and R. M. Honea (1968) Billingsleyite, a new silver sulfosalt. *Am. Mineral.* 53, 1791-1798.

_____, W. H. Newhouse, and R. F. Jarrell (1942) Spatial distribution of minor elements in single crystals. *Am. Mineral.* 27, 726-745.

Frueh, Alfred J. (1955) The crystal structure of stromeyerite, AgCuS: a possible defect structure. *Z. Kristallogr.* 106, 299-307.

_____, (1958) The crystallography of silver sulfide Ag_2S. *Z. Kristallogr.* 110, 136-144.

_____ (1961) The use of zone theory in problems of sulfide mineralogy, Part III; Polymorphism of Ag_2Te and Ag_2S. *Am. Mineral.* 46, 654-700.

Fujii, T. (1970) Unmixing in the system, sphalerite and chalcopyrite. In, T. Tatsumi, Ed., *Volcanism and Ore Genesis*. Univ. of Tokyo Press, Tokyo, p. 357-366.

Gaines, R. V. (1957) Luzonite, famatinite and some related minerals. *Am. Mineral.* 42, 766-779.

Gamble, F. R. (1974) Ionicity, atomic radii and structure in the layered dichalcogenides of group IVb, Vb, and VIb transition metals. *J. Solid State Chem.* 9, 358-367.

Gammon, J. B. (1966) Some observations on minerals in the system CoAsS-FeAsS. *Norsk Geol. Tidsskř.* 46, 405-426.

Garvin, P. L. (1973) Phase relations in the Pb-Sb-S Systems. *Neues Jahrb. Mineral. Abh.* 118, 235-267.

Geller, S. (1962) Refinement of the crystal structure of Co_9S_8. *Acta Crystallogr.* 15, 1195-1198.

_____ (1966) Pressure-induced phases of sulfur. *Science,* 152, 644-646.

_____, and M. D. Lind (1969) Indexing of the ψ-sulfur fiber pattern. *Acta Crystallogr.* B25, 2166-2167.

_____, and J. H. Wernick (1959) Ternary semiconducting compounds with sodium chloride-like structure: $AgSbS_2$, $AgSbTe_2$, $AgBiS_2$, $AgBiSe_2$. *Acta Crystallogr.* 12, 46-54.

Genkin, A. (1971) Some replacement phenomena in copper-nickel sulphide ores. *Mineral. Deposita* 6, 348-355.

_____, and I. V. Murav'eva (1963) Indite and dzhalindite, indium minerals. *Zap. Vses. Mineral. Obshch.* 92, 445-457.

Gerard, A. P. Imbert, H. Prange, F. Varret, and M. Wintenberger (1971) Fe^{2+} impurities, isolated and in pairs, in ZnS and CdS studies by the Mössbauer effect. *J. Phys. Chem. Solids* 32, 2091-2100.

Giese, R. F., and P. F. Kerr (1965) The crystal structures of ordered and disordered cobaltite. *Am. Mineral.* 50, 1002-1014.

Giletti, B. J., R. A. Yund, and T. J. Lin (1968) Sulfur vapor pressure of pyrite-pyrrhotite (abstr.). *Econ. Geol.* 63, 702.

Glutz, A. C. (1967) The Bi_2Te_3-Bi_2S_3 system and the synthesis of the mineral tetradymite. *Am. Mineral.* 52, 161-170.

Godovikov, A. A., and N. A. Il'yasheva (1971) Phase diagram of a bismuth-bismuth sulfide-bismuth selenide system. *Mater. Genet. Eksp. Mineral.* 6, 5-14.

_____, K. V. Kochetkova, and V. G. Lavrent'ev (1970) Study of the bismuth sulfotellurides of the Sokhondo deposit. *Geol. Geofiz.* 11, 123-127. [*Am. Mineral.* 56, 1839-1840 (1971)]

_____, and S. N. Nenasheva (1969a) The $AgSbS_2$-PbS system above 480°C. *Dokl. Acad. Sci. USSR Earth Sci. Sect.* 185, 76-79.

Godovikov, A. A., and S. N. Nenasheva (1969b) $Ag_3PbSb_3S_7$, a new phase of the $AgSbS_2$-PbS system. *Dokl. Akad. Nauk, SSSR* 184, 151.

_____, and A. B. Ptitsyn (1968) Syntheses of Cu-Bi-sulfides under hydrothermal conditions. *Eksp. Issled. Mineral.*, 29-41. [*Chem. Abstr.* 73, 104990 (1969)]

Gomes de Mesquita, A. H. (1967) Refinement of the crystal structure of SiC type 6H. *Acta Crystallogr.* 23, 610-617.

Goodenough, J. B. (1967) Description of transition metal compounds: Application to several sulfides. Propriétés thermodynamiques physiques et structurales des dérivés semi-metalliques. *Centre Nat. Sci. No.* 157, 263-292.

_____ (1969) Descriptions of outer *d* electrons in thiospinels. *J. Phys. Chem. Solids*, 30, 261-280.

_____ (1972) Energy bands in TX_2 compounds with pyrite, marcasite and arseno-pyrite structures. *J. Solid State Chem.* 5, 144-152.

Graeser, S. (1964) Über Funde der neven rhombohedrichan MoS_2 - Modifikation (Molybdanit - 3R) und von Tungstenit in den Alpen *Schweiz. Mineral. Petrogr. Mitt.* 44, 121-128.

_____ (1967) Ein Vorkommen von Lorandite ($TlAsS_2$) in der Schweiz. *Contrib. Mineral. Petrol.* 16, 45-50.

Graf, R. B. (1968a) The system Ag_3AuS_2-Ag_2S. *Am. Mineral.* 53, 496-500.

_____ (1968b) Phase transformation in the system Cu_2S-Ag_2S. *J. Electrochem. Soc.* 115, 433-434.

Graham, A. R. (1969) Quantitative determination of hexagonal and monoclinic pyrrhotites by *X*-ray diffraction. *Can. Mineral.* 10, 4-24.

Graterol, M., and A. J. Naldrett (1971) Mineralogy of the Marbridge no. 3 and no. 4 nickel-iron sulfide deposits. *Econ. Geol.* 66, 866-900.

Gray, H. B. (1965) *Electrons and Chemical Bonding.* W. A. Benjamin, New York.

Grice, J. D., and R. B. Ferguson (1974) Crystal structure refinement of millerite (β-NiS). *Can. Mineral.* 12, 248-252.

Gronvold, F., and H. Haraldsen (1952) On the phase relations of synthetic and natural pyrrhotites. *Acta Chem. Scand.* 6, 1452-1469.

_____, _____, and A. Kjekshus (1960) On the sulfides, selenides and tellurides of platinum. *Acta Chem. Scand.* 14, 1879-1893.

Grover, B., G. Kullerud, and G. H. Moh (1973) Phasen gleichgewichtsbeziehungen im ternaren system Fe-Mo-S in Relation zu naturlichen Mineralien und Erzlagerstatten. *Neues Jahrb. Mineral. Monatsch.*

_____, and G. H. Moh (1966) Experimentelle untersuchungen des quaternaren systems Kupfer-Eisen Molybdan-Schwefel. *Referat 44 Jahrestagung der Dentschen Mineralogischen Gesellschaft, Munchen.*

_____, _____ (1969) Phasen gleichgewichtsbeziehungen im system Cu-Mo-S in Relation zu naturlichen Mineralien. *Neues Jahrb. Mineral. Monatsch.*

Gruner, J. W. (1929a) Structural reasons for oriented intergrowths in some minerals. *Am. Mineral.* 14, 227-237.

_____ (1929b) Structures of sulfides and sulfosalts. *Am. Mineral.* 14, 470-481.

Guillermo, Romas R., and B. J. Wuensch (1973) The crystal structure of getchellite, $AsSbS_3$. *Acta. Crystallogr.* B29, 2536-2541.

Gustavson, L. B. (1963) Phase equilibria in the system Cu-Fe-As-S. *Econ. Geol.* 58, 667-701.

Haas, J. L., Jr., and R. W. Potter, II (1974) Internally consistent thermodynamic functions for sulfides and sulfosalts obtained using PVT data for sulfur (abstr.). *Geol. Soc. Am. Abstr. Programs,* (in press).

Hägg, G., and I. Sucksdorff (1933) Die Kristallstruktur von Troilite und Magnetkies. *Z. Phys. Chem.* 22B, 444-452.

Hahn, H., B. Harder, and W. Brockmuller (1965) Untersuchungen der ternären Chalcogenide X. Versuche zur Umsetzung von Titan sulfiden mit sulfiden Zweiwretiger Übergangsmetalle. *Z. Anorg. Allg. Chem.* 288, 260-268.

_____, and W. Klinger (1950) Über die Kristallstruktur einiger ternäre Sulfide, die sich vom Indium III Sulfid ableiten. *Z. Anorg, Allg. Chem.* 263, 177-190.

Hall, H. T. (1966) *The Systems Ag-Sb-S, Ag-As-S, and Ag-Bi-S: Phase Relations and Mineralogical Significance.* Ph.D. dissertation, Brown University.

_____ (1967) The pearceite and polybasite series. *Am. Mineral.* 52, 1311-1321.

_____ (1968) Synthesis of two new silver sulfosalts. *Econ. Geol.* 63,289-291.

Hall, L. H. (1969) *Group Theory and Symmetry in Chemistry.* McGraw-Hill, New York.

Hall, S. R., and E. J. Gabe (1972) The crystal structure of talnakhite, $Cu_{18}Fe_{16}S_{32}$. *Am. Mineral.* 57, 268-280.

_____, and J. F. Rowland (1973) The crystal structure of synthetic mooihoekite, $Cu_9Fe_9S_{16}$. *Acta Crystallogr.* B29, 2365-2372.

_____, and J. M. Stewart (1973a) The crystal structure of argentian pentlandite $(FeNi)_8AgS_8$, compared with the refined structure of pentlandite $(Fe,Ni)_9S_8$. *Can. Mineral.* 12, 169-177.

_____, _____ (1973b) The crystal structure refinement of chalcopyrite, $CuFeS_2$. *Acta Crystallogr.* B29, 579-585.

Hansen, M., and K. Anderko (1958) *Constitution of Binary Alloys.* McGraw-Hill, New York.

Harada, K., O. Sakamoto, K. Nakao, and K. Nagashima (1970) Bournonite from Daikoku, Chichibu mine, Saitama, Japan. *Mineral. J.* 6, 186-188.

Harker, David (1936) The application of the three-dimensional Patterson method and the crystal structures of proustite, Ag_3AsS_3, and pyrargyrite, Ag_3SbS_3. *J. Chem. Phys.* 4, 381-390.

Harris, D. C., and D. K. Owens (1972) A stannite-kesterite exsolution from British Columbia. *Can. Mineral.* 11, 531-534.

_____, and R. I. Thorpe (1969) New observations on matildite. *Can. Mineral.* 9, 655-662.

Hayase, K. (1955) Minerals of bismuthinite-stibnite series with special reference to horobetsuite from the Horobetsu Mine. *Mineral. J. Japan,* 1, 188-197.

Heier, K. (1953) Clausthalite and selenium-bearing galena in Norway. *Norsk Geol. Tidskr.* 32, 228-231.

Helgeson, H. C. (1969) Thermodynamics of hydrothermal systems at elevated temperatures and pressures. *Am. J. Sci.* 267, 729-804.

Hellner, Erwin (1958) A structural scheme for sulfide minerals. *J. Geol.* 66, 503-525.

_____, and H. Burzlaff (1964) Die Struktur des Smithit $AgAsS_2$. *Naturwissenschaften,* 51, 35-36.

_____, and Günter Leineweber (1956) Über komplex zusammengesetzte sulfidische Erze. *Z. Kristallogr.* 107, 150-154.

Hill, V. G. (1968) The hydrothermal corrosion and recrystallization of noble metals. *J. Electrochem. Soc.* 115, 720-721.

Hiller, J. E., and K. Probsthain (1956) Thermische und röntgenographische Untersuchungen an Kupferkres. *Z. Kristallogr.* 108, 108-129.

Hoda, S. N., and L. L. Y. Chang (1972) Phase relations in the systems Ag_2S-PbS-Sb_2S_3 and Ag_2S-PbS-Bi_2S_3 (abstr.). *Geol. Soc. Am. Abstr. Programs,* 4, 539-540.

Hofmann, Wilhelm (1933a) Die Struktur der Minerale der Antimonitgruppe. *Z. Kristallogr.* 86, 225-245.

_____ (1933b) Strukturelle und morphologische Zusammenhange bei Erzen vom Formeltyp ABC_2. I Die Stuktur von Wolfsbergit $CuSbS_2$ und deren Beziehungen zu der Struktur von Antimonit Sb_2S_3. *Z. Kristallogr.* 84, 177-203.

Höll, R., and K. Weber-Diefenbach (1973) Tungstenit-Molybdanit-Mischphasen in der Scheelit lagerstatte Felbartal (Hohe Tauern, Österreich). *Neues Jahrb. Mineral. Monatsch.* 1, 27-34.

Hollingsworth, C. A. (1967) *Vectors, Matrices, and Group Theory for Scientists and Engineers.* McGraw-Hill, New York.

Hollister, L. S. (1970) Origin, mechanism, and consequences of compositional sector-zoning in staurolite. *Am. Mineral.* 55, 742-766.

_____, and A. E. Bence (1967) Staurolite: Sectoral compositional variations. *Science* 158, 1053-1056.

Hrušková, J. (1969) The crystal structure of samsonite, $2Ag_2S \cdot MnS \cdot Sb_2S_3$. *Acta Crystallogr.* B25, 1004-1006.

Hulliger, F. (1968) Crystal chemistry of chalcogenides and prictides of the transition elements. *Structure and Bonding* 4, 83-229.

Hurlbut, C. S. (1957) The wurtzite-greenockite series. *Am. Mineral.* 42, 184-190.

Iitaka, Y., and W. Nowacki (1961) A refinement of the pseudo crystal structure of scleroclase $PbAs_2S_4$. *Acta Crystallogr.* 14, 1291-1292.

_____, _____ (1962) A redetermination of the crystal structure of galeno-bismutite, $PbBi_2S_4$. *Acta. Crystallogr.* 15, 691-698.

Ito, Tetsuzo, and W. Nowacki (1974a) The crystal structure of freieslebenite, $PbAgSbS_3$. *Z. Kristallogr.* 139, 85-102.

_____, _____ (1974b) The crystal structure of jordanite, $Pb_{28}As_{12}S_{46}$. *Z. Kristallogr.* 139, 161-185.

Jagodzinski, H. (1954a) Der Symmetrieeinfluss auf den allgemeinen Lösungsansatz eindimensionaler Fehlordnungs probleme. *Acta Crystallogr.* 7, 17-25.

_____ (1954b) Fehlordnungserscheinungen und ihr Zusammenhang mit der Polytypie des SiC. *Neues Jahrb. Mineral.* 10, 49-65.

_____ (1954c) Polytypism in SiC. *Acta Crystallogr.* 7, 300.

Jambor, J. L. (1967a) New Lead sulfantimonides from Madoc, Ontario. Part I. *Can. Mineral.* 9, 7-24.

_____ (1967b) New lead sulfantimonides from Madoc, Ontario. Part 2. Mineral descriptions. *Can. Mineral.* 9, 191-213.

_____ (1967c) New lead and sulfantimonides from Madoc, Ontario. Part 3. Syntheses, parageneses, origin. *Can. Mineral.* 9, 505-521.

_____ (1969a) Dadsonite (minerals Q and QM), a new lead sulfantimonide. *Mineral. Mag.* 37, 437-441.

_____ (1969b) Sulphosalts of the plagionite group. *Mineral. Mag.* 37, 442-446.

Jankovic, S. (1953) The structures of systems $ZnS-CuFeS_2$ in the ores of Saplja Stijena (Crna Gora). *Yugoslav. Geol. Inst. Geol. Vesnite,* 9-10, 255-271.

Janosi, A. (1964) La structure du sulfure cuivreux quadratiques. *Acta Crystallogr.* 17, 311-312.

Jellinek, F. (1957) The structures of the chromium sulphides. *Acta Crystallogr.* 10, 620-628.

_____ (1970) Sulfides. In, *Inorganic Sulfur Chemistry,* Ed. G. Nickless. Elsevier, Amsterdam, 669-748.

Jenkins, Judith K. (1969) *The Crystal Structure of Emplectite, CuBiS₂*. S. B. Thesis, Department of Metallurgy and Materials Science, Massachusetts Institute of Technology.

Jensen, E. (1942) Pyrrhotite: Melting relations and composition. *Am. J. Sci.* 240, 695-709.

_____ (1947a) Melting relations of chalcocite. *Norsk Videns - Akad. Oslo Avh. I. Mat.-Natur. Kl.* 6, 14 p.

_____ (1947b) The system silver sulfide - antimony trisulfide: a thermal study. *Avh. Norske Videns - Akac. Oslo I Mat. Naturv. Kl.* 2, 23 p.

Juza, R. A. Rabenau, and G. Pascher (1956) Über fests Lösungen in den systemen ZnS/MnS, ZnSe/MnSe, ZnTe/MnTe. *Zeit anorg. und allg. Chemie.* 285, 61-69.

Kajiwara, Y. C. (1969) Fukuchilite, Cu₃FeS₃, a new mineral from the Hanawa Mine, Akita Projecture, Japan. *Mineral. J. Japan* 5, 399-416.

Kalbskopf, R. (1971) Die Koordination des Quecksilbers im Schwazit. *Tschermaks Mineral.. Petrogr. Mitt.* 16, 173-175.

_____ (1972) Strukturverfeinerung des Freibergits. *Tschermaks Mineral. Petrogr. Mitt.* 18, 147-155.

Karakhanova, M. I., A. S. Pushinkin, and A. V. Novoselova (1966) Phase diagram of the tin-sulfur system. *Izv Adak Navk. SSSR Neorg. Mater.* 2, 991-996. [*Chem. Abstr.* 65, 14509-10]

Karup-Moller, S. (1972) New data on pavonite, gustavite and some related sulphosalt minerals. *Neues Jahrb. Mineral. Monatsh.,* 19-38.

_____ (1973) New data on schirmerite. *Can. Mineral.* 11, 952-957.

_____, and E. Makovicky (1974) Skinnerite, Cu₃SbS₃, a new sulfosalt from the Ilimaussag alkaline intrusion South Greeland. *Am. Mineral.* 59, 889-895.

Kato, A. (1959) Ikunolite, a new bismuth mineral from the Ikuno Mine, Japan. *Mineral. J. Japan* 2, 397-407.

_____, K. Sakurai, and K. Ohsumi (1970) Wakabayashilite (As,Sb)₁₁S₁₈. In, M. Nambu, Ed., *Introduction to Japanese Minerals,* Geol. Surv. Japan.

Kawada, Isao, and Erwin Hellner (1970) Zur Struktur von Cycloheptaschwefel. *Angew. Chemie* 82, 320.

_____, _____ (1971) Die Kristallstruktur der Pseudozelle (sub-cell) von Andorit VI (Ramdohrit). *Neues Jahrb. Mineral. Monatsh.,* 551-560.

Keighin, C. W., and R. M. Honea (1969) System Ag-Sb-S from 600 to 200°C. *Mineral. Deposita* 4, 153-171.

Keil, K., and R. Brett (1974) Heiderite, (Fe,Cr)₁₊ₓ(Ti,Fe)₂S₄, a new mineral in the Bustee eustatite achondrite. *Am. Mineral.* 59, 465-470.

_____, and K. G. Snetsinger (1967) Niningerite, a new meteorite sulfide. *Science,* 155, 451-453.

Kerr, P. F. (1945) Cattierite and vaesite: new Co-Ni minerals from the Belgium Congo. *Am. Mineral.* 30, 483-497.

Kingston, P. W. (1970) On alloclasite, a Co-Fe sulpharsenide. *Can. Mineral.* 10, 838-846.

Kirkinskiy, V. A., A. P. Ryaposov, and V. G. Yakushev (1967) Phase diagram for arsenic tri-sulfide up to 20 kilobars. *Isv. Akad. Nauk SSSR,* 3, 1931-1933.

Kissin, S. A. (1974) *Phase Relations in a Portion of the Fe-S System.* Ph.D. Thesis, University of Toronto, Canada.

_____, and S. D. Scott (1972) Phase relations of intermediate pyrrhotites (abstr.). *Econ. Geol.* 67, 1007.

Kitikaze, A. (1968) *The PbS-Sb₂-S₃ system.* M.S. Thesis, Yamaguchi University.

Klemm, D. D. (1962a) Synthese untersuchungen in der Mischkristallreihe FeAsS-CoAsS und ihre Beziehungen zu den Mineralian Arsenkies, Danait, Ghukodat und Glanzkobalt, *Vortr. DMG - Tagung, Wurzberg.*

_____ (1962b) Untersuchungen über die Mischkristall bildung im Dreieck diagramm $FeS_2-CoS_2-NiS_2$ und ihre Beziehungen zum Aufbau der naturlichen "Bravoite" *Neues Jahrb. Mineral. Monatsh.*

_____ (1965) Synthesen und Analysen in den Dreickdiagrammen FeAsS-CoAsS-NiAsS und $FeS_2-CoS_2-NiS_2$. *Neues Jahrb. Mineral. Abh.* 103, 205-255.

Klominsky, J., M. Rieder, C. Kieft, and L. Mraz (1971) Heyrovskyite, $6(Pb_{0.86}Bi_{0.08}(Ag,Cu)_{0.04})S \cdot Bi_2S_3$ from Hurky, Czechoslovakin, New Mineral of Genetic Interest. *Mineral. Deposita* 6, 133-147.

Knop, O. and M. A. Ibrahim (1961) Chalcogenides of the transition elements. II. Existence of the π phase in the M_9S_8 section of the system Fe-Co-Ni-S. *Can. J. Chem.* 39, 297-317.

Knowles, Charles R. (1964) A redetermination of the structure of miargyrite, $AgSbS_2$. *Acta Crystallogr.* 17, 847-851.

_____ (1965) A refinement of the structure of the structure of lorandite, $TlAsS_2$ (abstr.). *Program Abstr. Am. Crystallogr. Assoc.* Gatlinbug, Tenn., 79.

Kocman, V., and E. W. Nuffield (1973) The crystal structure of wittichenite, Cu_3BiS_3. *Acta Crystallogr.* B29, 2528-2535.

Kodera, M., V. Kupcik, and E. Makovicky (1970) Hadrushite - a new sulfosalt. *Mineral. Mag.* 37, 641-648.

Kohatsu, Iwao, and B. J. Wuensch (1971) The crystal structure of aikinite, $PbCuBiS_3$. *Acta Crystallogr.* B27, 1245-1252.

_____, _____ (1973a) The crystal structure of nuffieldite, $Pb_2Cu(Pb,Bi)Bi_2S_7$. *Z. Kristallogr.* 138, 343-365.

_____, _____ (1973b) The crystal structure of gladite, $PbCuBi_5S_9$ (abstr.). *Am. Mineral.* 58, 1098.

Kohatsu, J. J., and B. J. Wuensch (1974) Prediction of structures in the homologous series $Pb_{3+2n}Sb_8S_{15+2n}$ (the plagionite group). *Acta Crystallogr.* B30 (in press).

Koulenker, V. A., I. P. Laputina, and L. N. Vyal'sov (1971) First find of minerals of galena clausthalite series in copper-nickel ores. *Geol. Rud. Mestorozhd.* 13, 98-101. [*Chem. Abstr.* 75, 23735f]

Kozielski, M. (1967) Growth of ZnS single crystals from the melt at 1850°C under argon pressure of 50 atm. *J. Crystal Growth,* 1, 293-296.

Kracek, F. C. (1946) Phase relations in the system sulfur-silver and the transitions in silver sulfide. *Trans. Am. Geophys. Union,* 27, 364-374.

Krebs, H. (1968) *Fundamentals of Inorganic Crystal Chemistry.* McGraw-Hill, London.

Krestovnikov, A. N., A. Y. Mendelevich, and V. M. Glazov (1968) Phase equilibria in the Cu_2S-Ag_2S system. *Inorg. Material.* 4, 1047-1048.

Kretschmar, U. (1973) *Phase Relations Involving Arsenopyrite in the System Fe-As-S and their Application.* Ph.D. Thesis, University of Toronto, Canada.

Kroger, F. A. (1939a) Formation of solid solutions in the system zinc sulfide - manganese sulfide. *Z. Kristallogr.* A100, 543-545.

_____ (1939b) Solid solutions in the ternary system ZuS-CdS-MnS. *Z. Kristallogr.* A102, 123-135.

_____ (1964) *The Chemistry of Imperfect Crystals.* North-Holland Publ. Co., Amsterdam, 1039 p.

_____, and H. J. Vink (1954) The origin of the fluorescence in selfactivated ZnS, CdS and ZnO. *J. Chem. Phys.* 22, 250-252.

Kulagov, E. A., T. L. Erstigneeva, and O. E. Yashko-Zakharova (1969) The new nickel sulfide godlevskite. *Geol. Rud. Mestoro Zhd.* 11, 115-121.

Kullerud, G. (1953) The FeS-ZnS system, a geological thermometer. *Norsk Geol. Tiddsskr.* 32, 61-147.

_____ (1963a) The Fe-Ni-S system. *Carnegie Inst. Wash. Year Book,* 62, 175-189.

_____ (1963b) Thermal stability of pentlandite. *Can. Mineral.* 7, 353-366.

_____ (1965a) Covellite stability relations in the Cu-S system. *Frei. Forsch.* C186, 145-160.

_____ (1965b) The mercury-sulfur system. *Carnegie Inst. Wash. Year Book,* 64, 193-195.

_____ (1967a) The Fe-Mo-S system. *Carnegie Inst. Wash. Year Book,* 65, 337-342.

_____ (1967b) Sulfide studies. In, P. H. Abelson, Ed., *Researches in Geochemistry,* Vol. 2. John Wiley and Sons, p. 286-321.

_____ (1969a) Cubic $\rightleftharpoons$ hexagonal inversions in some M_3S_4-type compounds. *Carnegie Inst. Wash. Year Book,* 67, 179-182.

_____ (1969b) The lead-sulfur system. *Am. J. Sci.* 267-A, 233-267.

_____ (1971) Experimental techniques in dry sulfide research. In, G. C. Ulmer, Ed., *Research Techniques for High Pressure and High Temperature.* Springer-Verlag, New York, pp. 288-315.

_____, and H. S. Yoder (1959) Pyrite stability relations in the Fe-S system. *Econ. Geol.* 54, 533-572.

_____, and R. A. Yund (1962) The Ni-S system and related minerals. *J. Petrol.* 3, 126-175.

_____, _____, and G. H. Moh (1969) Phase relations in the Cu-Fe-S, Cu-Ni-S and Fe-Ni-S system. In, H.D.B. Wilson, Ed., Magmatic Ore Deposits, *Econ. Geol. Monogr.* 4, 323-343.

Kulpe, S. (1961) Die Kristallstrukur des Lautit. *Fortschr. Mineral.* 39, 332 (1961).

Kupĉík, V. (1965) Struktur des Emplektits CuBiS$_2$. *Program Abstr. Meet. Crystallogr. Sec. German Mineral. Soc., Marburg,* 16-17 (Abstract).

_____ (1967) Die Kristallstruktur des Kermesits, Sb$_2$S$_2$O. *Naturwissenschaften,* 54, 114.

_____, and E. Makovický (1968) Die Kristallstruktur des Minerals (Pb,Ag,Bi)Cu$_4$Bi$_5$S$_{11}$, *Neues Jahrb. Mineral. Monatsh.,* 236-237.

_____, and Ludmila Veselá-Nováková (1970) Zur Kristallstruktur des Bismuthinits, Bi$_2$S$_3$. *Tschermaks Mineral. Petrogr. Mitt.* 14, 55-59.

Kurash, V. V., G. A. Kulikov, and Y. S. Makarov (1973) Mg-Fe isomorphism in the MgS-FeS system. *Geochem. Int.* 10, 956.

Kushima, I., and N. Asano (1953) The phase diagram of the system nickel sulfide-copper sulfide-iron sulfide: I. Binary systems. *J. Min. Inst. Japan* 69, 297-300.

Kutoglu, A. (1968) Die Struktur des Pyrostilpnits (Feuerblende) Ag$_3$SbS$_3$. *Neues Jahrb. Mineral. Monatsh.,* 145-160.

_____ (1969) Rönt genographicsche und thermische Untersuchungen in Quasibinaren system PbS-As$_2$S$_3$. *Neues Jahrb. Mineral. Monatsh.* 2, 68-72.

_____, and R. Allmann (1972) Strukturverfeinerung des Patronits, V(S$_2$)$_2$. *Neues Jahrb. Mineral. Monatsh.,* 339-345.

Kuznetsov, V. G. and K. Ch'ih-Fa (1964) X-ray study of the stannous sulfide-lead sulfide system. *Ah. Neorgan. Khim.* 9, 1201-1206.

_____, M. A. Sokolova, K. K. Palkina, and Z. V. Popova (1965) System cobalt-sulfur. *IZV. Akad. Nauk. SSR Neogram. Matenaly* 1, 675-689.

Lafitte, M. (1959) Etude thermodynamique - des monosultures de nickel et de cobalt. *Bull. Soc. Ehim. Fr.* 7-8, 1223-1233.

Lange, W., and H. Schlegel (1951) Die Zustandbilder des Systeme Eisen-Antimon-Schwefel und Kobalt-Antimon-Schwefel. *Metallkunda* 42, 257-268.

Larimer, J. W. (1968) An experimental investigation of old hamite, CaS and the petrologic significance of oldhamite in meteorites. *Geochim. Cosmochim. Acta* 32, 965-982.

Laudise, R. A., E. D. Kolb, and J. P. DeNeufville (1965) Hydrothermal solubility and growth of sphalerite. *Am. Mineral.* 50, 382-391.

_____, J. B. Mullin, and B. Mutaftschiev, Eds. (1972) Crystal growth 1971. Proceedings of the third international conference on crystal growth, Marseilles, France, 5-9 July, 1971. *J. Crystal Growth,* 13/14, 1-876.

Lawson, A. C. (1972) Lattice instabilities in superconducting ternary molybdenum sulfides. *Mat. Res. Bull.* 7, 773-776.

Learned, R. E. (1966) *The solubilities of quartz, quartz-cinnabar, and cinnabar-stibnite in sodium sulfide solutions and their implications for ore genesis.* Ph.D. Thesis, University of California, Riverside.

_____, G. Tunnell and F. W. Dickson (1974) Equilibria of cinnbar , stibnite and saturated solutions in the system $HgS-Sb_2S_3-Na_2S-H_2O$ from 150° to 250°C at 100 bars with implications concerning ore genesis. *U.S. Geol. Surv. J. Res.* 2, 457-466.

Le Bihan, M.-Th. (1962) Étude structurale de quelques sulfures de plomb et d'arsenic naturels du gisement de Binn. *Bull. Soc. franc. Mineral. Cristallogr.* 85, 15-47.

_____ (1963) Étude structurale de quelques sulfures de plomb et d'arsenic naturels du gisement de Binn. *Mineral. Soc. Amer. Spec. Pap.* 1, 149-152.

Lee, J. Y. (1972) Experimental investigation on stannite-sphalerite solid solution series. *Neues Jahrb. Mineral. Monatsh.* 556-559.

Leegaard, T. and T. Rosenqvist (1964) Der Zersetsungsdruck und die hoheren Sulfides von Kobalt und Nickel. *Z. Anorg. allgem. Chem.* 328, 294-298.

Leonard, B. F., G. A. Desborough and N. J. Page (1969) Ore microscopy and chemical composition of some lavrites. *Am. Mineral.* 54, 1330-1346.

Lewis, G. N.,and M. Randall (1961) *Thermodynamics,* 2nd ed. Revised by K. S. Pitzer and L. Brewer, McGraw-Hill, Inc., New York.

Lindqvist, M., D. Lundqvist, and A. Westgren (1936) The crystal structure of Co_9S_8 and of pentlandite $(Ni,Fe)_9S_8$. *Kem. Tidskr.* 48, 156-160.

Lowenhaupt, D. E., and D. K. Smith (1974) The high-temperature crystal structure of Ag_2S-II (abstr.). *Program Abstr., Summer Meet. Am. Crystallogr. Assoc., University Park, Pennsylvania,* p. 265.

Lundqvist, D. (1947a) *X*-ray studies on the binary system Ni-S. *Ark. Kemi. Mineral. Geol.* 24A, no. 21, 12 p.

_____ (1947b) *X*-ray studies in the ternary system Fe-Ni-S. *Ark. Kemi. Mineral. Geol.* 24A, no. 22.

Luquet, H., F. Guastavino, J. Bougnot, and J. C. Vaissiere (1972) Étude du systeme Cu-S dans le domaine $Cu_{1.78}S-Cu_{2.1}S$ par analyse thermique differentielle. *Mat. Res. Bull.* 7, 955-962.

Malik, O. P., and R. A. Arnold (1971) Evidence for an invariant equilibrium at 1005±3°C in the Fe-Ni-S system (abstr.). *Geol. Assoc. Can. - Mineral. Assoc. Can.,* Sudbury, p. 40.

Manning, P. G. (1967) Absorption spectra of Fe (III) in octahedral sites in sphalerite. *Can. Mineral.* 9, 57-64.

Mardix, S., E. Alexander, O. Brafman, and I. T. Stinberger (1967) Polytype families in zinc sulphide crystals. *Acta Crystallogr.* 22, 808-812.

Mariano, A. N., and E. P. Warekois (1963) High Pressure phases of some compounds of groups II - VI. *Science,* 142, 672-673.

Markham, N. L. (1962) Plumbian ikunolite from Kingsgate, New South Wales. *Am. Mineral.* 47, 1431-1434.

Marumo, F. (1967) The crystal structure of nowackiite, $Cu_6Zn_3As_4S_{12}$. *Z. Kristallogr.* 124, 352-368.

_____, and W. Nowacki (1964) The crystal structure of lautite and of sinnerite, a new mineral from the Lengenbach Quarry. *Schweiz. Mineral. Petrogr. Mitt.* 44, 439-454.

_____, _____ (1965) The crystal structure of rathite I. *Z. Kristallogr.* 122, 433-456.

_____, _____ (1967a) A refinement of the crystal structure of luzonite. *Z. Kristallogr.* 124, 1-8.

_____, _____ (1967b) The crystal structure of dufrenoysite, $Pb_{16}As_{16}S_{40}$. *Z. Kristallogr.* 124, 409-419.

_____, _____ (1967c) The crystal structure of hatchite, $PbTlAgAs_2S_5$. *Z. Kristallogr.* 125, 249-265.

Maske, S., and B. J. Skinner (1971) Studies of the sulfosalts of copper. I. Phases and phase relations in the system Cu-As-S. *Econ. Geol.* 66, 901-918.

Matsumoto, Takeo, and Werner Nowacki (1969) The crystal structure of trenchmannite, $AgAsS_2$. *Z. Kristallogr.* 129, 163-177.

Matzat, E. (1972) Die Kristallstruktur des Wittichenits, Cu_3BiS_3. *Tschermaks Mineral. Petrogr. Mitt.* 18, 312-316.

Maurel, C. (1973) Mechanism of hydrothermal sphalerite ⇄ galena replacement at 300°C. *Econ. Geol.* 66, 665-670.

McClure, D. S. (1957) The distribution of transition metal cations in spinels. *J. Phys. Chem. Solids* 3, 311-17.

McKinstry, H. (1963) Mineral assemblages in sulfide ores: the system Cu-Fe-As-S. *Econ. Geol.* 58, 483-450.

McWeeny, R. (1963) *Symmetry.* Pergamon Press, Oxford.

Menzer, G. (1926) Über die Kristallstruktur von Linneit, ein schliesslich Polydymit und sychnodymit. *Z. Kristallogr.* 64, 506-507.

Merwin, H. E., and R. H. Lombard (1937) The system Cu-Fe-S. *Econ. Geol.* 32, 203-284.

Miehe, G. (1971) Crystal structure of kobellite. *Nature,* 231, 133-134.

Miller, R. O., F. Dachille, and R. Roy (1966) High Pressure phase equilibrium studies of CdS and MnS by static and dynamic methods. *J. Appl. Phys.* 37, 4913-4918.

Mills, K. C. (1974) *Thermodynamic Data for Inorganic Sulphides, Selerides, and Tellurides.* Butterworths, London, 845 p.

Misra, K. C., and M. E. Fleet (1973a) The chemical composition of synthetic and natural pentlandite assemblages. *Econ. Geol.* 68, 518-539.

_____, _____ (1973b) Unit cell parameters of monosulfide, pentlandite and taenite solid solutions within the Fe-Ni-S system. *Mat. Res. Bull.* 8, 669-678.

_____, _____ (1974) Chemical composition and stability of violarite. *Econ. Geol.* 69, 391-403.

Moh, G. H. (1964) Blaubleibender covellite. *Carnegie Inst. Wash. Year Book* 63, 208-209.

Moh, G. H. (1969) The tin-sulfur system and related minerals. *Neues Jahrb. Mineral. Abh.* 111, 227-263.

_____ (1972) Vergleicheude experimentelle Untersuchungen am molybdan und wolframhaltigen sulfid systemen. *Fortschr. Mineral.* 50, 65-67.

_____ (1973) Das Cu-W-S system und seine Mineralien sowie ein neues Tungstenit vorkommen in Kipushi/Katanga. *Mineral. Deposita* 8, 291-300.

_____, and F. Berndt (1964) Two new natural tin sulfides Sn_2S_3 and SnS_2. *Neues Jahrb. Mineral. Monatsh.*, 94-95.

_____, and L. A. Taylor (1971) Laboratory techniques in experimental sulfide Petrology. *Neues Jahrb. Mineral. Monatsh.*, 450-459.

Moore, D. E., and F. W. Dixon (1973) Phases of the system $Sb_2S_3-As_2S_3$. *Trans. Am. Geophys. Union* 54, 1223-1224.

Moore, Paul B. (1967) A classification of sulfosalt structures derived from the structure of aikinite. *Am. Mineral.* 52, 1874-1876.

Mootz, D., and H. Puhl (1967) Crystal structure of Sn_2S_3. *Acta Crystallogr.* 23, 471-476.

Morehead, F. F. (1963) A Dember effect study of shifts in the stoichiometry of ZnS. *J. Electrochem. Soc.* 110, 285-288.

_____, and A. B. Fowler (1962) The Dember effect in ZnS-type materials. *J. Electrochem. Soc.* 109, 688-695.

Morimoto, N. (1954) Paper on orpiment. *Mineral. J. Japan* 1, 160-169.

_____ (1962) Djurleite, a new copper sulphide mineral. *Mineral. J. Japan* 3, 338-344.

_____ (1964) Structures of two polymorphic forms of Cu_5FeS_4. *Acta Crystallogr.* 17, 351-360.

_____ (1970) Crystal-chemical studies of the Cu-Fe-S system. In, T. Tatsumi, Ed., *Volcanism and Ore Genesis.* Univ. of Tokyo Press, p. 323-328.

_____, and Lloyd A. Clark (1961) Arsenopyrite crystal-chemical relations. *Am. Mineral.* 46, 1448-1469.

_____, and K. Koto (1970) Phase relations of the Cu-S system at low temperatures: stability of anilite. *Am. Mineral.* 55, 106-117.

_____, _____, and Y. Shimazaki (1969) Anilite, Cu_7S_5, a new mineral. *Am. Mineral.* 54, 1256-1268.

_____, and G. Kullerud (1962) The Mo-S system. *Carnegie Inst. Wash. Year Book* 61, 143-144.

_____, _____ (1963) Polymorphism in digenite. *Am. Mineral.* 48, 110-123.

_____, _____ (1965) Pentlandite: thermal expansion. *Carnegie Inst. Wash. Year Book* 64, 204-205.

_____, _____ (1966) Polymorphism on the $Cu_5FeS_4-Cu_9S_5$ join. *Z. Kristallogr.* 123, 235-254.

_____, H. Nakazawa, M. Tokonami, and K. Nishiguchi (1971) Pyrrhotites: Structure type and composition. *Soc. Mineral. Geol. Japan, Spec. Issue 2, Proc. IMA-IAGOD Meet.'70*, 15-21.

Morozova, N. K., M. M. Veselkova, A. F. Botnev, and K. V. Shalimova, (1969) Influence of point defects on the crystalline structure of zinc sulfide powders. *Sov. Phys. Crystallogr.* 14, 74-78.

Mukaiyama, H., and E. Izawa (1966) Phase relations of pyrrhotite. *J. Min. Inst. Kyushu* 34, 194-213 (in Japanese).

_____, _____ (1970) Phase relations in the Cu-Fe-S system: The copper-deficient part. In, T. Tatsumi, Ed., *Volcanism and Ore Genesis.* Univ. Tokyo Press, p. 339-355.

Mukherjee, B. (1969) Crystallography of pyrrhotite. *Acta Crystallogr.* B25, 673-676.

Mullen, D. J. E., and W. Nowacki (1972) Refinement of the crystal structures of realgar, AsS and orpiment, As_2S_3. *Z. Kristallogr.* 136, 48-65.

Munson, R. A. (1966) Synthesis of copper disulfide. *Inorg. Chem.* 5, 1296-1297.

Nafziger, R. H., G. C. Ulmer, and E. Woermann (1971) Gaseous buffering for the control of oxygen fugacity at one atmosphere. In, C. G. Ulmer, Ed., *Research Techniques for High Pressure and High Temperature,* Springer-Verlag, New York, p. 9-41.

Nagao, I. (1955) Chemical studies on the hot springs of Nasu, Nippon. *Kageke Zasshi* 76, 1071-1073.

Nakazawa, H., and N. Morimoto (1971) Phase relations and superstructures of pyrrhotite, $Fe_{1-x}S$. *Mater Res. Bull.* 6, 345-358.

_____, T. Osaka, and K. Sakaguchi (1973) A new cubic iron sulfide prepared by vacuum deposition. *Nature Phy. Sci.* 242, 13-14.

Naldrett, A. J., J. R. Craig, and G. Kullerud (1967) The central portion of the Fe-Ni-S system and its bearing on pentlandite exsolution in iron-nickel sulfide ores. *Econ. Geol.* 62, 826-847.

_____, E. Gasparrini, R. Buchan, and J. E. Muir (1972) Godlevskite (β-$N_{17}S_6$) from the Texmont Mine, Ontario. *Can. Mineral.* 11, 879-885.

_____, and G. Kullerud (1967) A study of the Strathcona Mine and its bearing on the origin of the nickel-copper ores of the Sudbury district, Ontario. *J. Petrol.* 8, 453-531.

Nedachi, M., T. Takeuchi, K. Yamaoka, and M. Taniguchi (1973) Bi-Ag-Pb-S minerals from Agenosawa Mine, Akita Projecture, Northeastern Japan. *Mineral. J. Japan* 9, 69-80.

Nekrasov, I. Y., M. P. Kulakov, and Z. N. Sokolovskaya (1974) The subsolidus relations in the system PbS-SnS. *Geokhimiya,* 80-88.

Neuhaus, A., and L. Cemic (1970) Structur und Mischbarkeit im System ZnS-FeS im Druckbereich bis 60 kbar. *Naturwissenschaften* 7, 354-355.

Nickel, E. H. (1965) A review of the properties of zinc sulphide. *Dep. Mines. Tech. Surv., Ottawa, Inf. Cicr.* 170.

_____ (1970) The application of ligand field concepts to an understanding of the structural stabilities and solid solution limits of sulfides and related minerals. *Chem. Geol.* 5, 233-241.

_____ (1972) Nickeliferous smythite from some Canadian occurrences. *Can. Mineral.* 11, 514-519.

_____ (1973) Violarite - a key mineral in the uspergene alteration of nickel sulphide ores. *Aust. I.M.M. Conf.,* West. Australia, May, 1973, 111-116.

Niizeki, N., and M. J. Buerger (1957a) The crystal structure of livingstonite, $HgSb_4S_8$. *Z. Kristallogr.* 109, 129-157.

_____, _____ (1957b) The crystal structure of jamesonite, $FePb_4Sb_6S_{14}$. *Z. Kristallogr.* 109, 161-183.

Nowacki, Werner (1969) Zur Klassifikation und Kristallchemie der Sulfosalze. *Schweiz. Mineral. Petrogr. Mitt.* 49, 109-156.

Nuffield, E. W. (1974) The crystal structure of fülöppite ($Pb_3Sb_8S_{15}$) (abstr.) *Program Abstr. Summer Meet. Am. Crystallogr. Assoc.,*University Park, Pennsylvania, p. 270.

Oftedal, Ivar (1932) Dis Kristallstruktur des Covelline (CuS). *Z. Kristallogr.* 83, 9-25.

Ohmasa, Masaaki (1973) The crystal structure of $Cu_{2+x}Bi_{6-x}S_9$ (x = 1.21). *Neues Jahrb. Mineral. Monatsh.,* 227-233.

Ohmasa, Masaaki, and Werner Nowacki (1970) A redetermination of the crystal structure of aikinite [$BiS_2|S|Cu^{IV}Pb^{VII}$]. Z. Kristallogr. 132, 71-86.

_____, _____ (1971) The crystal structure of vrbaite $Hg_3Tl_4As_8Sb_2S_{20}$. Z. Kristallogr. 134, 360-380.

_____, _____ (1973) The crystal structure of synthetic $CuBi_5S_8$. Z. Kristallogr. 137, 422-432.

Ontoev, D. A. (1964) Peculiarities of bismuth mineralization in some tungsten deposits of Eastern Transbaikaba. Trudy Mineral. Muz. 15, 134-153.

Orgel, L. E. (1966) An Introduction to Transition-Method Chemistry. 2nd ed. John Wiley, New York.

Otto, H. H., and H. Strunz (1968) Zur Kristallchemic synthetischer Blei - Wismut - Spiessglanze. Neues Jahrb. Mineral. Abh. 108, 1-19.

Ozawa, Tohru, and Hoshio Takéuchi (1972) The crystal structure of $Cu_4Bi_4S_9$ and its relation to simple sulfide structures (abstr.). Acta Crystallogr. A28, p. S70.

Pabst, A. (1959) The pyrite-marcasite relation - a belated comment. Am. Mineral. 44, 685-688.

Pankratz, L. B., and E. G. King (1965) High-temperature heat contents and entropies of two zinc sulfides and four solid solutions of zinc and iron sulfides. U.S. Bur. Mines, Rep. Invest. 6708.

Parravano, N. and P. DeCesaris (1912) The system Sb_2S_3-SuS. Atti Acad Lincei 21, 535-540. [Chem. Abstr. 6, 2044]

Parthé, Erwin (1964) Crystal Chemistry of Tetrahedral Structures. Gordon and Breach, New York.

Pauling, Linus (1965) The nature of the chemical bonds in sulvanite, Cu_3VS_4. Tschermaks Mineral. Petrogr. Mitt. 10, 379-384.

_____, and L. O. Brockway (1932) The crystal structure of chalcopyrite $CuFeS_2$. Z. Kristallogr. 82, 188-194.

_____, and Ralph Hultgren (1933) The crystal structure of sulvanite, $CuVS_4$. Z. Kristallogr. 84, 204-212.

_____, and E. W. Neumann (1934) The crystal structure of binnite $(Cu,Fe)_{12}As_4S_{13}$, and the chemical composition and structure of minerals of the tetrahedrite group. Z. Kristallogr. 88, 54-62.

_____, and Sidney Weinbaum (1934) The crystal structure of enargite, Cu_3AsS_4. Z. Kristallogr. 88, 48-53.

Pauwels, L. J. (1970) Energy level diagrams of the "NiAs," "pyrite," and "spinel"-type sulfides of Fe, Co and Ni. Bull. Soc. Chim. Belges 79, 549-566.

Peacock, M. A. (1947) On heazlewoodite and the artificial compound Ni_3S_2. Univ. Toronto Studies, Geol. 51, 59-69.

_____, and McAndrew, J. (1950) On parkerite and shandite and the crystal structure of $Ni_3Pb_2S_2$. Am. Mineral. 35, 425-439.

Pearson, A. D. (1964) Sulphide, selenide and telluride glasses. In, Modern Aspects of the Vitreous State, Vol. 3. Butterworth, Inc., London.

_____, and M. J. Buerger (1956) Confirmation of the crystal structure of pentlandite. Am. Mineral. 41, 804-805.

Pearson, W. D. (1972) The Crystal Chemistry and Physics of Metals and Alloys. Wiley-Interscience, New York.

Petruk, W. (1973) Tin sulphides from the deposit of Brunswick Tin Mines Limited. Can. Mineral. 12, 46-54.

Petterd, W. F. (1896) Catalogue of the Minerals of Tazmania. Lunceston, 47 p.

Pettit, F. S. (1964) Thermodynamic and electrical investigators on molten antimony sulfide. *J. Chem. Phys.* 38, 9-20.

Petz, J., R. F. Kruh, and G. C. Amstutz (1961) X-ray diffraction study of PbS-As$_2$S$_3$
glasses. *J. Chem. Phys.* p. 526-529.

Platonov, A. N., and A. S. Marfunin (1968) Optical absorption spectra of
sphalerites. *Geochem. Int.* 5, 245-260.

Plovnick, R. H., M. Vlasse, and A. Wold (1968) Preparation and structural properties of some ternary chalcogenides of titanium. *Inorg. Chem.* 7, 127-129.

Potter, R. W. (1973) *The Systematics of Polymorphism in Binary Sulfides: I. Phase
Equilibria in the System Mercury-Sulfur. II. Polymorphism in Binary Sulfides.*
Ph.D. Thesis, Pennsylvania State University.

_____, and H. L. Barnes (1971) Mercuric sulfide stoichiometry and phase rela-
tions (abstr.). *Geol. Soc. Am. Abstr. Programs* 3, 674.

Preisinger, A. (1952) Über die Kristallstruktur des Julienit, Na$_2$Co(NCS)$_4$·8H$_2$O.
Tschermaks Mineral. Petrogr. Mitt. 3, 376-380.

Radtke, A. S., C. M. Taylor, and C. Heropoulas (1973) Antimony-bearing orpiment,
Corlin gold deposit, Nevada. *J. Res. U.S. Geol. Surv.* 1, 85-87.

_____, _____, R. C. Erd, and F. W. Dickson (1974) Occurrence of lorandite,
TlAsS$_2$, at the Corlin gold deposit, Nevada. *Econ. Geol.* 69, 121-124.

_____, _____, F. W. Dickson, and C. Heropoulos(1974) Thallium bearing orpiment,
Corlin Gold deposit, Nevada. *J. Res. U.S. Geol. Surv.* 2, 341-342.

Rajamani, V., and C. T. Prewitt (1973) Crystal chemistry of natural pentlandites.
Can. Mineral. 12, 178-187.

_____, _____ (1974a) The crystal structure of millerite. *Can. Mineral.* 12,
253-257.

_____, _____ (1974b) Thermal expansion of the pentlandite structure. *Am.
Mineral.* (in press)

Rakcheev, A. D., and L. V. Chernyshev (1968) Energy of activation and chemical
composition of pyrites as dependent on the conditions of their synthesis.
Dokl. Akad. Nauk. SSSR, 183, 1184-1187.

Ramsdell, L. S. (1947) Studies on silicon carbide. *Am. Mineral.* 32, 64-82.

Rau, H. (1965) Thermodynamische Messungen an SnS. *Ber. Bunsengs.* 69, 731-736.

_____ (1967) Defect equilibrium in cubic high temperature copper sulfide
(digenite). *J. Phys. Chem. Solids,* 28, 903-916.

_____, and A. Rabenau (1967) Crystal syntheses and growth in strong acid
solutions under hydrothermal conditions. *Solid State Commun.* 5, 331-332.

Ribar, B., Charles Nicca, and W. Nowacki (1969) Dreidimensionale Verfeinerung
der Kristallstruktur von Dufrenoysit, Pb$_8$As$_8$S$_{20}$. *Z. Kristallogr.* 130, 15-40.

_____, and W. Nowacki (1969) Neubestimmung der Kristallstruktur von Gratonit,
Pb$_9$AsS$_{15}$. *Z. Kristallogr.* 128, 321-338.

_____, _____ (1970) Die Kristallstruktur von Stephanit, [SbS$_3$|S|Ag$_5^{III}$].
Acta Crystallogr. B25, 201-207.

Rice, Francis Owen, and Jerome Ditter (1953) Green sulfur, a new allotropic form.
J. Am. Chem. Soc. 75, 6066-6067.

_____, and Calvin Sparrow (1953) Purple sulfur, a new allotropic form. *J. Am.
Chem. Soc.* 75, 848-850.

Richardson, F. D., and J. H. E. Jeffes (1952) The thermodynamics of substances
of interest in iron and steel making III sulphides. *J. Iron Steel Inst.* 171,
165-175.

Rickard, D. T. (1968) Synthesis of smythite - rhombohedral Fe$_3$S$_4$. *Nature* 218,
356.

Rickard, D. T. (1972) Covellite formation in low temperature aqueous solution. *Mineral. Deposita* 7, 180-188.

Riley, J. F. (1965) An intermediate member of the binary system FeS_2 (pyrite) - CoS_2 (cattierite). *Am. Mineral.* 50, 1083-1086.

_____ (1968) The cobaltiferous pyrite series. *Am. Mineral.* 53, 293-295.

_____ (1974) The tetrahedrite-freibergite series with reference to the Mount Isa Pb-Zn-Ag ore body. *Mineral. Deposita* 9, 117-124.

Ripley, L. G. (1971) Crystal growth part II: The growth of zinc sulphide crystals. *Mines Branch Res. Rep. R236,* Dep. Energy, Mines and Resources, Ottawa.

Rising, B. A. (1973) *Phase Relations Among Pyrite, Marcasite and Pyrrhotite Below 300°C.* Ph.D. Thesis, Pennsylvania State University, University Park, Pennsylvania.

Robie, R. A., and D. R. Waldbaum (1968) Thermodynamic properties of minerals and related substances at 298.15°K (25.0°C) and one atmosphere (1.013 bars) pressure and at higher temperatures. *U.S. Geol. Surv. Bull.* 1259, 256 p.

Robinson, S. C. (1948) Synthesis of lead sulphantimonides. *Econ. Geol.* 43, 293-312.

Roedder, E., and E. J. Dwornik (1968) Sphalerite color banding: lack of correlation between color and iron content, Pine Point, Northwest Territories, Canada. *Am. Mineral.* 53, 1523-1529.

Roland, G. W. (1968a) The system Pb-As-S. Composition and stability of jordanite. *Mineral. Deposita* 3, 249-260.

_____ (1968b) Synthetic trechmannite. *Am. Mineral.* 53, 1208-1214.

_____ (1970) Phase relations below 575°C in the system Ag-As-S. *Econ. Geol.* 65, 241-252.

Rooymans, C. J. M. (1963) A phase transformation in the wurtzite and zinc blende lattice under pressure. *J. Inorg. Nucl. Chem.* 25, 253-255.

Rösch, Heinrich (1963) Zur Kristallstruktur des Gratonite-$9PbS \cdot 2As_2S_3$. *Neues Jahrb. Mineral.* 99, 307-337.

_____, and E. Hellner (1959) Hydrothermale untersuchungen am system $PbS-As_2S_3$. *Naturwissenshaften,* 46, 72.

Roseboom, E. H. (1962). Djurleite, $Cu_{1.96}S$, a new mineral. *Am. Mineral.* 47, 1181-1183.

_____ (1966) An investigation of the system Cu-S and some natural copper sulfides between 25° and 700°C. *Econ. Geol.* 61, 641-672.

Rosenqvist, T. (1954) A thermodynamic study of the iron, cobalt and nickel sulfides. *J. Iron Steel Inst.* 176, 37-57.

Rowland, J. F., and S. R. Hall (1973) The crystal structure of haycockite, $Cu_4Fe_5S_8$. *Am. Mineral.* 58, 1110 (abstract).

Roy, R., A. J. Majunder, and C. W. Hulbe (1959) The Ag_2S and Ag_2Se transitions as geologic thermometers. *Econ. Geol.* 54, 1278-1280.

Roy-Choudhury, K. (1974) Experimental investigations on the solid solution series between Cu_2SnS_3 and Cu_2ZnSnS_4 (kesterite). *Neues Jahrb. Mineral. Monatsh.* (in press)

Rucklidge, J. C., and E. L. Gasparrini (1969) *Specifications of a Computer Program for Processing Electron Microprobe Analytical Data: EMPADR VII.* Department of Geology, University of Toronto.

Sadanaga, Ryoichi, and Sigeho Sueno (1967) X-ray study on the $\alpha-\beta$ transition of Ag_2S. *Mineral. J. Japan* 5, 124-148.

Sakharova, M. S. (1966) The basic problems of isomorphism and origin of tetrahedrite-tennantite ores. *Geology of Ore Deposits,* 8, 23-40 [*Econ. Geol.* 64, 120, 1969]

Salanci, B. (1965) Untersuchungen am System Bi_2S_3-PbS. *Neues Jahrb. Mineral. Monatsh.*, 384-388.

_____, and G. H. Moh (1969) Die experimentelle Untersuchung des pseudobinären Schnittes PbS-Bi_2S_3 innerholb des Pb-Bi-S systems in Beziehung zu natürlichen Blei - Bismut - Sulfosalzen. *Neues Jahrb. Mineral. Abh.* 112, 63-95.

_____, _____ (1970) The pseudobinary join galena-antimonite, PbS-Sb_2S_3. *Neues Jahrb. Mineral. Monatsh.* 11, 524-528.

Sands, Donald E. (1965) The crystal structure of monoclinic (β) sulfur. *J. Am. Chem. Soc.* 87, 1395.

Sato, M. (1971) Electrochemical measurements and control of oxygen fugacity and other gaseous fugacities with solid electrolyte sensors. In, G. C. Ulmer, Ed., *Research Techniques for High Pressure and High Temperature*. Springer-Verlag, New York, p. 43-99.

Scanlon, W. W. (1963) The physical properties of semiconducting sulfides, selenides, and tellurides. *Mineral. Soc. Am. Spec. Pap.* 1, 135-143.

Scavnicar, S. (1960) The crystal structure of stibnite. A redetermination of atomic positions. *Z. Kristallogr.* 114, 85-97.

Schenck, R., and P. von der Forst (1939) Gleichgewichtsstudien an erzbildenchen Sulfiden II. *Z. anorg. allg. Chem.* 2H, 145-157.

_____, I. Hoffman, W. Knepper, and H. Vogler (1939) Goeichgewichsstudien über erzbildende Sulfide, I. *Z. anorg. allg. Chem.* 240, 178-197.

Schlegel, H., and A. Schiller (1952) Die Schmelzund Kristallisations - gleischgewichte in System Cu-Fe-S und ihre Bedeutung für Kupfergewinnung. *Freiberg. Forsch.* Sec. B, No. 2, 1-32.

Schneeberg, E. P. (1973) Sulfur fugacity measurements with the electrochemical cell Ag|AgI|Ag_{2+x}S,f_{S_2}. *Econ. Geol.* 68, 507-517.

Schneer, C. J. (1955) Polymorphism in one dimension. *Acta Crystallogr.* 8, 279-285.

Schüller, A., and E. Wholmann (1955) Betektinite ein neues Blei-Kupfer-Sulfid aus dem Mansfelder Rücken. *Geologie* 4, 535-555.

Schwarcz, H. P., and S. D. Scott (1974) Pressures of formation of two iron meteorites based on iron content of sphalerites (abstr.). *Trans. Am. Geophys. Union,* 55, 333.

Schwarz, E. J., and D. C. Harris (1970) Phases in natural pyrrhotite and the effect of heating on their magnetic properties and composition. *J. Geomag. Geoelec.* 22, 463-470.

Scott, S. D. (1968) *Stoichiometry and Phase Changes in Zinc Sulfide*. Ph.D. Thesis, The Pennsylvania State University, State College, Pennsylvania.

_____ (1971) Mössbauer spectra of synthetic iron-bearing sphalerite. *Can. Mineral.* 10, 882-885.

_____ (1973) Experimental calibration of the sphalerite geobarometer. *Econ. Geol.* 68, 466-474.

_____ (1974) Sphalerite geobarometry of regionally metamorphosed terrains (abstr.). *Geol. Soc. Am. Abstr. Progr.* 6, (in press).

_____, and H. L. Barnes (1967) Nonstoichiometric phase changes in sphalerite and wurtzite (abstr.). *Can. Mineral.* 9, 306.

_____, _____ (1969) Hydrothermal growth of single crystals of cinnabar (red HgS). *Mat. Res. Bull.* 4, 897-904.

_____, _____ (1971) Sphalerite geothermometry and geobarometry. *Econ. Geol.* 66, 653-669.

_____, _____ (1972) Sphalerite-wurtzite equilibria and stoichiometry. *Geochim. Cosmochim. Acta* 36, 1275-1295.

Scott, S. D. and E. Gasparrini (1973) Argentian pentlandite $(FeNi)_8AgS_8$, from Bird River, Manitoba. *Can. Mineral.* 12, 165-168.

_____, and S. A. Kissin (1973) Sphalerite composition in the Zn-Fe-S system below 300°C. *Econ. Geol.* 68, 475-479.

_____, A. J. Naldrett, and E. Gasparrini (1972) FeS activities in the $Fe_{1-x}S-Ni_{1-x}S$ (mss) solid solution at 930°C (abstr.). *Econ. Geol.* 67, 1010.

_____, _____, _____ (1974) Regular solution model in the $Fe_{1-x}-Ni_{1-x}S$ (mss) solid solution. *Int. Mineral. Assoc. Ninth Gen. Meet.*, Berlin.

Serebryanaya, N. R. (1966) On the tetragonal modification of chalcocite. *Geochem. Int.* 3, 687-688.

Shalimova, K. V., and N. K. Morozova (1965) Effect of excess zinc on the crystal structure of ZnS. *Soc. Phys. - Crystallogr.* 9, 471-472.

Shannon, R. D. (1971) Relationships between covalency, interatomic distances, and magnetic properties in halides and chalcogenides. *Chem. Commun.* 881.

_____, and C. T. Prewitt (1968) Effective ionic radii in oxides and fluorides. *Acta Crystallogr.* B25, 925-945.

_____ and H. Vincent (1974) Relationships between covalency, interatomic distances, and magnetic properties in halides and chalcogenides. *Struct. Bonding* (in press).

Shewman, R. W., and L. A. Clark (1970) Pentlandite phase relations in the Fe-Ni-S system and notes on the monosulfide solid solution. *Can. J. Earth Sci.* 7, 67-85.

Shibata, Z. (1928) The equilibrium diagram of the iron sulfide-manganese sulfide system. *Tech. Rep. Tohoku Imp.Univ.* 7, 279-289.

Shimazaki, H. (1971) Thermochemical stability of bravoite. *Econ. Geol.* 66, 1080-1082.

_____, and L. A. Clark (1970) Synthetic FeS_2-CuS_2 solid solution and fukuchilite-like minerals. *Can. Mineral.* 10, 648-664.

Short, M. N. (1940) Microscopic determination of the ore minerals. *U.S. Geol. Surv. Bull.* 914, 2nd ed.

_____, and E. V. Shannon (1930) Violarite and other rare nickel sulfides. *Am. Mineral.* 15, 1-17.

Silverman, M. S. (1964) High-temperature high-pressure synthesis of a new bismuth sulfide. *Inorg. Chem.* 3, 1041.

Simkovich, G. (1966) Some effects of point imperfections in ionic type crystals. *Mineral. Ind.* 35, 1-6.

Skinner, Brian J. (1960) Assemblage enargite-famatinite, a possible geologic thermometer, (abstr.). *Geol. Soc. Am. Bull.* 71, 1975.

_____ (1966) The system Cu-Ag-S. *Econ. Geol.* 61, 1-26.

_____ (1970) Stability of the tetragonal polymorph of Cu_2S. *Econ. Geol.* 65, 724-730.

_____, and P. B. Barton, Jr. (1960) The substitution of oxygen for sulfur in wurtzite and sphalerite. *Am. Mineral.* 45, 612-625.

_____, F. R. Boyd, and J. L. England (1964) A high-pressure polymorph of chalcocite, Cu_2S. *Trans. Am. Geophys. Union* 45, 121-122.

_____, R. C. Erd, and F. S. Grimaldi (1964) Greigite, the thio-spinel of iron; a new mineral. *Am. Mineral.* 49, 543-555.

_____, J. L. Jambor, and M. Ross (1966) Mckinstryite, a new copper-silver-sulfide. *Econ. Geol.* 61, 1383-1389.

_____, and F. D. Luce (1971) Solid solutions of the type (Ca,Mg,Mn,Fe)S and their use as geothermometers for the enstatite chondrites. *Am. Mineral.* 56, 1269-1296.

Skinner, Brian J., Frederick D. Luce, and Emil Mackovicky (1972) Studies of the sulfosalts of copper III. Phases and phase relations in the system Cu-Sb-S. *Econ. Geol.* 67, 924-938.

‗‗‗‗‗‗, and D. L. Peck (1969) An immiscible sulfide melt from Hawaii. In, H. D. B. Wilson, Ed., Magmatic Ore Deposits, *Econ. Geol. Monogr.* 4, 310-322.

‗‗‗‗‗‗, D. E. White, H. J. Rose, and R. E. Mays (1967) Sulfides associated with the Salton Sea geothermal brine. *Econ. Geol.* 62, 316-330.

Smith, F. G. (1955) Structure of zinc sulfide minerals. *Am. Mineral.* 40, 658-675.

‗‗‗‗‗‗ (1967) Machine plotting of liquidus data of binary and ternary salt systems. *Can. Mineral* 9, 180-190.

‗‗‗‗‗‗, M. I. Smith, and D. F. Watson (1974) ALKHAL, a file of information on salt systems with an alkali halide component. *Can. J. Earth Sci.* 11, 719-721.

Snetsinger, K. G. (1971) Erlichmanite (OsS_2), a new mineral. *Am. Mineral.* 56, 1501-1506.

Somanchi, S. (1963) *Subsolidus Phase Relations in the systems Ag-Sb and Ag-Sb-S.* M.S. Thesis, McGill University.

Springer, G. (1969) Naturally occurring compositions in the solid solution series Bi_2S_3-Sb_2S_3. *Mineral. Mag.* 37, 295-296.

‗‗‗‗‗‗ (1971) The synthetic solid-solution series Bi_2S_3-$BiCuPbS_3$ (Bismuthinite-Aikinite). *Neues Jahrb. Mineral. Monatsh.* 19-24.

‗‗‗‗‗‗ (1972) The pseudobinary system Cu_2FeSnS_4-Cu_2ZnSuS_4 and its mineralogical significance. *Can. Mineral.* 11, 535-541.

‗‗‗‗‗‗, and J. H. G. LaFlammo (1971) The system Bi_2S_3-Sb_2S_3. *Can. Mineral.* 10, 847-853.

‗‗‗‗‗‗, D. Schachner-Korn, and J. V. P. Long (1964) Metastable solid solution relations in the system FeS_2-CoS_2-NiS_2. *Econ. Geol.* 59, 475-491.

Stanton, R. L. (1972) *Ore Petrology.* McGraw-Hill, Inc.

Staples, L. W. (1972) Sulfosalts. In, R. W. Fairbridge, *Encyclopedia of Geochemistry and Environmental Sciences IVA.* Van Nostrand Reinhold, p. 1141-1142.

Stemprok, M. (1967) The Bi-Mo-S system. *Carnegie Inst. Wash. Year Book,* 65, 336-337.

‗‗‗‗‗‗ (1971) The Fe-W-S system and its geological application. *Mineral. Deposita* 6, 302-312.

‗‗‗‗‗‗, and G. H. Moh (1969) Phase relations in the lead-tin-sulfur system at 630°. *Vestr. Ustred. Ustavu Geol.* 44, 359-381.

Stout, George H., and Lyle H. Jensen (1968) *X-ray Structure Determination.* The Macmillan Company, Collier-Macmillan Ltd, London.

Straumanis, M. E., G. C. Amstutz, and S. Chan (1964) Synthesis and X-ray investigations within the System FeS_2-CoS_2. *Neues Jahrb. Mineral. Monatsh.* 101, 127-141.

Strunz, H., B. H. Geier, and E. Seeliger (1958) Gallit, $CuGaS_2$, das erste selbstandige Gallium-Mineral und seine Paragenese in Tsumeb. *Neues Jahrb. Mineral. Monatsh.* 85-96.

Stubbles, J. R., and C. S. Birchenall (1959) A redetermination of the lead-lead sulfide equilibrium between 585°C and 920°C. *AIMME Trans.* 215, 535-538.

Sugaki, A., K. Hirofumi, and A. Kitekaze (1969) A synthesized phases of the system PbS-Sb_2S_3. *Japan Assoc. Mineral. Petrol. Econ. Geol.* 61, 172.

‗‗‗‗‗‗, and A. Kitakaze (1972) Chemical composition of synthetic alabandite solid solution and its phase relations in the system Fe-Mn-S. *Proc. 6th Int. Conf. X-ray Optics and Microanalysis.* G. Shinoda, K. Kohra, and T. Ichinokawa, Eds., p. 755-760.

Sugaki, A., and H. Shima (1965) BiSbS$_3$ (Horobetsuite). *Mem. Fac. Eng. Yamaguchi Univ.* 16, 116-118.

_____, _____ (1966) Studies on the pyrrhotite group minerals. I. Dry synthesis of monoclinic pyrrhotite. *Jap. Assoc. Mineral. Petrol. Econ. Geol.* 55, 242-253.

_____, _____ (1970) The phase equilibrium study of the Cu-Bi-S system (abstr.). *Collected Abstr., IMA-IAGOD Meet.*, Tokyo, Kyoto, p. 220.

_____, _____ (1972) Phase relations of the Cu$_2$S-Bi$_2$S$_3$ system. *Tech. Rep. Yamaguchi Univ.*, 45-70.

_____, _____, and A. Kitakaze (1972) Synthetic sulfide minerals (IV). *Tech. Rep. Yamaguchi Univ.* 1, 71-77.

Suhr, N. (1965) The Ag$_2$S-Cu$_2$S system. *Econ. Geol.* 50, 347-250.

Sutarno, O. Knop, and K. I. G. Reid (1967) Chalcogenides of the transition elements V. Crystal structures of the disulfides and ditellurides of ruthenium and osmium. *Can. J. Chem.* 45, 1391-1400.

Szymanski, J. T., S. R. Hall, and L. J. Cabri (1973) Twinning and disorder in cubic CuFe$_2$S$_3$. *Am. Crystallogr. Assoc. Winter Meet., January, 1973. Gainesville, Fla. Program Abstr.*, P3, pg. 89.

Takagi, J., and Y. Takeuchi (1972) The crystal structure of lillianite. *Acta Crystallogr.* B28, 649-651.

Takeno, S., and A. H. Clark (1967) Observations on tetragonal (Fe,Ni,Co)$_{1+x}$S, mackinawite. *J. Sci. Hiroshima Univ., Ser. C.* 5, 287-293.

Takeuchi, Y. (1957) The absolute structure of ullmanite, NiSbS. *Mineral. J. Japan* 2, 90-102.

_____ (1970) On the crystal chemistry of sulphides and sulphosalts. In, *Volcanism and Ore Genesis*, Ed., T. Tatsumi. Univ. Tokyo Press, p. 395-419.

_____, and N. Haga (1969) On the crystal structures of seligmannite, PbCuAsS$_3$, and related minerals. *Z. Kristallogr.* 130, 254-260.

_____, and W. Nowacki (1964) Detailed crystal structure of rhombohedral MoS$_2$ and systematic deduction of possible polytypes of molybdenite. *Schweiz. Mineral. Petrogr. Mitt.* 44, 105-120.

_____, and R. Sadanaga (1969) Structural principles and classification of sulfosalts. *Z. Kristallogr.* 130, 346-368.

_____, S. Ghose, and W. Nowacki (1965) The crystal structure of hutchinsonite, (Tl,Pb)$_2$As$_5$S$_9$. *Z. Kristallogr.* 121, 321-347.

_____, M. Ohmasa and W. Nowacki (1968) The crystal structure of wallisite, PbTlCuAs$_2$S$_5$, the Cu analogue of hatchite, PbTlAgAs$_2$S$_5$. *Z. Kristallogr.* 127, 349-365.

Tatsuka, K., and N. Morimoto (1973) Composition variation and polymorphism of tetrahedrite in the Cu-Sb-S system below 400°C. *Am. Mineral.* 58, 425-434.

Taylor, L. A. (1969) The significance of twinning in Ag$_2$S. *Am. Mineral.* 54, 961-963.

_____ (1970a) Low-temperature phase relations in the Fe-S system. *Carnegie Inst. Wash. Year Book* 68, 259-270.

_____ (1970b) Smythite, Fe$_{3+x}$S$_4$, and associated minerals from the Silverfields Mine, Cobalt, Ontario. *Am. Mineral.* 55, 1650-1658.

_____ (1970c) The System Ag-Fe-S: Phase Equilibria and Mineral Assemblages. *Mineral. Deposita* 5, 41-58.

_____ (1970d) The System Ag-Fe-S: Phase relations between 1200° and 700°C. *Metall. Trans.* 1, 2523-2529.

_____ (1971) Oxidation of pyrrhotites and the formation of anomalous pyrrhotite. *Carnegie Inst. Wash. Year Book* 70, 287-289.

Taylor, L. A., and L. W. Finger (1971) Structure refinement and composition. *Carnegie Inst. Wash. Year Book,* 69, 319-323.

_____, and G. Kullerud (1971) Pyrite-type compounds. *Carnegie Inst. Wash. Year Book* 69, 322-325.

_____, _____ (1972) Phase equilibria associated with the stability of copper disulfide. *Neues Jahrb. Mineral. Monatsh.* 458-463.

_____, and K. L. Williams (1972) Smythite, $(Fe,Ni)_9S_{11}$ - a redefinition. *Am. Mineral.* 57, 1571-1577.

Togari, K. (1961) Mineralogical studies on Japanese sphalerite. *J. Fac. Sci. Hokkaido Univ., Ser. 4., Geol. Mineral.* 10, 703-733.

Tokonami, M., K. Nishiguchi, and N. Morimoto (1972) Crystal structure of a monoclinic pyrrhotite (Fe_7S_8). *Am. Mineral.* 57, 1066-1080.

Toulmin, P. (1960) Effect of Cu on sphalerite phase equilibria - a preliminary report. *Geol. Soc. Am. Abstr.,* p. 1993.

_____ (1963) Proustite-pyrargyrite solid solutions. *Am. Mineral.* 48, 725-736.

_____, and P. B. Barton, Jr. (1964) A thermodynamic study of pyrite and pyrrhotite. *Geochim. Cosmochim. Acta,* 28, 641-671.

Trahan, J., R. G. Goodrich, and S. F. Watkins (1970) X-ray diffraction measurements on metallic and semiconducting hexagonal NiS. *Phys. Rev.* B2, 2859-2862.

Trail, R. J., and R. W. Boyle (1955) Hawleyite, isometric cadmium sulfide, a new mineral. *Am. Mineral.* 40, 555-559.

Trojer, F. J. (1966) Refinement of the structure of sulvanite. *Am. Mineral.* 51, 890-894.

Tuinstra, F. (1966) The structure of fibrous sulfur. *Acta Crystallogr.* 20, 341-349.

Tunnell, G. (1964) Chemical processes in the formation of mercury ores and ores of mercury and antimony. *Geochim. Cosmochim. Acta* 28, 1019-1037.

Uchida, I. (1964) Fluorescence of pure zinc sulfide in correlation with deviation from stoichiometry. *J. Phys. Soc. Japan* 19, 670-674.

Uda, M. (1967) The structure of synthetic Fe_3S_4 and the nature of transition to FeS. *Z. Anorg. Allg. Chem.* 350, 105-109.

Urazov, G. G., K. A. Bol'shakov, P. I. Federov, and I. I. Vasilevskaya (1960) The ternary bismuth-iron-sulfur system (a contribution to the theory of precipitation smelting of bismuth). *Russ. J. Inorg. Chem.* 5, 303-307.

Vaasjoki, O. (1971) On the striation developed in cubanite by heating, a criterion for thermometamorphism. *Mineral. Deposita* 6, 103-110.

_____, T. A. Hakli, and M. Tonitti (1974) The effect of cobalt on the thermal stability of pentlandite. *Econ. Geol.* 69, 549-551.

Valverde, D. N. (1968) Phase diagram of the Cu-Ag-S system at 300°C. *Z. Phys. Chem. (Frankfurt),* 62, 218-220.

Vand, V. (1951a) Application of dislocation theory to the polytypism of silicon carbide. *Phil. Mag.* 42, 1384-1386.

_____ (1951b) Polytypism arising from screw dislocation. *Nature,* 168, 783.

van der Berg, C. B. (1972) Phase diagram of the semiconducting uniaxial ferroelectric $Fe_{1-x}S$. *Ferroelectrics* 4, 117-120.

van Hook, J. J. (1960) The ternary system $Ag_2S-Bi_2S_3-PbS$. *Econ. Geol.* 55:759-788.

Vaughan, D. J. (1969) Zonal variation in bravoite. *Am. Mineral.* 54, 1075-1083.

_____, R. G. Burns, and V. M. Burns (1971) Geochemistry and bonding of thiospinel minerals. *Geochim. Cosmochim. Acta* 35, 365-381.

Verma, A. R., and P. Krishna (1966) *Polymorphism and Polytypism in Crystals.* John Wiley and Sons, Inc., New York.

Viaene, W. (1968) Le Systeme Fe-Ge-S a 700°C. *C. R. Acad. Sci. Paris* 266, 1543-1545.

_____ (1972) The Fe-Ge-S system: phase equilibria. *Neues Jahrb. Mineral. Monatsh.* 23-35.

_____, and G. Kullerud (1971) The Ti-S and Fe-Ti-S Systems. *Carnegie Inst. Wash. Year Book* 70, 297-299.

_____, _____, and L. A. Taylor (1971) Phase equilibria studies in the system Ti-S and Fe-Ti-S (abstr.). *Geol. Soc. Am. Abstr. Programs,* 3, 739.

Vogel, R. (1953) Über das System Blei silber-schwefal. *Z. Metallkd.* 44, 133-135.

_____ (1956) The system bismuth sulfide-copper sulfide and the ternary system Bi-Bi$_2$S$_3$-Cu$_2$S-Cu. *Z. Metallkd.* 47, 694-699.

_____ (1968) Zur Entstehung des Daubréelith im meteorischen Eisen. *Neues Jahrb. Mineral. Monatsh.* 453-463.

_____, and W. Gilde (1949) Über das System Zinn-Antimon-Schwefel. *Z. Metallkd.* 40, 121-126.

_____, and R. Neumann (1950) Über daubréelith. *Neues Jahrb. Mineral. Monatsh.* 175-190.

_____, and F. G. Hillner (1953) Das Zustendsschaubild Eisen-Eisensulfide-Kobaltsulfid-Kobalt. *Arch. Eisenhuttenwessen* 24, 133-141.

_____, and H. H. Weizenkorn (1961) Über das Drei-staff system Eisen-Schwefel-Wolfram. *Arch. Eisenhuttenwessen* 32, 413-420.

Vuorelainen, Y., T. A. Hakli, and H. Papunen (1972) Argentian pentlandite from some Finnish sulfide deposits. *Am. Mineral.* 57, 137-145.

Walia, D. S., and L. L. Y. Chang (1973) Investigations in the systems PbS-Sb$_2$S$_3$-As$_2$S$_3$ and PbS-Bi$_2$S$_3$-As$_2$S$_3$. *Can. Mineral.* 12, 113-119.

Wang, N. (1973) A study of the phases on the pseudobinary join PbS-Sb$_2$S$_3$. *Neues Jahrb. Mineral. Monatsh.* 2, 79-81.

_____, (1974) The three ternary phases in the system Cu-Sn-S. *Neues Jahrb. Mineral. Monatsh.* (in press).

Watanabe, Y. (1974) The crystal structure of monoclinic γ-sulphur. *Acta Crystallogr.* B30, 1396-1401.

Wehmeier, F. H., R. A. Laudise, and J. W. Shieves (1968) The system Ag$_2$S-As$_2$S$_3$ and the growth of crystals of proustite, smithite and pyrargyrite. *Mat. Res. Bull.* 3, 767-778.

Weissberg, G. A. (1965) Getchellite, AsSbS$_3$, a new mineral from Humboldt County, Nevada. *Am. Mineral.* 50, 1817-1826.

Weitz, G., and E. Hellner (1960) Über komplex zusammengesetzte sulfidische Erze VII. Zur Kristallstruktur des Cosalits, Pb$_2$Bi$_2$S$_5$. *Z. Kristallogr.* 113, 385-402.

Welin, E. (1966) Notes on the mineralogy of Sweden 5. Bismuth-bearing sulphosalts from Gladhammar, a revision. *Ark. Mineral. Geol.* 4, 377-386.

Werner, A. (1965) Investigations on the system Cu-Ag-S. *Z. Phys. Chem. (Frankfurt),* 47, 267-285.

Wernick, J. H. (1960) Constitution of the AgSbS$_2$-PbS, AgBiS$_2$-PbS, AgBiS$_2$-AgBiSe$_2$ systems. *Am. Mineral.* 45, 591-611.

_____, S. Geller, and K. E. Benson (1958a) New semiconductors. *J. Phys. Chem. Solids,* 4, 154-155.

_____, _____, _____ (1958b) Synthetic proustite Ag$_3$AsS$_3$. *Anal. Chem.* 30, 303.

Westgren, A. (1938) Die Kristallstruktur von Ni_3S_2. Z. Anorg. Chem. 239, 82-84.

Wickman, F. E. (1953) The crystal structure of aikinite, $CuPbBiS_3$. Ark. Mineral. Geol. 1, 501-507.

Wiegers, G. A., and F. Jellinek (1970) The system titanium-sulfur II. The structure of Ti_3S_4 and Ti_4S_5. Solid State Chem. 1, 519-525.

Wiggins, L. B. (1974) A Reconnaissance Investigation of Chalcopyrite-Sphalerite Relationships in the Cu-Fe-Zn-S System. M.S. Thesis, Virginia Polytechnic Institute and State University, Blacksburg.

Williams, K. L., and G. Kullerud (1970) The Ni-Sb-S system. Carnegie Inst. Wash. Year Book 68, 270-273.

Williamson, D. P., and N. W. Grimes (1974) An X-ray diffraction investigation of sulphide spinels. J. Phys. & Appl. Phys. 7, 1-6.

Wright, H. D., W. M. Bernard, and J. B. Halbig (1965) Solid solutions in the systems ZnS-ZnSe and PbS-PbSe at 300°C and above. Am. Mineral. 50, 1802-1815.

Wuensch, B. J. (1964) The crystal structure of tetrahedrite, $Cu_{12}Sb_4S_{13}$. Z. Kristallogr. 119, 437-453.

_____ (1972) Sulfur. 16A Crystal Chemistry. In, Handbook of Geochemistry, Vol. II/3. Springer-Verlag, Berlin.

_____, and M. J. Buerger (1963) The crystal structure of chalcocite, Cu_2S. Mineral. Soc. Am. Spec. Pap. 1, 164-170.

_____, and W. Nowacki (1966) The substructure of the sulfosalt jordanite. Schweiz. Mineral. Petrogr. Mitt. 46, 89-96.

_____, _____ (1967) The crystal structure of marrite, $PbAgAsS_3$. Z. Kristallogr. 125, 459-488.

_____, Y. Takeuchi, and W. Nowacki (1966) Refinement of the crystal structure of binnite, $Cu_{12}As_4S_{13}$. Z. Kristallogr. 123, 1-20.

Yamaguchi, S., and H. Wada (1973) Fe_2S_3 of the spinel type structure with lattice defect. Kristall und Technik, 8, 1017-1019.

Yamaoka, S., and B. Okai (1970) Preparation of barium tin sulfide, strontium tin sulfide and lead tin sulfide at high pressures. Mat. Res. Bull. 5, 789-794.

Ying-chen, J., and T. Yu-jen (1973) Isomorphous system $RuS_2-O_5S_2-IrS_2$ and the mineral system PdS-PtS. Geochemia, 4, 262-270.

Yu, W. D., and P. J. Gidisse (1971) High pressure polymorphism in CdS, CdSe, and CdTe. Mat. Res. Bull. 6, 621-638.

Yund, R. A. (1962) The system Ni-As-Si phase relations and mineralogic significance. Am. J. Sci. 260, 761-782.

_____ (1963) Crystal data for synthetic $Cu_{5.5x}S_{6.5x}$ (idaite). Am. Mineral. 48, 672-676.

_____, and H. T. Hall (1968) The miscibility gap between FeS and $Fe_{1-x}S$. Mat. Res. Bull. 3, 779-784.

_____, _____ (1969) Hexagonal and monoclinic pyrrhotites. Econ. Geol. 64, 420-423.

_____, _____ (1970) Kinetics and mechanism of pyrite exsolution from pyrrhotite. J. Petrol. 11, 381-404.

_____, and G. Kullerud (1966) Thermal stability of assemblages in the Cu-Fe-S system. J. Petrol. 7, 454-488.

Zelikman, A. N., and L. V. Belyaerskaya (1956) The melting point of molybdenite. Zhus Neorg. Khim. 1, 2239-2244.

Zemann, A., and J. Zemann (1959) Zur Kenntnis der Kristallstruktur von Lorandit $TlAsS_2$. Acta Crystallogr. 12, 1002-1006.

Zhdanov, G. S. (1945) The numerical symbol of close packing of spheres and its application in the theory of close packings. *C. R. Acad. Sci. USSR*, 48, 39–42.

Zoka, H., L. A. Taylor, and S. Takeno (1973) Compositional variations in natural mackinawite and the results of heating experiments. *J. Sci. Hiroshima Univ., Ser. C*, 7, 37–53.

Zoltai, T. (1974) *Systematics of Sulfide Structures*. Department of Geology and Geophysics, University of Minnesota, Minneapolis, Minn.

ACKNOWLEDGMENTS

The authors and editor are grateful to the following for permission to reproduce figures used in this book:

Dr. H. L. Barnes, Editor of *Geochemistry of Hydrothermal Ore Deposits**

Dr. F. D. Bloss, Editor of the *American Mineralogist*

Dr. L. G. Berry, Editor of the *Canadian Mineralogist*

Dr. B. J. Skinner, Editor of *Economic Geology*

Dr. D. M. Shaw, Executive Editor of *Geochimica et Cosmochimica Acta*

Dr. H. S. Yoder, Jr., Director of the Geophysical Laboratory

Dr. E. H. Wainwright, Editor of *Minerals Science and Engineering*

Mrs. M. Kelso, Permissions Editor of W. H. Freeman and Company**

Dr. V. H. Borsodi, Senior Editor, Springer-Verlag New York, Inc.

*The authors and editor of these short course notes regrettably omitted proper acknowledgment, in the first printing, of the excellent book *Geochemistry of Hydrothermal Ore Deposits* edited by H. L. Barnes and published by Holt Rinehart and Winston, We gratefully acknowledge this book as a primary source of information and illustrations and apologize to the authors and editor for previously failing to give proper credit.

**For the use of their Figure 217 (p. 232) from *The Interpretation of Geological Phase Diagrams* by Ernest G. Ehlers, W. H. Freeman and Company. Copyright 1972.

SUPPLEMENTAL REFERENCES (Second Printing)

Birnie, R. W. (1974) Phase relations in the galena-rich portion of the PbS-As_2S_3-Sb_2S_3 system. *Geol. Soc. Am. Abstr. Progr. 6*, 655.

Cabri, L. J. (1967) A new copper-iron sulfide. *Econ. Geol. 62*, 910-925.

_____, D. C. Harris, J. H. Stewart, and J. F. Rowland (1970) Willyamite redefined. *Aust. I.M.M. Proc.* no. 233, 95-100.

Caye, R., P. Picot, R. Pierrot and F. Permingeat (1967) New data on the mercury content of vrbaite. *Bull. Soc. fr. Mineral. Cristallogr. 90*, 185-191.

Chen, T. T. and L. L. Y. Chang (1974) Investigations in the system Ag_2S-Cu_2S-Bi_2S_3 and Ag_2S-Cu_2S-Sb_2S_3. *Canad. Mineral. 12*, 404-410.

Drabek, M. and M. Stemprok (1974) The system Sn-S-O and its geological application. *Neues Jahrb. Mineral. Abh. 122*, 90-118.

Harris, D. C. and E. H. Nickel (1972) Pentlandite compositions and associations in some mineral deposits. *Canad. Mineral. 11*, 861-878.

Hoda, S. N. and L. L. Y. Chang (1974) Phase relations in the systems PbS-Cu_2S-Bi_2S_3 and PbS-Cu_2S-Sb_2S_3. *Geol. Soc. Am. Abstr. Progr. 6*, 1042.

MacLean, W. H., L. J. Cabri, and J. E. Gill (1972) Exsolution products in heated chalcopyrite. *Canad. J. Earth Sci. 9*, 1305-1317.

Miehe, G. (1971) Crystal structure of kobellite. *Nature, Physical Sciences 231*, 133-134.

Moh, G. H. (1971) Blue-remaining covellite and its relations to phases in the sulfur rich portion of the copper sulfur system at low temperatures. *Min. Soc. Japan Spec. Paper 1*, 226-232.

Morimoto, N., A. Gyobu, K. Tsukuma, and K. Koto (1975) Superstructure and non-stoichiometry of intermediate pyrrhotite. *Am. Mineral. 60*, 240-248.

Mumme, W. G. (1975) Crystal structure of the krupkaite from the Juno Mine at Tennant Creek, Northern Territory, Australia. *Am. Mineral. 60*, 300-308.

Nowacki, W. (1968) Über Hatchite, Lengenbachit, und Vrbait. *Neues Jahrb. Mineral. Monat.*, 69-75.

Petruk, W., D. R. Owens, J. M. Stewart, and E. J. Murray (1974) Observations on acauthite, aguilarite, and naumannite. *Canad. Mineral. 12*, 365-369.

Potter, R. W. (1974) The low temperature phase relations in the system Cu-S derived from an electrochemical investigation. *Geol. Soc. Am. Abstr. Progr. 6*, 915-916.

Vaughan, D. J., and J. R. Craig (1974) The crystal chemistry and magnetic properties of iron in the monosulfide solid solution of the Fe-Ni-S system. *Am. Mineral. 59*, 889-895.

Wu, I. J., U. Petersen, and H. D. Holland (1974) Tetrahedrite-temmantite at Casapalca, Peru, and its application to mineral exploration. *Geol. Soc. Am. Abstr. Progr. 6*, 1010-1011.

Zak, L., U. Synecek, and J. Hybler (1974) Krupkaite, $CuPbBi_3S_6$, a new mineral of the bismuthinite-aikinite group. *Neues Jahrb. Mineral. Monatsh.*, 533-541.

CS-15   Add Ag-Mn-Sb-S. Samsonite = $Ag_4MnSb_2S_6$.

CS-17   Second Co-Ni-Fe-S should be "Co-Ni-Sb-S." Toward should be "between."

CS-18   Cu-Fe-Sn-S. Mawsonite formula should be $Cu_5Fe_2SnS_8$. Add reference
         Lee et al. (1975) Econ. Geol. 70, 834-843.

CS-18   Pb-Sb-Sn-S. Add reference Makovicky (1974) N. Jh. Min. Mh., 235-256.

CS-19   Cu-Fe-Sn-Zn-S. Add "stannoidite $Cu_5(Fe,Zn)_2SnS_8$" and Kato (1969) Nat.
         Sci. Museum (Tokyo) Bull. 12, 165-172.

CS-26   9 lines from bottom. Reflection should be "reflections."

CS-28   Monoclinic Pyrrhotite. Ferromagnetic should be "ferrimagnetic."

CS-29   Line 8. Ferromagnetic should be "ferrimagnetic."

CS-29   9 lines from bottom. . . .of the $Fe_{n-1}S_n$ clan should be "similar to the
         $Fe_{n-1}S_n$ clan."

CS-33   3 lines from bottom. Choose should be "chose."

CS-36   Equation (CS-4). αFeS should be "a_{FeS}."

CS-51   Effect of Pressure. . . . Line 1-3 should be "The partial molar volume
         of FeS in sphalerite is large compared to that in pyrrhotite, so the ΔV
         for reaction Fe-rich sphalerite → Fe-poor sphalerite + pyrrhotite is
         negative."

CS-59   Table CS-6. Anilite formula should be "Cu_7S_4."

         See supplemental reference, Moh (1971), for information
         on blaubleibender covellites.

         Add "1971" to reference (15).

CS-61   8 lines from bottom. Cu_7S_5 should be "Cu_7S_4."

CS-64   Top of page. Add heading "The Cu-Fe-S System."

CS-64   Line 6. Schiller should be "Schüller."

CS-64   6 lines from bottom. Idaite formula should be Cu_3FeS_4.

CS-65   Table CS-7. Idaite formula should be Cu_3FeS_4.

         Reference chalcopyrite thermal stability "(11)" and
         structure as "(22)."

         Reference mooihoekite structure as "(21)."

CS-66   Table CS-7. CS-1 should be "CS-7."

         Reference haycockite structure as "(20)."

         Add references "(20) Rowland and Hall (1973)
                        (21) Hall and Rowland (1973)
                        (22) Hall and Stewart (1973b)."

CS-72   Lines 4,5. Should be ". . .twinning normal to [111] of the cubic iss
         was common. . . ."

CS-72   Lines 12-14. Should be ". . .cubanite compositions a mixed x-ray
         pattern which may be misindexed as tetragonal and
         probably accounts for the. . . ."

CS-72   3 lines from bottom. Idaite formula should be Cu_3FeS_4.

CS-77   Table CS-8. High temperature $Ni_{3\pm x}S_2$ is cubic $F\bar{4}3m$, a = 5.21.
         (Liné and Huber, 1963, Comptes Rendus, 256, 2, 3118).

         Vaesite formula is NiS_2 and its upper stability is 1022°C.

CS-80   The 5 equations should read:

$$3Ni + S = Ni_3S_2 \qquad\qquad \Delta G = -79,240 + 39.01T$$
$$2Ni_3S_2 + S_2 = 6NiS \qquad\qquad \Delta G = -76,360 + 57.21T$$
$$7/2\ Ni_3S_2 + S_2 = 3/2\ Ni_7S_6 \qquad \Delta G = -49,260 + 19.91T$$
$$2Ni_7S_6 + S_2 = 14\ NiS \qquad\qquad \Delta G = -50,600 + 30.53T$$
$$2NiS + S_2 = 2NiS_2 \qquad\qquad \Delta G = -49,880 + 46.67T$$

CS-82   Table CS-9. Pentlandite structure, add reference (11): Hall and Stewart (1973a).

CS-87   Last line. Add reference (Hall and Stewart, 1973a).

CS-91   4 lines from bottom. Combimation should be "combination."

CS-91   Last line. Sulfbismathinides should be "sulfbismuthinides."

R-18   Leonard *et al.* reference. Lavrites should be "laurites."

R-20,21   The page numbers are switched.

R-25   Schlegel and Schiller should be "Schlegel and Schüller."

System	Compound	Mineral Name	References
Ag-S			Petruk *et al.* (1974)* Lowenhaupt & Smith (1974) Sadanaga *et al.* (1967)
As-S			Mullen & Nowacki (1972)
Bi-S			Kupcik & Vesela-Novakova (1970)
Co-S			Geller (1962)
Cu-S			Potter (1974)*
Fe-S			Morimoto *et al.* (1974)
Hg-S			Auvray & Genet (1973)
Sb-S			Scavnicar (1960) Bayliss & Nowacki (1972)
V-S			Allman *et al.* (1964) Kutoglu & Allman (1972)
Ag-As-S			Matsumoto & Nowacki (1969) Hellner & Burzlaff (1964) Engel & Nowacki (1966,1968)
Ag-Sb-S			Kutoglu (1968) Engel & Nowacki (1965) Ribar & Nowacki (1970)
Ag-Se-S	$Ag_2(Se,S)$	aguilarite	Petruk *et al.* (1974)
As-Cu-S			Kulpe (1961)
As-Ni-S			Bayliss & Stephenson (1967,1968)
As-Pb-S			Engel & Nowacki (1969,1974) Ito & Nowacki (1974) Iitaka & Nowacki (1961) Ribar & Nowacki (1969)
As-Sb-S			Guillermo & Wuensch (1973)
As-Tl-S			Zemann & Zemann (1959) Knowles (1965)
Bi-Cu-S			Ohmasa (1973) Kocman & Nuffield (1973) Ozawa & Takeuchi (1972)
Bi-Pb-S			Iitaka & Nowacki (1962) Takagi & Takeuchi (1972)
Cu-V-S			Pauling (1965)
Fe-Sn-S			Drabek and Stemprok (1974)*
Hg-Sb-S			Niizeki & Buerger (1957a)
Ni-Sb-S			Takeuchi (1957)
Pb-Sb-S			Cho & Wuensch (1970,1974) Jawdor (1969a,b) Kohatsu & Wuensch (1974) Nuffield (1974)
Ag-Bi-Cu-S			Chen & Chang (1974)*
Ag-Cu-Sb-S			Chen & Chang (1974)*
Ag-Mn-Sb-S			Hruskova (1969)
Ag-Pb-Sb-S			Ito & Nowacki (1974)

System	Compound	Mineral Name	References
As–Pb–Sb–S			Birnie (1974)*
As–Cu–Pb–S			Edenharter & Nowacki (1970) Takeuchi & Haga (1969)
As–Cu–Sb–S			Wu *et al.* (1974)*
As–Cu–Zn–S	$Cu_6Zn_3As_4S_{12}$	nowackiite	Marumo (1967)
As–Pb–Tl–S	$(Tl,Pb)_2As_5S_9$	hutchinsonite	Takeuchi *et al.* (1965)
Bi–Cu–Pb–S	$PbCuBi_3S_6$	krupkaite	Kohatsu & Wuensch (1971,1973a,1973b) Ohmasa & Nowacki (1970,1973) Hoda & Chang (1974)* Mumme (1974)*
Cu–Fe–Pb–S			Dornberger-Shiff & Hohne (1959)
Cu–Fe–Sn–S			Drabek & Stemprok (1974)*
Cu–Pb–Sb–S			Edenharter & Nowacki (1970) Euler & Hellner (1960) Hoda & Chang (1974)*
Fe–Pb–Sb–S	$FePb_4Sb_6S_{14}$	jamesonite	Niizeki & Buerger (1957b)
Ag–As–Pb–Tl–S	$PbTlAgAs_2S_5$	hatchite	Marumo & Nowacki (1967c)
As–Cu–Pb–Tl–S	$PbTlCuAs_2S_5$	wellisite	Tukeuchi *et al.* (1968)
As–Hg–Sb–Tl–S	$Hg_3Tl_4As_8Sb_2S_{20}$	vrbaite	Uhmasa & Nowacki (1971) Nowacki (1968)* Caye *et al.* (1967)*